BIOCHEMICAL MESSENGERS

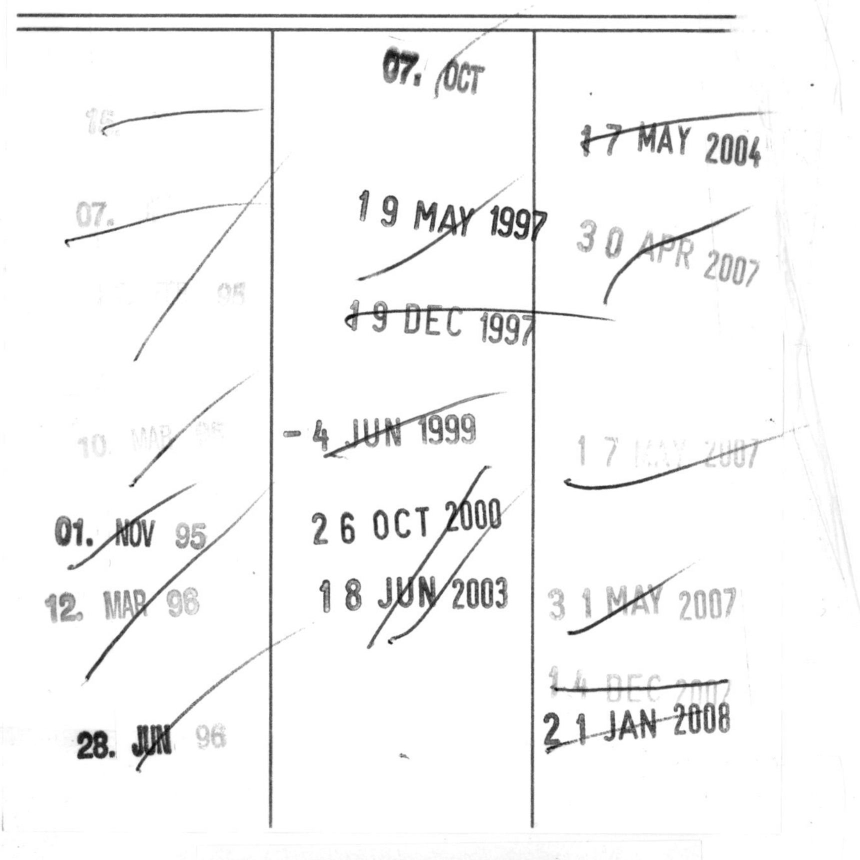

Front cover photograph

Calcium concentrations in cultured cells incubated with the α-adrenergic agonist phenylephrine. The cells have been loaded with the indicator fura-2, for which the fluorescence excitation spectrum changes in response to binding of calcium (Section 4.1.2). In this pseudocolour representation displayed on a computer, areas of high concentration (0.8–2 μM) are shown in reds and purples, areas of low concentration (<0.6 μM) in yellows and greens. The left-hand cell has responded with an increase in cytosolic calcium: surprisingly, the apparently identical cell on the right has not. Courtesy of Kelly Ambler, Martin Poenie, Roger Tsien and Palmer Taylor.

BIOCHEMICAL MESSENGERS

Hormones, neurotransmitters and growth factors

D.G. Hardie
Reader in Biochemistry
University of Dundee

CHAPMAN & HALL
London · New York · Tokyo · Melbourne · Madras

UK	Chapman & Hall, 2–6 Boundary Row, London SE1 8HN
USA	Chapman & Hall, 29 West 35th Street, New York NY10001
JAPAN	Chapman & Hall Japan, Thomson Publishing Japan, Hirakawacho Nemoto Building, 7F, 1-7-11 Hirakawa-cho, Chiyoda-ku, Tokyo 102
AUSTRALIA	Chapman & Hall Australia, Thomas Nelson Australia, 102 Dodds Street, South Melbourne, Victoria 3205
INDIA	Chapman & Hall India, R. Seshadri, 32 Second Main Road, CIT East, Madras 600 035

First edition 1991

Printed and bound in Great Britain at the University Press, Cambridge

ISBN 0 412 30340 X (HB)
0 412 30350 7 (PB)

British Library Cataloguing in Publication Data

Hardie, D.G.
Biochemical messengers.
I. Title
574.87

ISBN 0–412–30340–X
ISBN 0–412–30350–7 pbk

Library of Congress Cataloging-in-Publication Data

Available

To Lyn,

for putting up with the hours

I spent at the word processor

CONTENTS

PREFACE

The central theme of this book is that systems of cell–cell signalling via nerves, hormones, local mediators and growth factors are not distinct phenomena, but branches of one general mechanism. These topics therefore can and should be discussed in an integrated manner, and the division of cell signalling studies into separate pigeonholes such as neuroscience, endocrinology or cancer biology is unnecessary, if not counterproductive. I also believe it to be unfortunate that there is not a collective term to describe neurotransmitters, hormones, local mediators and growth factors, other than clumsy phrases such as "extracellular signal molecule". The lack of a short and distinctive word for these entities genuinely hampers people from thinking about them in an integrated way. Having decided that it was presumptuous to invent a new term, I have chosen in this book the term **first messenger** to cover all types of extracellular signal molecule, because of the widespread acceptance of the term **second messenger** to represent the intracellular signal molecules that are produced in response to many of them. I have given the book the title "**biochemical messengers**", which is a global term to cover both first and second messengers.

The impetus for writing the book came, as must often be the case, when I had to put together a course on cell–cell signalling for biochemistry students at the University of Dundee. Having carried out research for some years in one aspect of this subject (mainly that covered by Chapter 8) I realized that the existing general texts did not cover the area particularly well. This is not meant to imply any criticism of the many excellent general textbooks of biochemistry, molecular biology, cell biology, physiology, pharmacology, neurobiology or cancer biology. The point is that cell–cell communication is a multidisciplinary topic which lies at the interface of all of these subjects. My hope is that this book will act as a bridge between the larger, general texts, and will be of interest to biologists coming to the topic from a variety of different backgrounds.

I am a biochemist/molecular biologist by training (and personally I do not believe in making a distinction between the two), so the emphasis throughout is on explaining events at the molecular level. However, my aim was to provide enough background for a pure biochemist to understand the phyiological basis of each system, and enough explanation of biochemical concepts for a person with a more physiological background to be able to follow the arguments. A certain level of basic biochemical knowledge is assumed. I have made extensive reference to drugs or toxins acting on each system, mainly because these are often important experimental tools, but I hope that pharmacologists will consequently also find the book useful.

I believe this is a good time to write such a book, because recent advances in the application of protein, DNA and electrophysiological technologies now allow a reasonable understanding of the mechanism of action of most first messengers at the molecular level. One problem, i.e. the mechanism of action of insulin and growth factors acting through protein (tyrosine) kinase receptors, remains elusive, although at the time of writing even that seems to be slowly falling into place. In other cases, e.g. the ligand-gated ion channels, steroid hormone receptors or Ca^{2+}-mobilizing hormones, the events of the last ten years have been nothing short of revolutionary.

A brief word about the logic of the chapter organization may be helpful. Chapters 1 and 2 introduce the general concepts of intercellular signalling, and also cover the topic of direct cell–cell communication by cytoplasmic bridges (e.g. gap junctions). The remaining chapters cover the mechanism of action of the first messengers, starting with their structure, synthesis and release, progressing to the target cell and then working from the membrane inwards. Chapter 4 is a slight digression to discuss mechanisms of nervous conduction, and the regulation of the ionic balance of cells. The latter is essential background for several of the later chapters. Chapter 9 discusses the regulation of cell growth and division – due to restrictions on length I have reluctantly omitted the recent exciting advances in understanding of the cell cycle, because it is not yet clear how these relate to the action of first messengers. Chapter 10 covers the special case of messengers acting via nuclear receptors.

Finally I must thank all of the colleagues who helped in the production of this book. Special thanks go to all of those who contributed photographs, this being acknowledged in the figure legends. I am also very grateful to the following persons who provided advice and read first drafts of the indicated chapters for me: Nigel Unwin and Howard Evans (1); David Nicholls (4 and 5); Colin Watts (5); Peter Downes (6 and 7); Philip Cohen (7 and 8); Chris Marshall (9); Olaf Pongs (10). The responsibility for the accuracy of the text rests, however, entirely with myself: I would be very grateful to hear of any errors or serious omissions.

Grahame Hardie
Dundee, November 1990

ONE

INTRODUCTION: CELL–CELL SIGNALLING BY CYTOPLASMIC BRIDGES

1.1 *Why do cells need to communicate?*

In unicellular organisms, such as bacteria or yeast, every cell has to be capable of the full panoply of biochemical processes that the organism can perform. In these organisms, growth and metabolism is usually controlled only by the availability of nutrients, with each cell competing for these nutrients with its neighbours of the same and other species. The only occasion when cooperation between cells may be evident is when nutrients become limiting, and sexual processes may become necessary for the formation of spores which can survive the leaner times ahead. In yeast, this involves a true form of intercellular communication via the mating factors, which are analogous to hormones. However this is one of the few well-documented cases of intercellular signalling between unicellular organisms, which generally appear to follow an anarchistic doctrine of 'every cell for itself'.

In multicellular organisms, by contrast, things are different. The labour is divided between differentiated cells which are specialized for particular purposes, and close cooperation is required between the various cell types in order for the organism to function efficiently. The various methods by which the individual cells signal their needs to each other form the subject matter of this book. This communication is required in order to integrate cellular activity in all of the basic processes of life, e.g. movement, metabolic activity, and growth. The necessity for such coordination can be understood by considering three simple examples:

Coordination of movement

Among the most primitive of multicellular organisms are green algae of the genus *Volvox* (Figure 1.1). An individual *Volvox* is a colony of up to 50,000 cells arranged around the surface of a sphere, with little differentiation of specialized cells evident except for those involved in sexual reproduction. However while each cell can be regarded in one sense as an independent entity, they do have to cooperate in order for the colony to swim through their aquatic environment. Swimming is achieved by beating of the cilia which are present on every cell. Since the cells are arranged around a sphere, this beating clearly has to be carried out in an ordered manner or the colony would expend a great deal of energy in getting nowhere! Beating of the cilia is initiated by ion movements, and adjacent cells are believed to beat their cilia in unison because their intracellular spaces

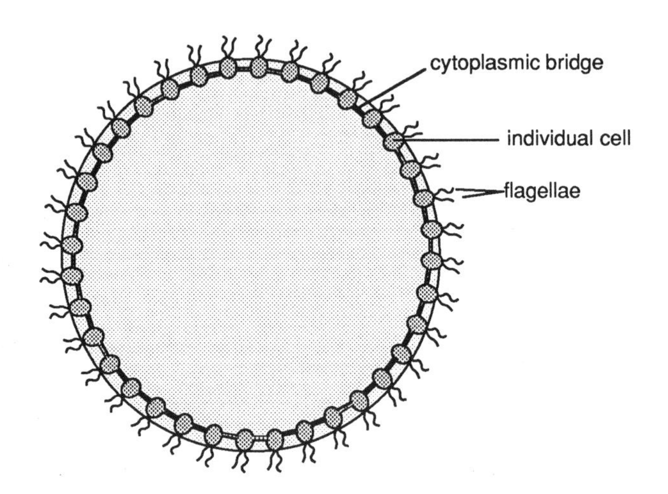

Fig. 1.1 Schematic structure of a member of the genus *Volvox*. The diagram shows a section through the colony, which is actually spherical, and may contain many more cells than shown here.

are connected by cytoplasmic bridges (see Section 1.2.1 for a detailed discussion of this type of intercellular signalling).

Coordination of metabolism

The majority of multicellular organisms are more differentiated than the colonial green algae, and different cells in each organism are often specialized for distinct metabolic processes. In mammals, adipose cells have the job of storing high energy fuel (fatty acids, in the form of triacylglycerols) which are required during periods of starvation or exercise. However the adipose cells themselves have a very low energy requirement, and the fatty acids are required instead by other cells (particularly muscle). There must therefore be a means for the rest of the body to signal these requirements to the adipose cells. This is achieved by changes in the concentration of chemical messengers such as insulin (Chapter 2).

Coordination of growth

Unicellular organisms grow in a medium in which nutrients are often limiting: when nutrients are available the cells normally grow and divide in an unrestricted way in order to outpace their competition (Figure 1.2). In most multicellular organisms the situation is reversed: the cells are constantly bathed in a nutrient-rich medium (the blood or extracellular fluid), but unrestricted growth would disrupt the structural and functional integrity of the organism, and the consequences of loss of growth control in higher animals can be appreciated from the destructive effects of cancer. In multicellular animals, cells do not grow and divide in the absence of the

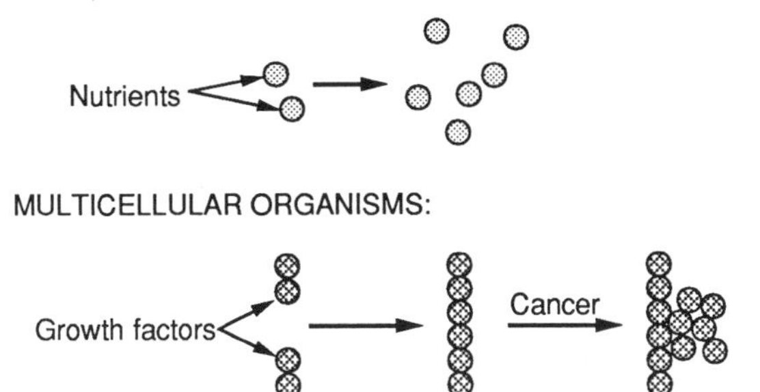

Fig. 1.2 Control of cell proliferation in unicellular and multicellular eukaryotes.

extracellular signals known as growth factors (Figure 1.2), despite the continuous presence of nutrient-rich medium (Chapter 9). Such growth and division does not normally occur in adult animals except for replacement of cells which have a limited lifetime (e.g. blood, skin and gut cells), or for repair of wounds, when growth factors appear to be released locally. Cancer results from a failure in this normally stringent control of cell proliferation, and will be discussed further in Chapter 9.

1.2 *Three basic types of intercellular signalling*

Signalling between cells of multicellular organisms appears to occur by one of three general mechanisms which are illustrated schematically in Figure 1.3. Firstly, cells which are in direct contact with each other can communicate via cytoplasmic bridges, which allow intracellular signal molecules to pass from one cell to the next without being secreted into the extracellular fluid. These bridges are found in even the simplest of multicellular organisms, as discussed in the previous section

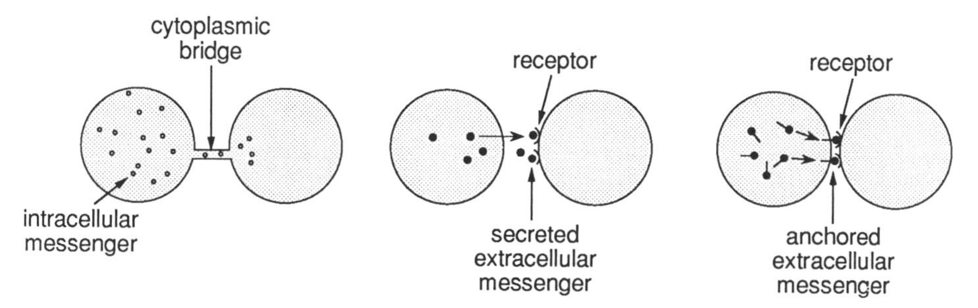

Fig. 1.3 Three possible mechanisms of cell–cell signalling.

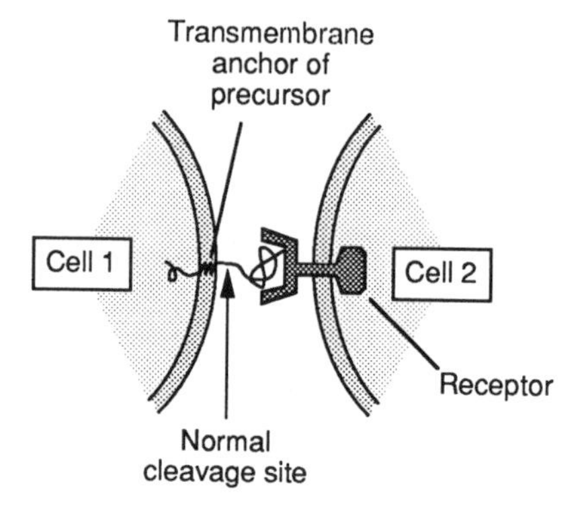

Fig. 1.4 Cell signalling via a membrane-anchored growth factor, transforming growth factor-α.

for *Volvox*. In higher animals the bridges are known as **gap junctions**, whose structure and function are discussed in detail in Section 1.3. Secondly, cells can communicate via extracellular signal molecules, or **first messengers**, which are secreted by one cell and can cause a response in another cell which may be at any distance from a few micrometres to many metres. These first messengers are introduced in Chapter 2, and form the major topic of discussion in the remaining chapters of the book. The third possibility, that integral surface molecules on one cell could signal, perhaps via a change in conformation, to receptor molecules on a neighbouring cell, has until recently been a theoretical possibility rather than an established mechanism. However it has been reported that the precursor form of transforming growth factor-α (TGF-α: Chapter 9), which is anchored in the membrane of the cell in which it is synthesized by a single transmembrane sequence, is biologically active on neighbouring cells. In order to demonstrate this, the cloned precursor DNA was expressed in very large amounts from high copy plasmids, and to prevent the release of the mature, soluble form of TGF-α, which normally occurs by proteolysis and would interfere with the experiment, the cleavage sites had been eliminated by genetic engineering (Figure 1.4). Whether these elegant but artificial experiments reflect a mechanism which occurs *in vivo* is still not clear. One problem is the question of how the signal would be switched off, since the normal pathway of degradation of a first messenger involves uptake into the target cell (Section 6.5).

1.3 *Intracellular signalling via cytoplasmic bridges*

1.3.1 *The structure of the gap junction*

When examining tissues of multicellular animals, electron microscopists noticed

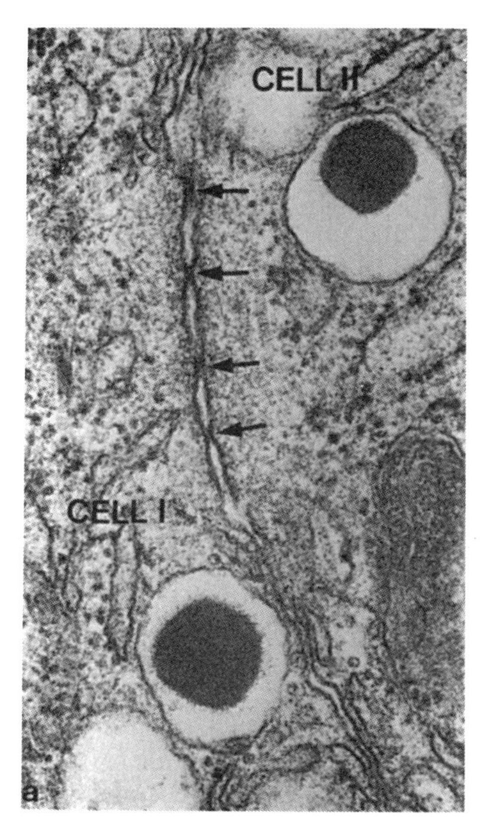

Fig. 1.5 Electron micrograph of the region of contact of two insulin-secreting β-cells in the pancreas. Arrows indicate gap junctions. Reproduced from Diabetes (1982) **31**, 538–565, courtesy of Lelio Orci.

that many different types of cells adhered together via two distinct types of structures which were named according to their appearance. **Tight junctions** were so-called because they form a zone of continuous electron density between cells. They are found particularly between cells of epithelial sheets, and are now believed to ensure that molecules which pass from one body compartment to another (e.g. from the gut to the blood in the intestine, or from the blood to the bile in the liver), can do so only **through** the cells rather than between them. This is essential to ensure that the epithelial sheets can function as selective, controllable permeability barriers between the body compartments. **Gap junctions** (Figure 1.5) were named because when seen in cross-section, they appeared as areas where the plasma membranes of the neighbouring cells were closely apposed, but with a very narrow (2–4 nm) gap of low electron-density remaining between them. When seen *en face* by freeze-fracture or negative-stain electron microscopy, gap

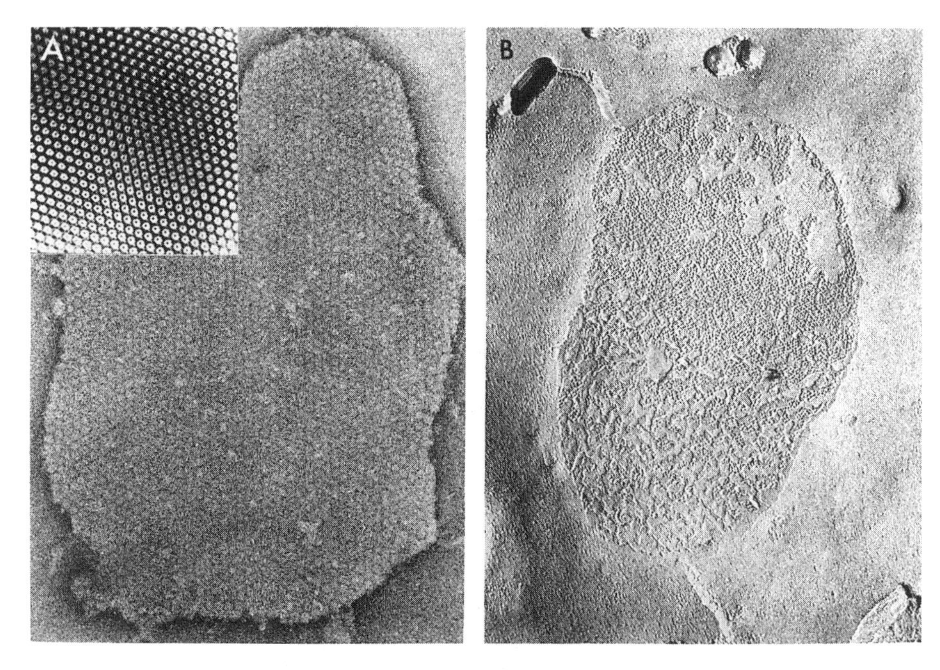

Fig. 1.6 (A) Negatively-stained and (B) freeze-fractured electron micrographs of gap junctions purified from rat liver and heart. The hexagonal packing of the connexons is particularly evident in the inset in (A), which has been optically filtered. Courtesy of Howard Evans.

junctions are seen to be composed of regular hexagonal arrays of tightly packed protein units or **connexons**, each of which shows a central 'pore' of diameter ~1.5 nm (Figure 1.6). The connexons of neighbouring cells are believed to line up so that the pores are in register, as shown schematically in Figure 1.7.

Gap junctions can be studied by electron microscopy using either negative-staining, or the newer technique of analysing frozen, unstained samples using low intensity electron beams. In either case, the image of any individual connexon is imperfect because of, for example, uneven staining or damage by the electron beam. However, because the connexons form such a regular hexagonal array, it is possible to carry out image enhancement by computing diffraction patterns from the electron micrographs. Any regular, repeating structures will produce diffraction peaks, while irregular, non-repeating structures will not, and are effectively filtered out. By analysing micrographs taken at different angles, and using techniques of Fourier transformation, it is possible to reconstruct a density map of the connexon. Essentially one is removing the noise from the image of any one connexon by averaging the signal from a large number of connexons. The maps (Figure 1.8) clearly reveal six subunits packed in symmetrical fashion around a central pore.

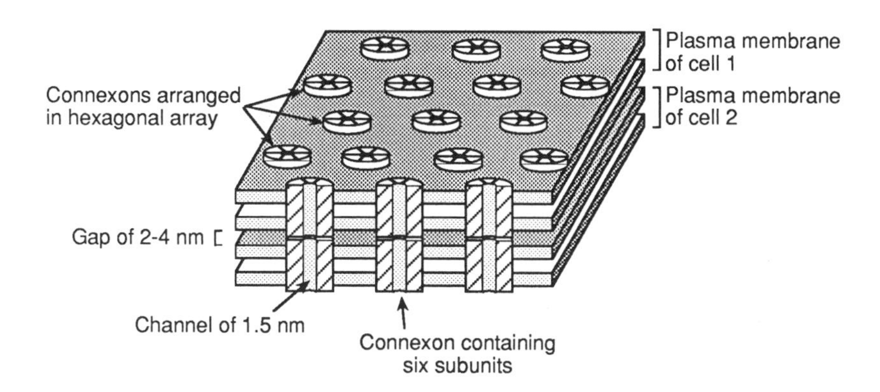

Fig. 1.7 Schematic diagram of part of a gap junction showing how hexagonal arrays of connexons from neighbouring cells line up, to form continuous pores between the cells.

Fig. 1.8 Reconstructed density map of connexons. The view is perpendicular to the plane of the membrane, and the six subunits are clearly seen. Reproduced from Unwin and Ennis (1984) Nature **307**, 609–613, courtesy of Nigel Unwin.

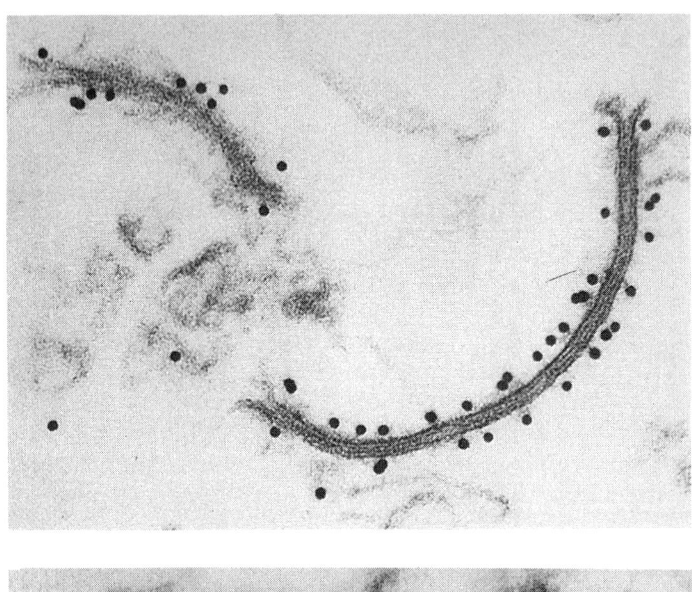

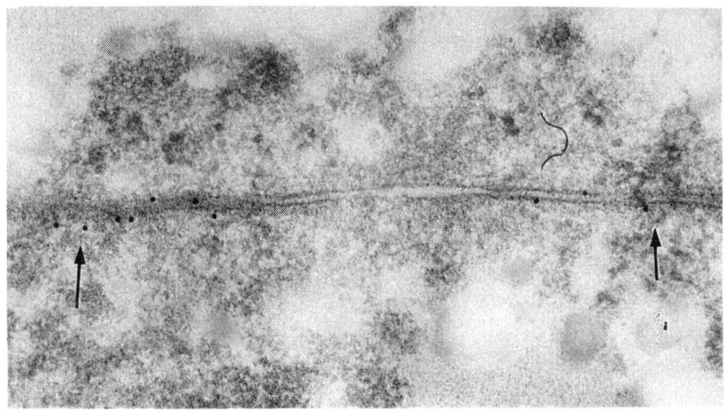

Fig. 1.9 Localization of connexin in isolated rat liver gap junctions (top), and gap junctions in intact rat liver cells (bottom), demonstrated using colloidal gold-labelled antibodies. The black dots are the gold particles. Courtesy of Howard Evans.

Reasonably pure preparations of gap junctions, which are recognizable by their characteristic appearance, have been isolated by subcellular fractionation. Although the nature of the protein components of the connexon has been an area of some controversy, there is general agreement that in rat liver a major component of these preparations is a polypeptide of 32 kDa known as **connexin**. Gap junctions from other cell types contain related connexins of different molecular mass. Antibodies against connexin immunolocalize by electron microscopy to gap junctions (Figure 1.9), suggesting that the polypeptide forms the subunits of the connexon.

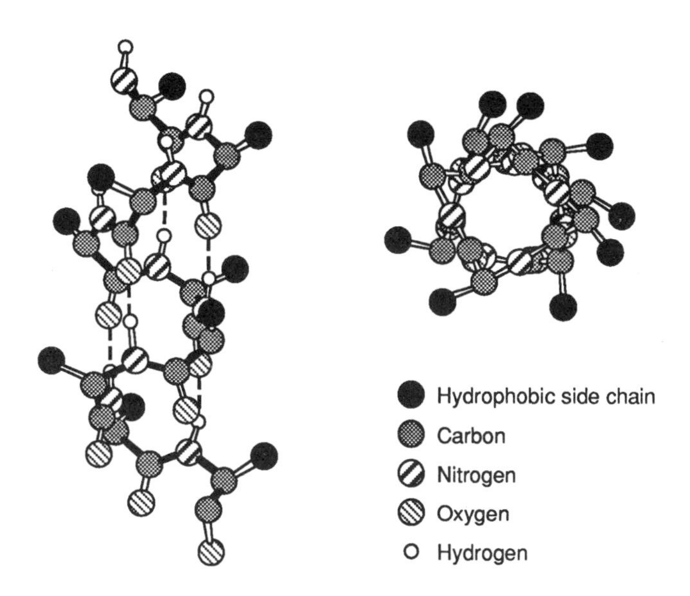

Fig. 1.10 Side view (left) and end-on view (right) of a simplified representation of a hydrophobic α-helix, showing how the hydrophilic backbone >NH and >CO groups form internal hydrogen bonds (dashed lines), while the hydrophobic side chains (black spheres) stick out towards the exterior. The covalent bonds in the backbone of the helix are shown in black.

cDNA encoding rat liver connexin has now been sequenced, which allows model building studies. Very few integral membrane proteins have had their three-dimensional structures solved by X-ray crystallography, but in those that have it is found that the transmembrane regions usually consist of α-helices about 20–22 residues long consisting mainly of hydrophobic residues. This can be rationalized when it is realized that by this arrangement the hydrophilic peptide backbone can form mutual hydrogen bonds within the core of the helix, while the hydrophobic side chains can interact with the membrane lipids (Figure 1.10). It is also possible to accommodate some hydrophilic side chains, particularly if they occur at intervals such that they all lie on the same side of the helix (a so-called **amphipathic helix** – Figure 1.11), where they can interact with another transmembrane helix, or line an aqueous channel in the membrane.

For a protein which has been sequenced but whose structure is not known at the three-dimensional level, it is possible to predict the positions of transmembrane regions by looking for stretches of about 20 hydrophobic amino acids, or amphipathic helices. This is usually carried out by a computer program which plots a **hydropathy profile** (Figure 1.12). Analysis of the sequence of rat liver connexin suggests that it contains four stretches of hydrophobic amino acids

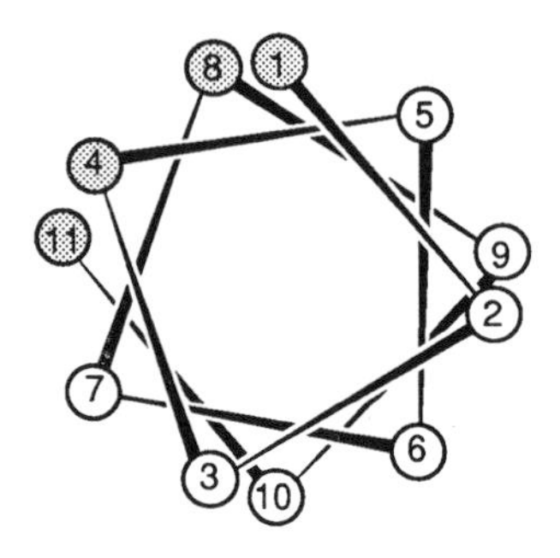

Fig. 1.11 A 'helical wheel' representation of an amphipathic helix. Each residue is represented by a sphere, and the view is along the helix. Side chains with a particular character (e.g. hydrophilic) will line one face of the helix if they occur with an average periodicity of 3.5 residues. In this example, grey spheres are shown at residues 1, 4, 8 and 11.

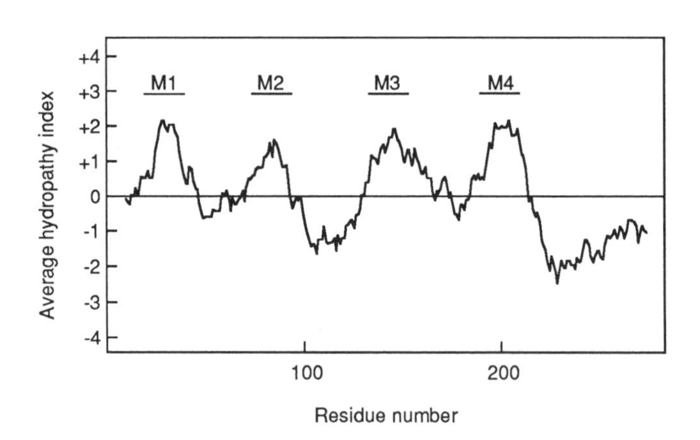

Fig. 1.12 Hydropathy plot on the amino acid sequence of rat liver connexin. Each amino acid residue has been assigned a "hydropathy index" between –4.5 and +4.5, based on its water-vapour partition coefficient and other parameters [Kyte and Doolittle (1982) J. Mol. Biol. **157**, 105–132]. The average hydropathy index of every segment of 19 residues is plotted, and the four putative transmembrane helices (M1 to M4) are clearly seen.

(M1 to M4) which would form transmembrane helices, with rather short loops protruding into the extracellular "gap" (Figure 1.13).

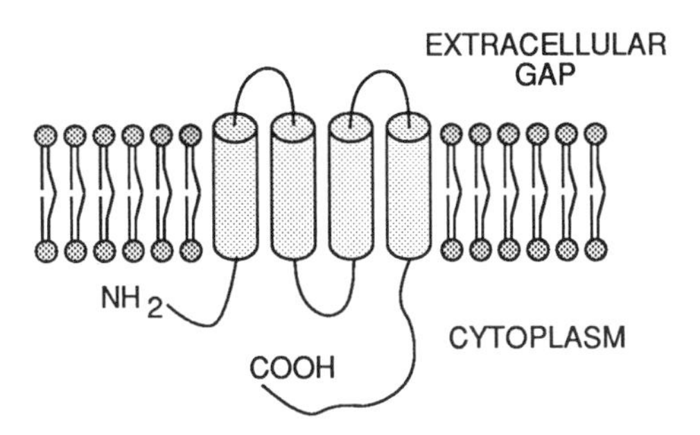

Fig. 1.13 Membrane topology of a single subunit of rat liver connexin based on the hydropathy plot in Fig. 1.12. It is likely that the four helices pack together into a compact structure.

1.3.2 *Gap junctions form cytoplasmic bridges between cells*

Early experiments in which microelectrodes were inserted into neighbouring nerve cells of crayfish showed that they were electrically connected by low resistance junctions, i.e. that small ions could readily pass from one cell into the next. Similar experiments were carried out in which fluorescent dyes were injected into cells and were observed to diffuse into neighbouring cells without passing into the extracellular space (Figure 1.14). Such connections were found between essentially all neighbouring cells within a tissue. Use of marker dyes with different molecular characteristics showed that this passage of molecules was relatively non-specific, but only occurred with molecules up to a molecular weight of about 1000. This suggested that it occurred through pores with a functional diameter of about

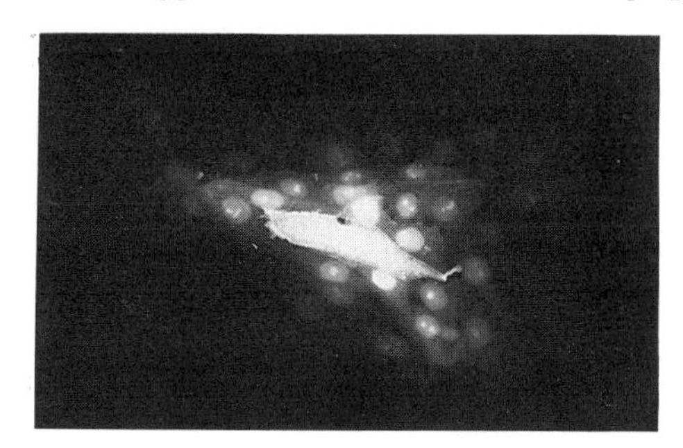

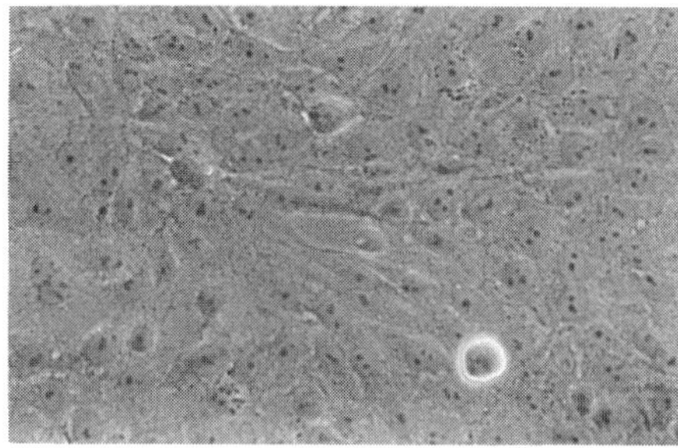

Fig. 1.14 Diffusion of fluorescent dye (Lucifer Yellow, molecular weight 450) between neighbouring cells of a rat liver parenchymal cell line via gap junctions (left). The long, highly fluorescent cell in the centre has been injected with the dye. The same field of cells is seen by phase contrast microscopy on the right. Courtesy of Ephraim Kam.

1.5 nm, consistent with the observed size of the central pores in the connexons of gap junctions.

Evidence has steadily accumulated suggesting that the gap junctions are indeed the functional pores between cells. When cultured cells, separated from their neighbours by proteolytic digestion, are reseeded onto a tissue culture dish they spontaneously form gap junctions when they grow together, and the time of appearance of the junctions correlates with the reappearance of functional pores as assessed by the passage of microinjected dyes. More recently, antibodies against connexin have been shown to block intercellular communication when injected into cells. The candidacy of connexin as the junctional protein is also supported by reconstitution studies. Functional channels form when the purified protein is inserted into lipid bilayers, and cell–cell channels also form when oocytes are injected with mRNA coding for the protein.

1.3.3 *Functions of gap junctions*

The permeability properties of gap junctions show that small molecules such as ions or metabolites can cross directly from one cell to the next, but macromolecules such as proteins or nucleic acids cannot. One can envisage that this form of intercellular communication would be of value in several ways:

1) *Movement of ions*

   Transmission of impulses along nerves is due to movement of ions across the plasma membrane (Chapter 4). Although most mammalian nerve cells communicate via secreted chemical messengers or **neurotransmitters** (Chapter 2), certain nerve cells, particularly in lower animals, do communicate directly by cytoplasmic bridges, and this would allow the impulse to spread in a particularly rapid and reliable manner. Another example is the contraction of myocardial cells in the atrium of the heart, which is initiated by an influx of ions (**depolarization**) into cells at a place called the sinu-atrial node, the natural pacemaker of the heart. This depolarization is then propagated via gap junctions throughout all of the cells of the atrial wall, leading to a wave of contraction. Similar coordination of ion movements may also help to explain the controlled beating of the cilia in *Volvox* (Section 1.1).

2) *Movement of metabolites*

   Although cells in multicellular organisms tend to be organized in such a manner that none are too distant from their source of nutrients (e.g. the bloodstream), diffusion of metabolites through gap junctions would ensure that they are shared between neighbouring cells. An elegant demonstration of this phenomenon in cultured cells came from the use of mutant cells that lacked the enzyme thymidine kinase, and were therefore unable to incorporate radioactive

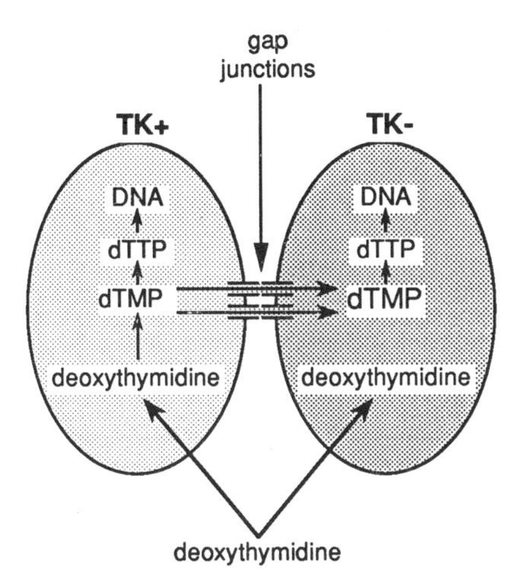

Fig. 1.15 Metabolic cooperation between cells via gap junctions, demonstrated by co-culture of thymidine kinase-deficient (TK–) and normal (TK+) cells. TK– cells cannot incorporate radioactive deoxythymidine into DNA (which is readily detectable by autoradiography). They can do so if deoxythymidine monophosphate (dTMP) is transferred from a neighbouring normal cell via gap junctions.

thymidine into DNA, a process that can be readily visualized in the light microscope by autoradiography. These cells could incorporate radioactive thymidine into DNA if they were seeded onto the same tissue culture dish as wild type cells containing thymidine kinase, but only when they were observed to form junctions with the normal cells (Figure 1.15). The normal cells would be able to form thymidine phosphates which could then pass into the mutant cells through gap junctions, and be utilized for DNA synthesis.

3) *Movement of second messengers and morphogens*

Second messengers are small signal molecules which are formed inside cells in response to extracellular signal molecules (or **first** messengers) acting on the outside of the cell. They are discussed in much more detail in Chapter 7. The classic example is cyclic AMP which, like most second messengers, is a small molecule well within the size range which could pass through the presumed pores in gap junctions. This raises the possibility that first messengers acting on one cell could cause increases in second messenger concentration in several neighbouring cells also. Experimental support for this idea came from experiments in which two mammalian cell types which produced intracellular cyclic AMP in response to different external messengers were grown together

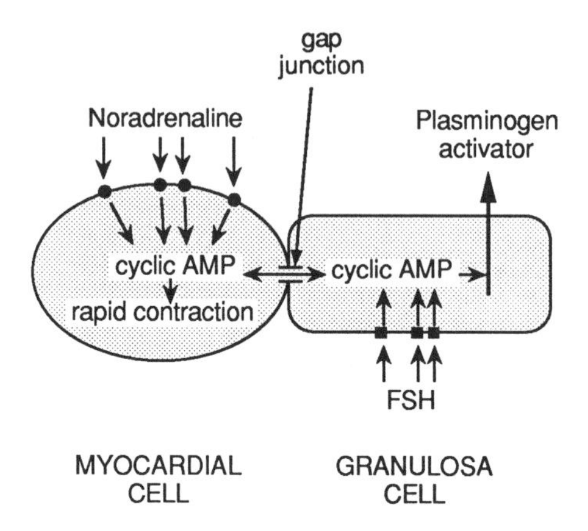

Fig. 1.16 Passage of cyclic AMP between cells, as demonstrated by co-culture of myocardial and ovarian granulosa cells. The culture produces plasminogen activator in response to noradrenaline, and the myocardial cells contract in response to follicle-stimulating hormone (FSH), but only when gap junctions have been formed.

in culture. The cell types used were: 1) myocardial cells, which are the muscle cells of the heart, and which beat more frequently in response to the first messenger noradrenaline; and 2) ovarian granulosa cells, which play an accessory role in the function of the ovary, and which produce a protein involved in lysis of blood clots (**plasminogen activator**) in response to a different messenger molecule, follicle stimulating hormone (FSH). These different responses to first messengers in myocardial and granulosa cells are mediated by the same intracellular second messenger, cyclic AMP. When seeded onto the same tissue culture dish, myocardial and granulosa cells formed functional junctions as assessed by their ability to transfer labelled nucleotides to each other. Intriguingly, addition of FSH to these co-cultures caused the myocardial cells to beat more frequently, while addition of noradrenaline caused release of plasminogen activator (Figure 1.16). This only occurred when the heterologous cells were actually in contact, and various control experiments suggested that it was due to transfer of cyclic AMP between cells rather than to fusion of cells or indirect transfer of electrical signals. These experiments provided elegant confirmation of the idea that second messengers can pass between cells via gap junctions.

There is also preliminary evidence that gap junctions are involved in the process of pattern formation during embryogenesis, in which cells differentiate according to their position within the embryo. Cells in embryos of all but the simplest animals

are found in distinct groups or **segments**, and the developmental fate of a particular cell depends on the segment in which it is located. It has been shown that while dye molecules diffuse freely between cells within a segment, diffusion across segment boundaries is restricted. Diffusion of **small** molecules does seem to occur relatively freely across boundaries, so the barrier to diffusion results from regulation of the permeability of gap junctions (see below) rather than their complete loss. Within one segment, cells do still develop in different ways, and it speculated that this is due to gradients of signal molecules (**morphogens**) set up by diffusion through gap junctions. Consistent with this idea, injection of antibodies against connexin into individual cells of early embryos blocks dye diffusion through gap junctions and also interferes with normal development. For example, incorporation of antibodies into one of the eight cells of an early embryo of the toad, *Xenopus*, led to development of tadpoles which showed abnormalities in a unilateral manner (Figure 1.17). This was despite the fact that the antibodies were probably only affecting the gap junctions briefly during the first few hours of development.

1.3.4 *Regulation of gap junctions*

As discussed above when considering segmental boundaries in embryos, there is now good evidence that intercellular communication via gap junctions can be regulated. Several factors have been shown to decrease the permeability of junctions, including a lowering in pH, and a rise in cytosolic Ca^{2+} ion concentration. Connexins have also been shown to be covalently modified by phosphorylation by e.g. cyclic AMP-dependent protein kinase (see Chapter 8), and there is evidence that cyclic AMP affects junctional conductance. A good example occurs in fish retina, where the **horizontal cells** receive nervous inputs from both the photoreceptor (light-detecting) cells, and from other cells which secrete the first messenger **dopamine**. The horizontal cells appear to be involved in controlling the lateral spread of inhibitory signals within the retina, and thus help to determine how many ganglion cells are stimulated by a beam of light (Figure 1.18). This lateral spread is brought about by ion movements through gap junctions. Dopamine acts on the horizontal cells to increase the intracellular messenger cyclic AMP, and hence causes phosphorylation of the gap junctions. This event, which can be mimicked experimentally by microinjection of purified protein kinase, causes closing of the junctions and limits the lateral spread of the signal.

How do these regulatory mechanisms cause opening and closing of the gap junction channel? Three-dimensional maps of connexons reconstructed from electron micrographs (Section 1.3.1) show that in the absence of Ca^{2+} the subunits are tilted with respect to their axis of symmetry perpendicular to the membrane. In the presence of Ca^{2+}, the subunits are more parallel to this axis, and it is easy to

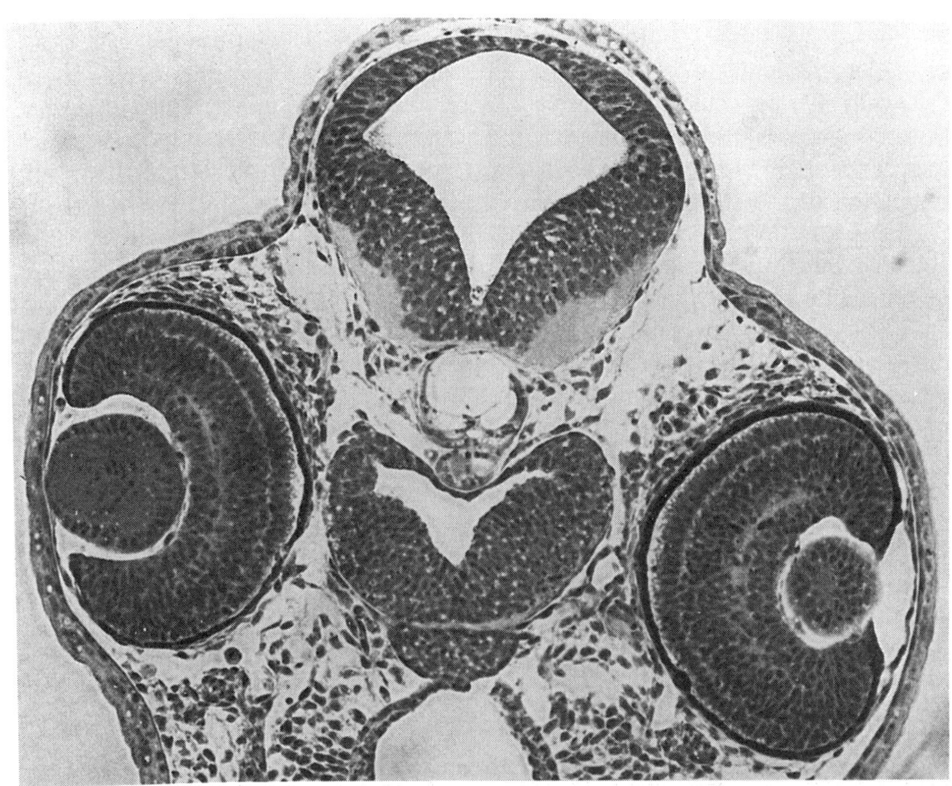

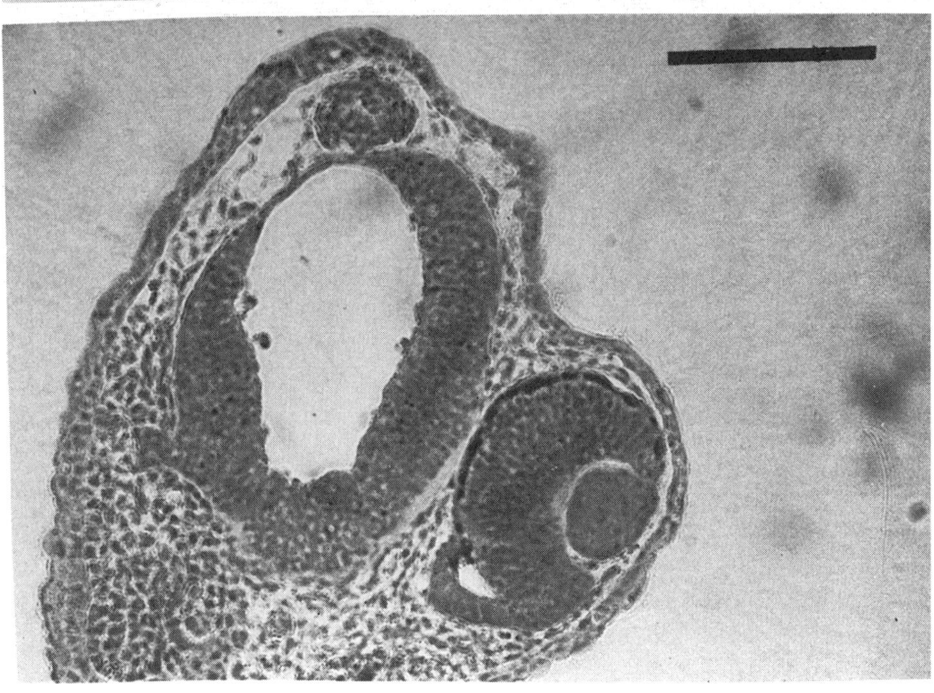

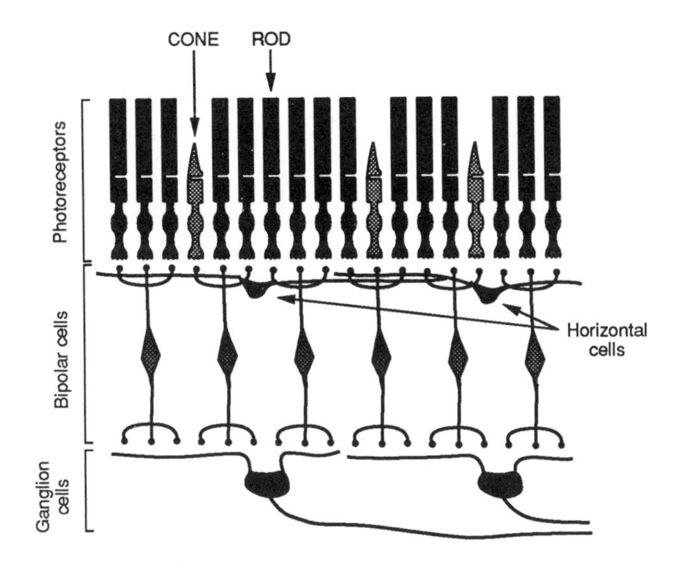

Fig. 1.18 Schematic diagram showing how horizontal cells provide connections between neighbouring groups of photoreceptor/bipolar cells in the retina.

visualize how this could cause narrowing of the channel (Figure 1.19). By this mechanism, very small cooperative changes in the interactions of the subunits could produce relatively large changes in the pore size.

One occasion when closure of gap junctions may be particularly important would be when an individual member of a cell layer dies. One of the first thing that happens when a cell dies is that there is a rapid rise in intracellular Ca^{2+} ion concentration, due probably to failure of the plasma membrane Ca^{2+} pump which normally maintains the concentration in the cell interior 10,000 times lower than that outside (Chapter 4). This rise in Ca^{2+} may cause closure of the gap junctions, preventing Ca^{2+} itself, and other potentially toxic substances, diffusing into healthy cells.

Fig. 1.17 (left) Micrograph showing the developmental effects of injecting antibody against connexin into one cell of an 8 cell stage frog embryo. The eye and brain on one side of the tadpole fail to develop (bottom): a control injected with pre-immune serum is shown at the top. Reproduced from Warner, Guthrie and Gilula (1984) Nature **311**, 127–131, courtesy of Anne Warner.

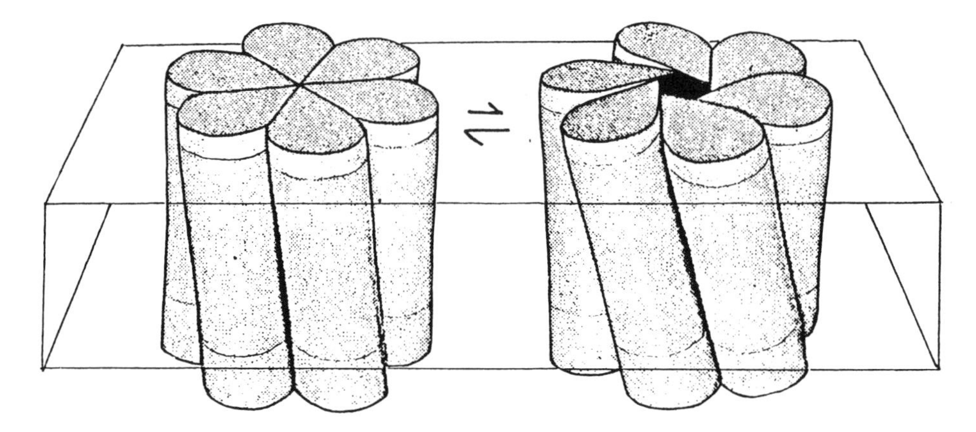

Fig. 1.19 Schematic models of the conformation of the connexons in the presence (left) and absence (right) of Ca^{2+}. Small changes in intersubunit contacts cause tilting of the subunits and opening and closing of the channel. Courtesy of Nigel Unwin.

SUMMARY

Individual cells in multicellular organisms need to communicate with each other in order to coordinate the basic processes of life, e.g. movement, metabolic activity and growth. This intercellular communication occurs in one of three ways, i.e. via cytoplasmic bridges (gap junctions), via secreted chemical messengers and, more speculatively, via messenger molecules anchored on the cell surface. Gap junctions are constructed from structures called connexons, which are assembled from six copies of an integral membrane polypeptide. The connexons from neighbouring cells line up, in gap junctions, to form continuous pores which allow diffusion of substances of molecular weight <1000 between cells. This may be important for direct transfer of ions (coordinating electrical activity), metabolites, second messengers and morphogens between cells. The permeability of gap junctions can be regulated through alterations in pore size that are probably produced by small cooperative changes in subunit interactions.

FURTHER READING

1. "Membrane-anchored growth factors work!" – minireview of studies using anchored TGF-α. Steele, R.E. (1989) Trends Biochem. Sci. **14**, 201–202.
2. "Gap junctions: towards a molecular structure" – review. Evans, W.H. (1988) BioEssays **8**, 3–6.

3. "The structure of proteins in biological membranes" – review of physical methods for structural analysis of membrane proteins. Unwin, N. and Henderson, R. (1984) Scientific American, February, pp. 56–66.
4. "Two configurations of a channel-forming membrane protein" – structure of gap junctions in the presence and absence of Ca^{2+}. Unwin, P.N.T. and Ennis, P.D. (1984) Nature **307**, 609–613.
5. "Molecular cloning of cDNA for rat liver gap junction protein". Paul, D.L. (1986) J. Cell Biol. **103**, 123–134.
6. "Retinal horizontal cell gap junctional conductance is modulated by dopamine through a cyclic AMP-dependent protein kinase". Lasater, E.M. (1987) Proc. Natl. Acad. Sci. USA **84**, 7319–7323.
7. "Evidence mounts for the role of gap junctions during development" – review. Green, C.R. (1988) BioEssays **8**, 7–10.
8. "Antibodies to gap-junctional protein selectively disrupt junctional communication in the early amphibian embryo". Warner, A.E., Guthrie, S.C. and Gilula, N.B. (1984) Nature **311**, 127–131.

TWO

THE FIRST MESSENGERS – HORMONES, NEUROTRANSMITTERS AND LOCAL MEDIATORS

Intercellular communication via gap junctions, or by the still hypothetical mechanism of anchored cell surface messengers (Chapter 1), is obviously limited to cells which are in contact with each other. The other major mechanism by which cells communicate, which forms the subject of the remainder of this book, is by secreted signal molecules or **first messengers**. This method has the advantage that communication can occur over a great range of distances (from tens of nanometres in the case of a neurotransmitter, to tens of metres in the case of a hormone acting in a blue whale!). Cells in higher animals signal to each other by probably hundreds of different first messengers, with the list of those that have been identified increasing year by year. They can be classified into several groups according to their structures, a topic which is discussed in Chapter 3. A somewhat more fundamental classification into three groups is necessary according to the distance over which the first messengers act, as shown in Figure 2.1. **Hormones** are released, usually by specialized secretory cells in **endocrine glands**, and are carried **by the circulation** (which in vertebrates is the bloodstream) to target cells where they have an effect. **Neurotransmitters** are released at specialized regions (nerve terminals) of nerve cells, and have to diffuse only across a small gap (the **synaptic cleft**) to the target cell. **Local mediators** are released by many cell types (which are not specialized for this purpose) and diffuse through the extracellular fluid to act on cells within the same local area. They are not found in significant amounts in the circulation. Regulation by local mediators is sometimes referred to as **paracrine** control (from the Greek *para-*, beside), to distinguish it from hormonal or **endocrine** control.

The three functional groups of first messengers therefore differ mainly by the medium through which they diffuse: the circulation, the synaptic cleft, or the extracellular fluid. However, classifying a particular first messenger into one of these three groups is not always straightforward, particularly for the distinction between hormones and local mediators. In any case a central theme of this book is the idea that these three groups are not fundamentally different, but that their actions are very similar, and can therefore be discussed in an integrated manner. A nerve cell can be thought of as a sort of endocrine cell which has grown a long process (the **axon**) so that its first messenger is released very close to the target cell, and no longer has to rely on being carried in the blood. The division between

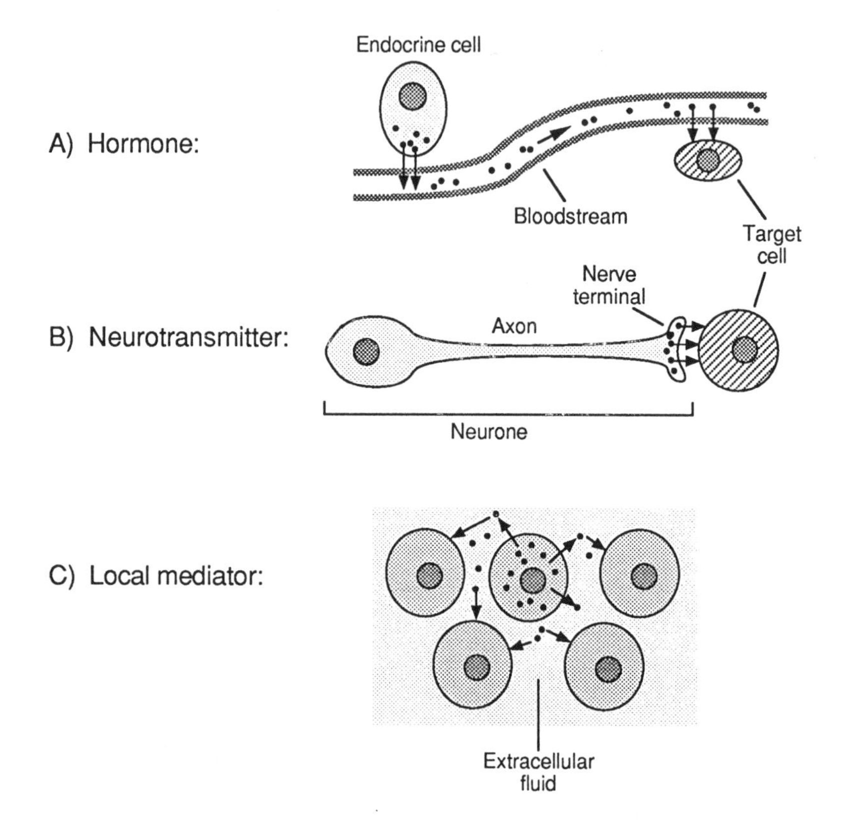

Fig. 2.1 The functional classification of first messengers into hormones, neurotransmitters and local mediators.

neurochemistry and endocrinology is therefore artificial and to some extent counter-productive. This point is also emphasized when one considers that some first messengers can act both as hormones and neurotransmitters: for example, noradrenaline (norepinephrine) is released as a neurotransmitter by sympathetic nerves, and also as a hormone by chromaffin cells in the adrenal medulla. This is also true of many of the peptides which are used as neurotransmitters in the brain, which are often found also in non-neural tissue.

For the remainder of this chapter I will further introduce the concepts of hormones, local mediators and neurotransmitters by considering specific examples. A comprehensive treatment of the complete range of first messengers and their physiological roles is beyond the scope of this book, and the reader who requires this is recommended to consult textbooks of physiology, endocrinology or neurobiology, some of which are given in *Further Reading* at the end of the chapter. However, I will try to provide sufficient background so that readers who

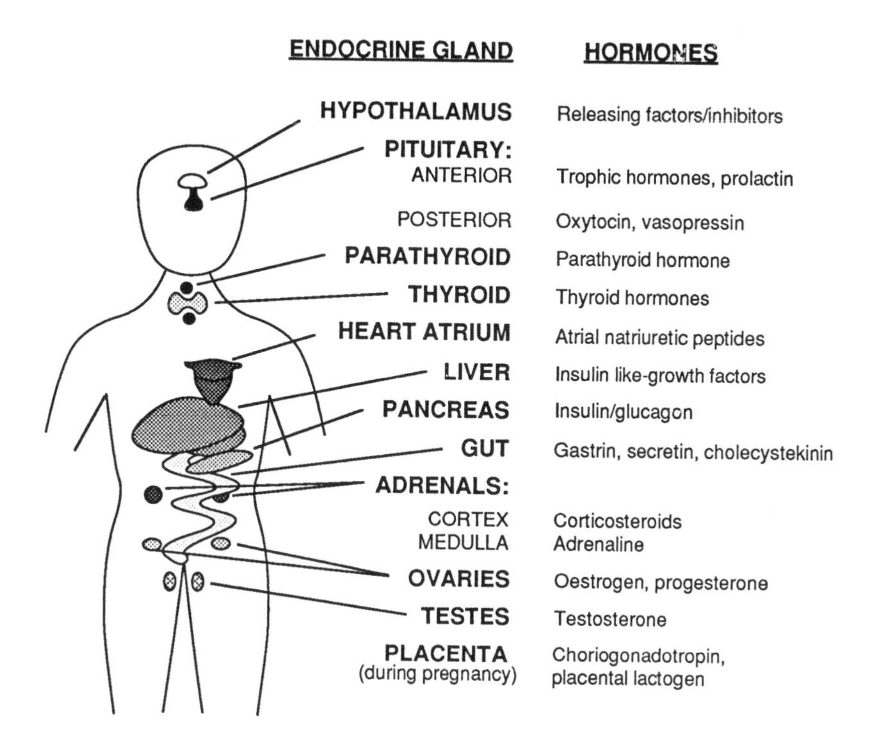

Fig. 2.2 Location of major endocrine organs in the body.

have not previously studied these branches of biology will be able to put the more biochemical and mechanistic treatment of later chapters into a physiological context.

2.1 *Hormones*

The term hormone was introduced by Starling in 1905 and comes from the Greek meaning "excite" or "arouse". Of the functional groups of first messengers, the hormones are the most familiar to the general public, due probably to the widespread pharmacological use and abuse of steroid hormones for diverse purposes, such as contraception and bodybuilding. As stated previously, a characteristic of hormones is that they are secreted into the bloodstream by endocrine cells which are usually specialized for that purpose (an **endocrine** gland secretes a substance into the body interior, as opposed to an **exocrine** gland, e.g. a sweat gland, which secretes substances to the exterior or into the gut). A schematic diagram of some of the major endocrine glands in the human is shown in Figure 2.2. I will now briefly review the physiological roles of the hormones released by each of these major endocrine glands.

2.1.1 *The major endocrine glands and their physiological roles*

The hypothalamus

The hypothalamus is the most important interface between the nervous system and the endocrine system. It secretes various **releasing hormones** and **inhibitors** which, as their names suggest, have antagonistic effects on release of hormones from the pituitary gland (see below). Secretion of these releasing hormones and inhibitors from the hypothalamus is triggered by nervous impulses initiated in the brain.

The pituitary gland

The pituitary gland or **hypophysis** is divided into two lobes, the anterior (**adenohypophysis**) and posterior (**neurohypophysis**). The anterior pituitary receives blood directly from the hypothalamus via blood vessels in the pituitary stalk, and in response to the releasing hormones from the hypothalamus secretes a variety of different hormones, in particular the so-called **trophic hormones**, which stimulate hormone release (and often growth) in endocrine glands elsewhere in the body. Examples of the trophic hormones are:

1) **Somatotropin** (growth hormone), which stimulates release of somatomedins, especially from the liver.
2) **Thyrotropin**, which stimulates growth of the thyroid gland and its release of thyroid hormones.
3) **Adrenocorticotropin**, which stimulates steroid hormone synthesis in the adrenal cortex.
4) **Follicle stimulating hormone** (FSH) and **lutropin** which regulate the production of sex steroids in the ovary and testis.
5) **Prolactin**, which stimulates growth of the mammary gland and milk production during lactation.

The posterior pituitary secretes two major hormones, **vasopressin** and **oxytocin**. Vasopressin is mainly concerned with regulation of blood volume and hence blood pressure. It causes contraction of smooth muscle lining blood vessels, and also promotes retention of water by the kidneys. Both effects increase blood pressure, and the release of vasopressin is under nervous control which is in turn regulated by changes in blood pressure. Oxytocin also causes contraction of smooth muscle, but its major targets are the smooth muscle in the uterus and mammary gland, which contract during childbirth and milk ejection respectively. Infusion of oxytocin is used by obstetricians to induce labour artificially in expectant mothers.

The thyroid and parathyroid glands

The thyroid gland produces the thyroid hormones T4 (**thyroxine**) and T3 (**tri-iodothyronine**) in response to thyrotropin from the pituitary. Thyroid hormones regulate gene expression and have the effect of increasing the metabolic rate in a wide variety of target cells. Abnormal overactivity of the thyroid gland (**hyperthyroidism**) is characterized by a nervous and irritable personality, while underactivity (**hypothyroidism**) is associated with a slow and placid nature.

The thyroid gland also produces **calcitonin**, while the closely associated parathyroid gland produces **parathyroid hormone**. These two hormones act antagonistically to regulate Ca^{2+} and phosphate levels and the formation of bone.

The liver

Although the liver clearly has many other functions and is not normally thought of as an endocrine gland, it does secrete the **insulin-like growth factors** (IGF-1 and IGF-2, also known as the somatomedins). This occurs in response to somatotropin (growth hormone), and the liver appears to be a major, although probably not the only, site of release of these hormones. Although receptors for these hormones are found on many cells and their functions are not completely understood, IGF-1 is thought to be responsible for promoting elongation of bones during growth. Intriguingly, plasma levels of IGF-1 in adolescent boys of the pygmy races are only one-third of those in controls from races of more normal stature.

The pancreas

The bulk of the pancreas consists of exocrine gland tissue which secretes digestive enzymes into the gut. However small spherical regions in the pancreas (the **Islets of Langerhans**) are endocrine glands which secrete hormones into the bloodstream. Around 75% of the cells in the Islets are β-cells, specialized for the secretion of **insulin**, while most of the remaining cells are α-cells, which secrete **glucagon** (Figure 2.3). The Islets receive blood direct from the gut via the **hepatic portal vein**, and insulin is released in response to high glucose and certain amino acids (i.e. after a carbohydrate or protein meal), while glucagon is released in response to low glucose, i.e. during fasting. Insulin and glucagon are a classical antagonistic pair of hormones (Table 2.1). In general insulin promotes uptake of glucose and amino acids and their conversion into stores of energy (glycogen, triacylglycerol and protein): it is the signal for "times of plenty". Glucagon opposes these actions of insulin, although it has no effects on muscle due to a lack of glucagon receptors on these cells: this is consistent with the fact that glycogen in muscle is required as fuel for muscle contraction rather than as a glucose reserve during starvation.

Insulin is of course among the most familiar of hormones to the layman due to the very high prevalence in Western society of the disease diabetes. The less

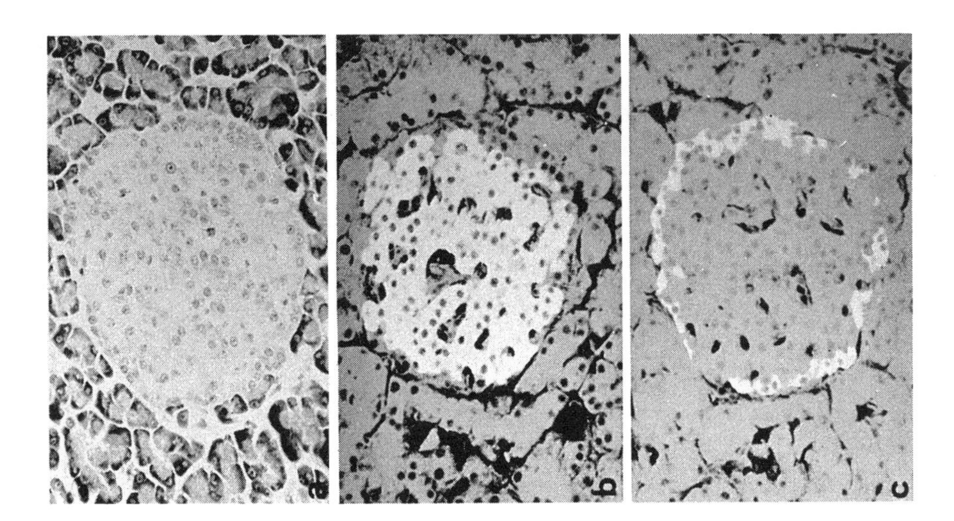

Fig. 2.3 Consecutive serial sections of an Islet of Langerhans from rat pancreas stained with haematoxylin-eosin (left), anti-insulin antiserum (centre) and anti-glucagon antiserum (right). The light areas in the two right hand panels are the location of the fluorescent label on the antibodies, and reveal that insulin is secreted by the β-cells in the centre of the Islet, while glucagon is secreted by the α-cells around the periphery of the Islet. The darkly staining cells outside the Islet are the exocrine pancreatic acinar cells which secrete digestive enzymes into the gut. Courtesy of Lelio Orci.

Table 2.1 Effects of insulin, glucagon and adrenaline on metabolism.

Pathway	Tissue	Insulin	Glucagon	Adrenaline
Fatty acid synthesis	Adipose	↑	↓	↓
Lipolysis	Adipose	↓ *	↑	↑
Glycogen synthesis	Liver	↑ *	↓	↓
Glycogen breakdown	Liver	↓ *	↑	↑
Gluconeogenesis	Liver	↓ *	↑	↑
Glycogen synthesis	Muscle	↑	–	↓
Glycogen breakdown	Muscle	–	–	↑
Protein synthesis	Muscle	↑	–	–

* In these cases insulin has little or no effect on its own, but antagonizes the effects of glucagon or adrenaline.

common, but more life-threatening, form of the disease (**insulin-dependent diabetes**) is believed to be caused by an autoimmune destruction of β-cells, leading to a severe deficiency of insulin, and if untreated results in fatal coma due to metabolic complications of the patient's inability to metabolize glucose. The advent in 1922 of diabetes treatment by injection with animal insulin preparations was one of the great "miracle cures" of medical science. While not in truth a complete cure, it has immeasurably improved the life expectancy and quality of life for diabetics, and is a outstanding example of the many ways in which basic research on biochemical messengers has led to practical benefit.

The gut

Although the main function of the gut is the digestion and absorption of food, it also contains specialized endocrine cells in the stomach and intestine which secrete peptide hormones into the bloodstream. Among these hormones is **gastrin**, which stimulates acid secretion in the stomach, **cholecystekinin**, which stimulates secretion of digestive enzymes by the exocrine pancreas and contraction of the gall bladder, and **secretin**, which stimulates the secretion of pancreatic fluid containing HCO_3^- (to neutralize stomach acid) into the small intestine. The gut also secretes a variety of other peptide messengers whose functions are less well understood.

The atrium of the heart

It was noticed many years ago that the heart muscle cells (myocytes) in the atria, but not the ventricles, contained granules similar to the secretory granules in endocrine cells. Moreover, a change in the number of these granules occurred in response to alteration of sodium and water balance. Subsequently, atrial extracts were shown to cause increased excretion of both sodium and water in the kidney. The agents responsible for this have now been purified, and shown to be a family of closely related peptides termed **atrial natriuretic peptides**. As well as the effects on the kidney described above, they cause relaxation of the smooth muscle lining blood vessels, and inhibition of release of the mineralocorticoid aldosterone (see below). All of these effects tend to lower blood pressure, and the peptides may be released in response to the distension of the atrium caused by high blood pressure.

The adrenal glands

Each of the two adrenal glands are composed of a central **medulla** and an outer **cortex**. Nerve cells from the sympathetic nervous system enter the medulla, and when these are triggered, the chromaffin cells release the catecholamine hormones, chiefly **adrenaline** (called **epinephrine** in the USA) with some **noradrenaline** (**norepinephrine**). This occurs particularly during stress or strenuous exercise, and adrenaline is the hormone that is released when an animal meets another animal whose intentions may not be friendly: the so-called "fight or flight" situation. Some of its effects are familiar to anyone who has been in a stressful situation: thumping heart (increase in the rate and strength of contraction of heart muscle), blushing (dilation of some blood vessels), sweating (stimulation of sweat glands), and dry mouth (inhibition of salivary glands). It also has numerous, less obvious, biochemical effects on metabolism (Table 2.1). It causes mobilization of stores of energy (glycogen and triacylglycerol) and inhibits biosynthetic pathways: the body is generally geared up for rapid action by both heart and skeletal muscle.

The adrenal cortex is specialized for the synthesis of steroid hormones. Steroid hormones act largely via induction or repression of the synthesis of proteins (Chapter 10) and their effects tend to be slower than those of hormones like insulin or glucagon. The adrenal steroids fall into two functional groups which are synthesized in different regions of the cortex, i.e. **glucocorticoids** (e.g. **cortisol**) and **mineralocorticoids** (e.g. **aldosterone**). Synthesis of glucocorticoids and growth of the cells that produce them are stimulated by adrenocorticotropin from the pituitary gland. Glucocorticoids have very diverse effects, but are particularly released during long-term stress or starvation. In general they have anabolic effects on liver and kidney, and catabolic effects on other tissues, e.g. they stimulate the mobilization of fatty acids from adipose tissue and of amino acids from muscle protein. They also have inhibitory effects on cells of the immune system, and have been much used to relieve the symptoms of inflammatory diseases such as rheumatoid arthritis, although not surprisingly there are numerous deleterious side-effects. Mineralocorticoids are released in response to low blood volume or pressure, and promote retention of water by stimulating Na^+ reabsorption by the kidney.

The testis, ovary and placenta

Specialized cells in the testis or ovary are responsible for the production of the sex steroid hormones. The Leydig cells in the testis produce the male sex hormone **testosterone** in response to stimulation by the pituitary hormone, lutropin: testosterone is responsible for the secondary sex changes that occur in males during and after puberty, e.g. growth of facial hair, increase in muscle bulk, and unfortunately for some, baldness! The situation in females is more complex. In humans hormonal levels fluctuate during the menstrual cycle (Figure 2.4). During the first half of the cycle follicle stimulating hormone (FSH) from the pituitary stimulates development of ovarian follicles and their release of **oestrogens**. These steroid hormones activate a process of rapid growth of the uterus and its blood supply. In mid-cycle a burst of release of lutropin from the pituitary stimulates a single mature follicle to release an egg, and the remainder of the follicle to develop into the **corpus luteum**, which releases both oestrogens and **progesterone**. These steroid hormones promote further changes in the uterus and also mammary tissue, preparing the way for pregnancy. If the egg is not fertilized, the corpus luteum regresses, steroid hormone synthesis decreases and the uterus atrophies, accompanied by bleeding (menstruation). If successful

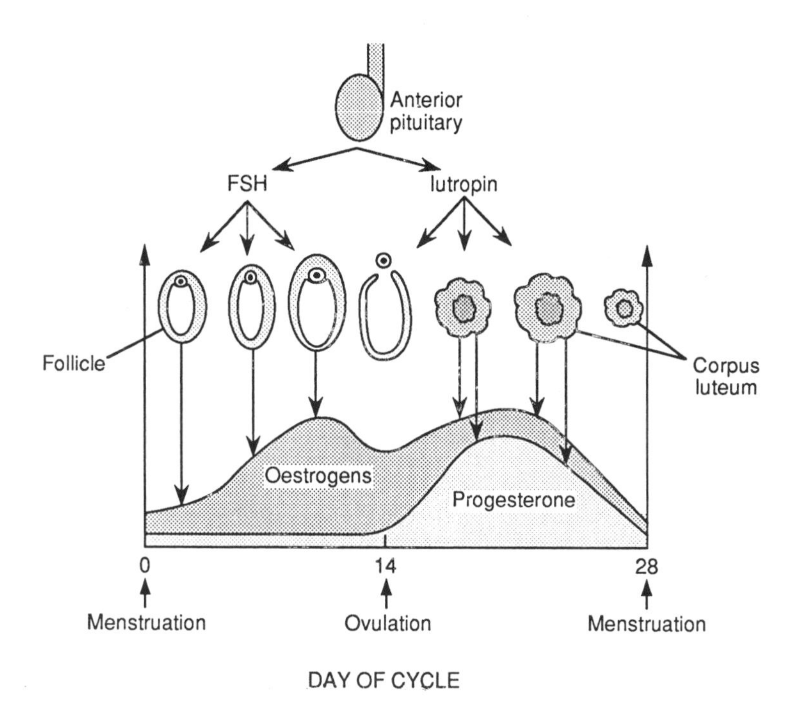

Fig. 2.4 Variation of hormone levels during the ovarian cycle in females.

fertilization does occur, the corpus luteum does not regress and continues to produce progesterone. As the embryo develops the placenta takes over some of the endocrine functions normally undertaken by the pituitary. It releases: (1) **choriogonadotropin**, which is very similar to lutropin, and prolongs progesterone production; and (2) **placental lactogen**, which is closely related to prolactin and further stimulates development of mammary tissue in preparation for birth of the infant.

Another effect of oestrogen and progesterone is to cause feedback inhibition of the release of their respective trophic hormones, FSH and lutropin. As well as playing a part in determining the cyclical nature of release of these hormones, this feedback control ensures that release of a new egg does not occur until all possibility of an existing pregnancy is finished. This forms the basis of the contraceptive actions of these steroid hormones when taken orally.

2.1.2 *The role of hormones in homeostasis*

The above survey of mammalian endocrine glands and their hormone products was

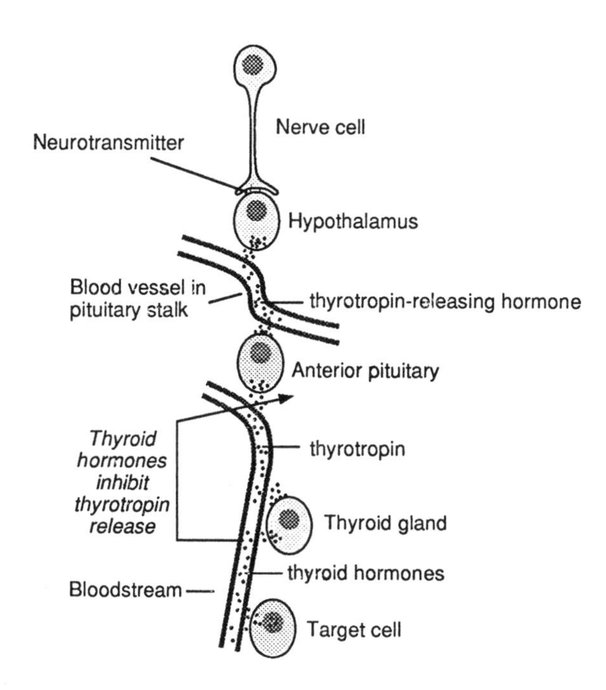

Fig. 2.5 The cascade of hormones triggering release of thyroid hormones, and feedback control on the same system.

necessarily brief, but contained sufficient detail to allow several generalizations about the roles of hormones:

1) Hormones are often concerned with **homeostasis**, i.e. the maintenance of a constant internal environment conducive to optimal function of cells. Examples of this can be seen in the tight control of blood glucose (insulin/glucagon), Ca^{2+} (parathyroid hormone/calcitonin) and Na^{+} (vasopressin/atrial natriuretic peptides/aldosterone).

2) Hormones often achieve homeostasis by working in antagonistic pairs. Insulin/glucagon and parathyroid hormone/calcitonin are prime examples of this.

3) In some cases hormone release is modulated directly by the metabolite which that hormone regulates (e.g. effect of glucose on release of insulin and glucagon). In other cases release is controlled by a complex cascade, or hierarchy, of messengers. An excellent example here is provided by the thyroid hormones (Figure 2.5). In response to a neurotransmitter released at a nerve ending in the hypothalamus, thyrotropin releasing hormone is secreted

into the blood vessel in the pituitary stalk. This in turn stimulates release of thyrotropin from the anterior pituitary, causing release of T3 and T4 from the thyroid gland itself.

4) The synthesis and release of hormones is often itself under homeostatic control. A common mechanism is a feedback inhibition of release of a pituitary trophic hormone by the hormone on which that trophic hormone acts, e.g. inhibition of thyrotropin release by thyroid hormones (Figure 2.5), adrenocorticotropin release by glucocorticoids, FSH release by oestrogen, and lutropin release by progesterone.

2.2 *Neurotransmitters*

As discussed in the introduction to this chapter, neurotransmitters are first messengers released by nerve cells or **neurones**, which can be thought of as being like endocrine cells which have a long extension or **axon** (which can be more than a metre in length) by means of which the first messenger is released very close to the target cell. Although this comparison may be a helpful concept, in truth the neurone is an extraordinary cell quite unlike any other. A distinguishing feature of the nerve cell, other than its great length, is of course its ability to conduct an electrical impulse or **action potential** along the axon. The details of this mechanism will be discussed in Chapter 4: for present purposes the action potential can be thought of as a unit of information sent electrically (analogous to a single binary digit or "bit" of information in a computer) which conveys down the axon the message that a certain amount of neurotransmitter should be released at the other end.

Neurones are of course found particularly in the brain and spinal cord, which together form the **central nervous system** (CNS). In addition the **peripheral nervous system** comprises the nervous tissue which connects body structures to the CNS. Within the central and peripheral nervous system, nerve cells can be subdivided into several classes (Figure 2.6). **Sensory neurones** send information from the various sense receptors back to the CNS. **Interneurones** connect neurones to other neurones, and provide the computational circuits that combine information from a large number of sensory neurones and convert it into appropriate signals for output. **Motor neurones** send this output to skeletal (voluntary) muscles, causing contraction. The **autonomic nervous system** sends output to control involuntary processes such as contraction of smooth muscle in the gut or blood vessels, or secretion by exocrine or endocrine glands. It is divided into two parts, the **sympathetic** and **parasympathetic** nervous systems, which often have antagonistic effects on the same target organ, e.g. contraction or dilation of the pupil in the eye. All nervous tissue also contains accessory cells, called collectively **glial** cells, which provide various service functions for the

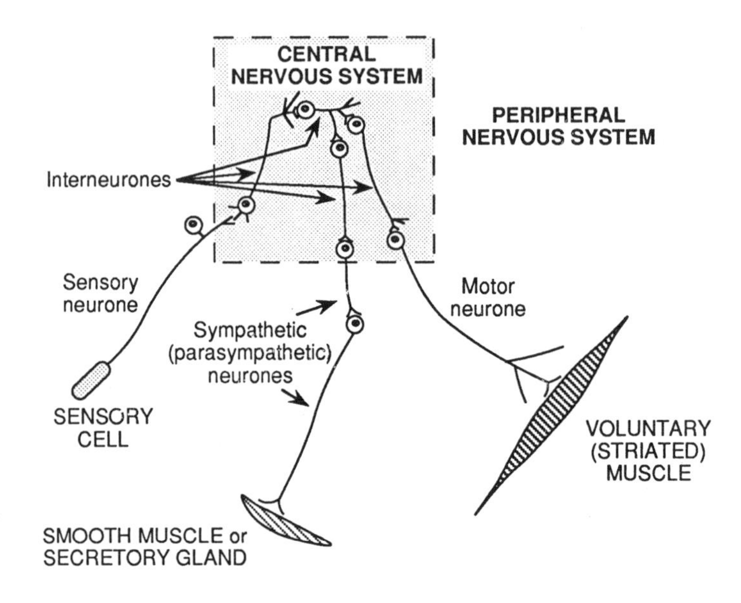

Fig. 2.6 A highly schematic view of the nervous system.

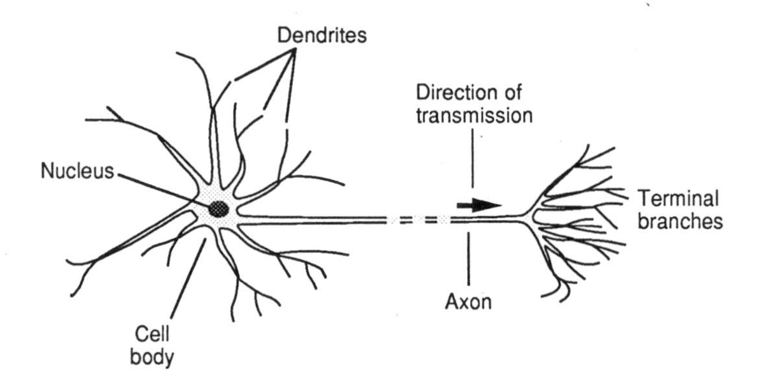

Fig. 2.7 Sketch showing principal features of typical neurone.

neurones themselves.

A schematic diagram of a typical vertebrate neurone is shown in Figure 2.7. The cell contains a single nucleus located in the cell body, which also contains the Golgi apparatus and most of the endoplasmic reticulum and ribosomes. Extending from the cell body, in addition to the axon, are many other fine processes or **dendrites**. These greatly increase the surface area over which the cell body may receive inputs from other neurones or, in the case of a sensory neurone, from sensory receptor cells. Although only a few are shown in the diagram, in reality

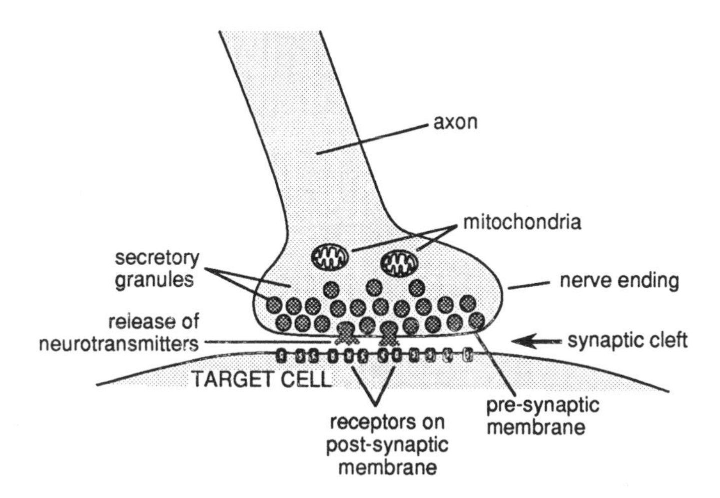

Fig. 2.8 Sketch showing principal features of a typical synapse.

there may be hundreds of branches in the dendrites, receiving inputs from an even larger number of other neurones. As a result of summation of the inputs received in the dendrites and the cell body itself, action potentials are initiated at the base of the axon and pass down it at speeds up to 100 metres per second. The axon normally has many terminal branches which form outputs (**synapses**) where communication with target cells occurs by means of chemical messengers (neurotransmitters).

2.2.1 *Structure of the synapse*

The target cell with which the neurone forms a synapse may be another neurone, a muscle cell, or perhaps a secretory cell. A particularly well-studied example of a synapse is that which forms between a motor nerve and a muscle cell: a **neuromuscular junction**. A schematic diagram of a typical synapse is shown in Figure 2.8. The axon of the nerve cell swells out into a bulbous or flattened **nerve terminal** which lies close to the target cell, separated from it only by the **synaptic cleft** (a gap of rather less than 100 nm, which contains extracellular matrix material). Nerve terminals contain a few mitochondria, plus a large number (probably many thousand) of **synaptic vesicles** which contain stored neurotransmitter. In the case of a neuromuscular junction, each vesicle contains of the order of 10^4 molecules of the neurotransmitter, acetylcholine. The vesicles lie in close proximity to the plasma membrane where it borders the synaptic cleft (**the pre-synaptic membrane**). When an action potential arrives at the nerve terminal, a few hundred of these vesicles fuse with the pre-synaptic membrane to release neurotransmitter into the synaptic cleft. Due to the short distance involved, the neurotransmitters diffuse very rapidly across the cleft and can build up to rather high concentrations. They bind to receptors located on the region of plasma membrane of the target cell which borders the synaptic cleft (the post-synaptic membrane) and activation of these receptors then mediates the effects on the target cell (Chapters 6 to 8).

2.2.2 *Some features of synaptic transmission*

It may be helpful at this point to summarize some of the features of nervous communication and to point out the similarities with, and differences from, intercellular signalling by hormones:

1) Although transmission of information **along** neurones is electrical and does not involve significant changes in chemical concentrations, communication **between** neurones is usually by means of chemical messengers (neurotransmitters) analogous to hormones.

2) While hormones must be carried between the endocrine cell and the target cell by the circulation, neurotransmitters only have to diffuse across the synaptic cleft (less than 100 nm). Synaptic transmission is therefore much more rapid than hormonal communication.

3) Hormones get greatly diluted when they enter the circulation, and so have to be able to affect the target cell at very low concentrations (usually in the nanomolar range). Because of the small volume of the synaptic cleft and their release from preformed vesicles, neurotransmitters build up very rapidly to high concentrations in the cleft (in the millimolar range).

4) Because a second action potential may arrive at the synapse a very short time after the first, it is essential that the neurotransmitter is removed very rapidly to prepare the synapse for the next signalling event. This is carried out either by enzymic breakdown in the synaptic cleft itself (which happens in the case of acetylcholine; Section 3.1.1) or by reuptake and metabolism by the pre-synaptic cell or by neighbouring glial cells. Breakdown of hormones need not be such a rapid event, and is usually mediated by the target cell (Section 6.5).

An obvious question to ask at this stage is why two nerve cells should signal to each other by means of chemical messengers. Since transmission along the nerves themselves is by electrical means, at first sight it might seem more logical for them to communicate electrically, which could be carried out by means of gap junctions. Some neurones (especially in invertebrates) do indeed communicate by that mechanism. Transmission by chemical messengers would seem to be not only slower and probably more costly (synthesis of neurotransmitters takes energy!) but also perhaps more unreliable. Several common mental or nervous disorders in humans, such as epilepsy, can be ascribed to derangement of chemical transmission at synapses.

While some would argue that this question is inappropriate because it is teleological (i.e. it makes the assumption that there has to be a reason why!), the answer in this case seems clear. It is at the synapses that **regulation** of nervous communication occurs. An action potential is indeed like a binary digit or "bit" of information in a computer: it can only have values of "0" or "1". The reason for this "all or none" nature will become clear when the action potential is discussed in detail in Chapter 4. However, the consequence is that only the frequency, and not the strength, of action potentials can be varied in a single neurone. Just as the silicon chips in computers contain logic gates which take decisions according to a summation of inputs of binary digits, so a single neurone may receive inputs (synapses) from hundreds of other neurones. Whether that neurone sends an action potential, and how frequently, is determined by a summation of those many inputs. An important point is that those inputs can be positive or negative in nature: while many neurotransmitters are **excitatory**, i.e. they promote the formation of action potentials in the target neurone, others are **inhibitory**, i.e. they antagonize the effects of excitatory transmitters acting on the same target cell.

Regulation can also occur at the synapse itself, since in some cases neurones form synapses not at the dendrites or cell body of the target cell but at the nerve terminal (Figure 2.9). In this way the regulatory neurone can modulate the amount of neurotransmitter released in response to a particular frequency of action potentials in its target neurone.

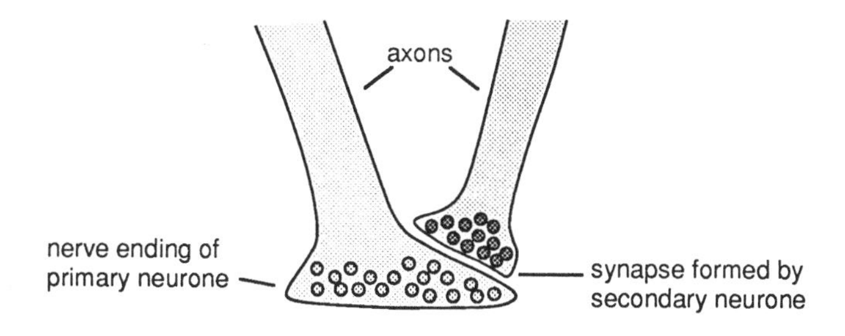

Fig. 2.9 A nerve terminal forming a synapse on the terminal of another nerve.

2.3 *Local mediators (paracrine control)*

The third functional class of first messenger is the type which is released into the intercellular fluid and affects neighbouring cells. Although the existence of this class is now well recognized, there is no agreed single word to describe such an entity, and terms such as local hormone or local mediator are used. Regulation by local mediators is referred to as **paracrine control** (from the Greek *para-*, beside) to distinguish it from **endocrine** control by hormones which are carried in the circulation. Local mediators differ from neurotransmitters in that they are released in a non-directional manner into the extracellular fluid, rather than into the defined space of the synaptic cleft.

Of the three major functional classes of second messenger, the local mediator is probably the least familiar to the layman, and perhaps the best way to understand the concept is to consider some specific examples.

2.3.1 *Nerve growth factor*

Growth factors, which regulate growth and division of cells in multicellular organisms and are discussed in detail in Chapter 9, are generally local in their action and therefore classed as local mediators. They must allow growth to occur in a strictly regulated manner, and in adults their function is probably mainly concerned with repair of wounds, and replacement of cells which turn over rapidly such as blood cells. It would clearly be harmful if they had general (**systemic**) effects throughout the body.

The first growth factor to be discovered was nerve growth factor (NGF), a fact that was recognized by the award of the Nobel prize for medicine to Rita Levi-Montalcini in 1986. NGF, a polypeptide of 118 residues (Section 3.3.1), is a rather atypical growth factor in that it promotes growth but **not** division of its target cells. Its target is neurones of the sympathetic nervous system, and it is absolutely required for the survival of these cells. This was shown by two types of experiment:

1) Injection of anti-NGF antibody into newborn mice caused the death of sympathetic nerves.

2) Sympathetic nerves can be maintained in culture in the absence of other cells but **only in the presence of NGF**.

As well as allowing survival of sympathetic neurones, NGF also directs their growth. This can be shown by culturing the cells in dishes containing compartments divided by barriers which the axons can grow through, but which are impermeable to NGF. Axons that penetrate a barrier into a compartment

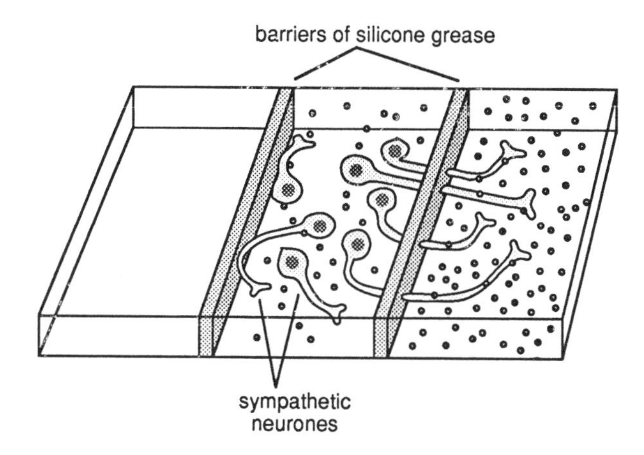

Fig. 2.10 Experiment showing how sympathetic neurones grow towards regions of higher concentration of nerve growth factor.

containing higher concentrations of NGF will grow much faster (Figure 2.10). Similar results can be obtained *in vivo:* if NGF is injected into the brain of newborn mice, sympathetic nerves grow into the central nervous system, where they are not normally found. These results suggest that NGF is secreted by target cells, and that a gradient of increasing NGF directs the axon of a sympathetic neurone to grow towards, and form a synapse with, the target cell. It is clearly important that NGF is released and acts locally, and does not build up to a uniform concentration throughout the body, which would presumably cause sympathetic neurones to grow everywhere! Like most other growth factors, NGF can therefore be regarded as a local mediator.

2.3.2 *Platelet-derived growth factor*

This is another of the classical growth factors. It was discovered during attempts to grow isolated mammalian cells in culture in defined medium. By a painstaking process of trial and error, the early tissue culturists worked out the requirements of different cells for nutrients and vitamins, but still needed to add an ill-defined mixture of protein components, in the form of serum which had been dialysed to remove the small molecules. Serum is made by taking whole blood, allowing it to clot, and taking the buff coloured fluid which remains. An early stage in blood-clotting involves the activation and aggregation of the blood components called platelets, and it became clear that something was being released from the platelets during clotting which stimulated cell growth. This factor (platelet-derived growth factor, or PDGF) has now been purified and shown to be a glycoprotein of 24,000

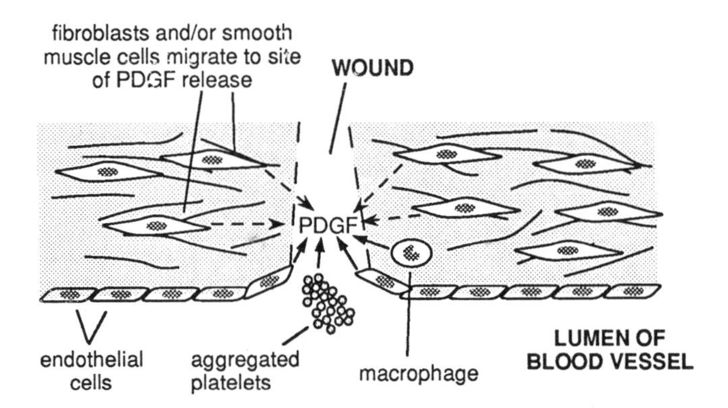

Fig. 2.11 Stimulation of migration and proliferation of fibroblasts by platelet-derived growth factor at the site of a wound in a blood vessel.

daltons. It stimulates growth and division of cells of connective tissue origin such as fibroblasts and smooth muscle cells. In addition, like NGF for sympathetic neurones, PDGF is **chemotactic**, i.e. it stimulates target cells to migrate along a gradient of increasing concentration of the factor.

PDGF is not found in detectable concentrations in blood and its effects are purely local. Indeed if it is injected into the bloodstream, it is rapidly cleared with a half-life of less than 2 minutes. PDGF is released locally at the site of a blood vessel injury, where the clotting pathway would be activated and platelets would aggregate. As well as activated platelets, PDGF is released by certain cells of the immune system (e.g. macrophages) which migrate to the site of an injury, and also possibly by damaged endothelial cells, these being the cells that line the blood vessel. This local release would stimulate the migration and proliferation of fibroblasts at the site of the injury (Figure 2.11). Fibroblasts secrete components of the extracellular matrix such as collagen, which other cell types of the blood vessel wall (smooth muscle and endothelial cells) require in order to divide and finish the repair. There is evidence that PDGF binds tightly to extracellular matrix components at the site of a wound, which may be another reason why its effects are local in nature.

Hence, PDGF is a local mediator thought to be involved in adults primarily in repair of damaged blood vessels. Derangements in the synthesis of PDGF can be involved in cancer of connective tissue (**sarcoma**), a topic discussed in detail in Chapter 9. Secretion of PDGF by macrophages in the artery wall may also account for the proliferation of smooth muscle cells that causes narrowing of arteries in **atherosclerosis**, the condition which leads to heart attacks and strokes.

2.3.3 *Histamine*

Histamine is derived from the amino acid histidine (Section 3.1.3) and is released by mast cells which are activated at the site of an injury or infection. One of its effects is to cause the blood vessel wall to become leaky. This accounts for the redness, swelling and pain at the site of a local infection, due to the fact that blood cells and fluid leaks into the tissue and cannot be returned via the circulation so readily. This of course happens for a good reason, because it allows components of the immune system (e.g. antibodies, complement, lymphocytes and phagocytic white cells) to gain easier access to the site of infection. That histamine is classed as a local mediator is particularly obvious in that its effects are confined to the site of the infection.

Histamine has many other effects, including secretion of fluid from the mucous membranes of the nose, and contraction of smooth muscle in the airways leading to the lung, and is involved in the allergic reactions to pollen and other air-borne antigens in individuals suffering from hay fever and asthma. Drugs which antagonize actions of histamine (anti-histamines) are familiar to the general public through treatment of these conditions. Histamine is also involved in control of acid secretion in the stomach, and the histamine antagonist ranitidine (Box 6.2), used in the treatment of stomach ulcers, is at the time of writing the world's biggest selling drug.

2.3.4 *Eicosanoids (prostaglandins)*

The eicosanoids, including the familiar sub-class of prostaglandins, are derivatives of the unsaturated, 20-carbon fatty acid arachidonic acid (the prefix *eicosa-* comes from the Greek for twenty). Prostaglandins received their name because they were originally detected in seminal fluid, which is secreted from the **prostate gland.** Although they can be detected in blood, their effects generally appear to be localized, and they are not hormones in the sense defined at the beginning of this chapter. Almost all cells are capable of synthesizing some eicosanoids, and they often affect the same cells in which they are synthesized: a so-called **autocrine** stimulation. Their effects are so numerous that it is impossible to summarize their roles in a few sentences. Contraction or relaxation of smooth muscle is a common response, and the contraction of smooth muscle in the airways during asthmatic attacks is caused by eicosanoids released from mast cells, as well as histamine (see above). They often operate in antagonistic pairs, e.g. thromboxane A_2 and prostacyclin. Thromboxane A_2 is released by platelets and triggers platelet aggregation, while prostacyclin is released by the endothelial cells which line blood vessels and has the opposite effect. This interplay controls the extent of a local blood clotting reaction. Prostaglandins are often released in response to hormones, and modulate the response of the target tissue to the hormone. Parturition (childbirth) is initiated by changes in the concentration of steroid hormones which in turn lead to release of oxytocin by the pituitary and prostaglandin $F_{2\alpha}$ ($PGF_{2\alpha}$) by the uterus itself. Oxytocin and $PGF_{2\alpha}$ both stimulate contraction of the smooth muscle in the uterus during childbirth. Injection of prostaglandins into the uterus is commonly used clinically to induce abortions.

Synthesis of prostaglandins is inhibited by the familiar drug aspirin, and this probably accounts for most of the numerous effects of the drug, including the relief of pain, since perception of pain appears to be modulated by prostaglandins. Prostaglandins also inhibit acid secretion in the stomach, and an unfortunate side effect of long-term use of aspirin is promotion of stomach ulcers.

2.3.5 *The purines*

The purines and purine nucleotides adenosine, AMP, ADP and ATP are familiar intracellular metabolites. However it has become clear in recent years that they are also found **outside** cells and at low concentrations have physiological effects on many cells. How they are released from cells is not clear, although it is known that ATP can be released by exocytosis of adrenal chromaffin granules (Section 5.2). Adenosine is released by homogeneous preparations of isolated cells such as adipocytes (fat cells), where it antagonizes the effects of the hormone adrenaline (epinephrine) on the same cells. In this case it is clearly acting as a local mediator, modulating the response to other first messengers.

2.3.6 *A morphogen – retinoic acid*

A morphogen is a special type of local mediator, gradients of which control development. Although several have been defined in lower eukaryotes, the only morphogen identified with reasonable certainty in higher animals is **retinoic acid**. The wings of birds contain three digits, equivalent to our middle three fingers and therefore labelled 2, 3 and 4. In the wing bud of developing embryos, if cells from a certain area on the posterior part of the wing (the **zone of polarizing activity**) are grafted on to the anterior side, extra digits are formed in the order 4,3,2,2,3,4 (Figure 2.12). This suggested that a factor was released at the zone of polarizing activity, and that a gradient of this factor determined the formation of the different digits. The factor has now been identified as all-*trans*-retinoic acid, which is formed by oxidation of retinol (vitamin A) (Figure 2.13). Application of retinoic acid to the anterior part of the wing has the same effect as grafting on cells from the posterior part, and it has also been shown that in the normal wing bud there is a gradient of retinoic acid from 50 $nmol.l^{-1}$ in the posterior part to 20 $nmol.l^{-1}$ in the anterior part.

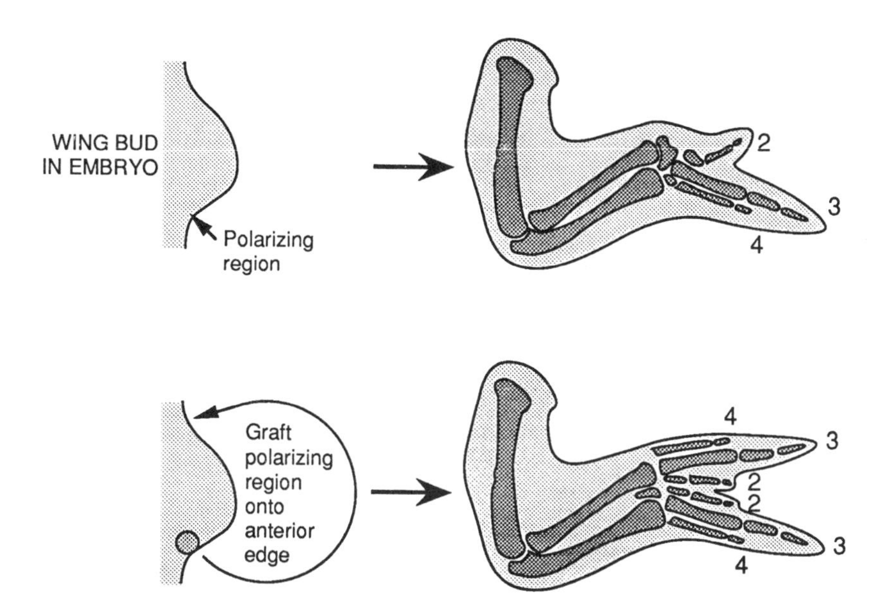

Fig. 2.12 Digits in wing bud of normal chick embryo and in embryo in which a region from the posterior of the wing bud (zone of polarizing activity) has been grafted onto the anterior side.

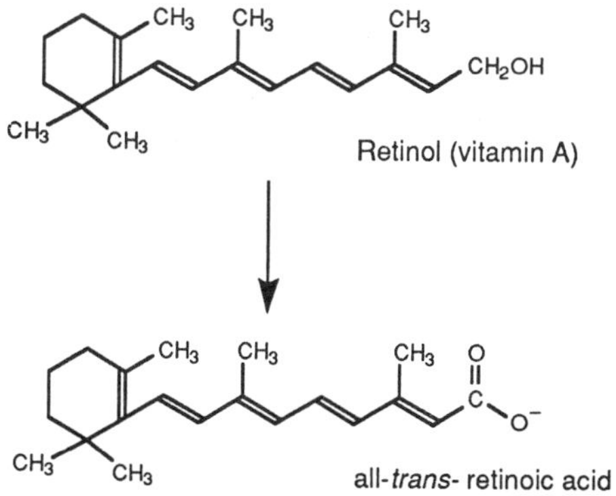

Retinol (vitamin A)

all-*trans*- retinoic acid

Fig. 2.13 Formation of retinoic acid from retinol (vitamin A).

2.3.7 *Immune system mediators*

The migration, division and differentiation of the cells of the immune system is controlled by a complex array of mediators, including antigens, chemotactic factors, interleukins, and interferons. Discussion of these is beyond the scope of this book and the reader is referred to some of the publications listed at the end of this chapter. In most cases the mechanism of action of these mediators is not clear, although this is an area of intense research activity, and it seems likely that many of them will turn out to act via mechanisms analogous to those described in Chapters 6 to 8. What does seem clear is that most of them are local in their effect, whether this is at the site of cell differentiation (e.g. bone marrow, thymus, spleen, lymph nodes) or at the site of an infection or injury. In most cases they should therefore be classified as local mediators.

SUMMARY

Secreted first messengers can be classified into three functional types: hormones, neurotransmitters and local mediators. Hormones are secreted into the bloodstream by cells in endocrine glands which are usually specialized for that purpose, and are carried large distances to the target cells. Neurotransmitters are secreted by nerve cells (neurones) at specialized regions on the cell surface (nerve terminals) into the very narrow space (the synaptic cleft) between the neurone and the target cell. Local mediators are secreted in a non-directional manner into the extracellular fluid by cells which are not specialized for that purpose, and their targets are neighbouring cells in the immediate locality.

Major endocrine organs are the hypothalamus/pituitary, thyroid/parathyroid, liver, pancreas, gut, atrium of the heart, adrenal glands, ovary/testis, and placenta. Hormones are often released in response to cascades of other hormones, and this release is in turn regulated by feedback networks.

Neurones are of several functional types: e.g. sensory neurones which take input from the sense organs back to the central nervous system, interneurones which connect other neurones and perform the complex summations of information which result in the final output, and motor neurones which take the output from the central nervous system to control voluntary muscles. In addition the sympathetic and parasympathetic nerves control the involuntary contraction of smooth muscle, and secretion by both endocrine and exocrine glands. An electrical impulse along a nerve is of an all-or-none nature, analogous to a binary digit of information in a computer. Communication between neurones occurs at synapses via chemical messengers (neurotransmitters), and it is at the synapses that regulation and "decision-making" occurs, in analogy to the logic gates in a computer.

Local mediators form a diverse group of first messengers which includes most growth factors. Nerve growth factor appears to direct the growth of the axon of a sympathetic nerve to its target cell. Platelet-derived growth factor stimulates proliferation of fibroblasts at the site of a blood vessel injury and thus is involved in the initiation of wound repair. Histamine causes the redness and swelling at a site of local infection and aids the access of immune system components to the site of infection. Eicosanoids, including the prostaglandins, are ubiquitous local mediators with very diverse effects. Purines such as adenosine often have physiological effects on the same cells which release them. Retinoic acid is a morphogen, gradients of which control development of digits of the wing in birds. The immune system is controlled by a complex set of local mediators including antigens, chemotactic factors, interleukins, and interferons.

FURTHER READING

1. "Hormones" – general text on the structure, metabolism and physiology of hormones. Norman, A.W. and Litwack, G. (1987) Academic Press, New York.
2. "Molecular aspects of atrial natriuretic peptides" – review. Imamura, T. and Miura, Y. (1986) BioEssays **5**, 66–69.
3. "Elements of Molecular Neurobiology" – introductory text on neurotransmitters. Smith, C.U.M. (1989) Wiley.
4. "The nerve growth factor: 35 years on" – review. Levi-Montalcini, R. (1987) Science **237**, 1154–1162.
5. "The biology of platelet-derived growth factor" – review. Ross, R., Raines, E.W. and Bowen-Pope, D.F. (1986) Cell **46**, 155–169.
6. "Identification and spatial distribution of retinoids in the developing chick limb bud" – retinoic acid as a morphogen. Thaller, C. and Eichele, G. (1987) Nature **327**, 625–628.
7. "Haemopoietic growth factors" – review of growth factors acting on the immune system. Whetton, A. and Dexter, T.M. (1986) Trends Biochem. Sci. **11**, 207–211.
8. "Hormones that stimulate the growth of blood cells" – review of growth factors acting on the immune system. Golde, D.W. and Gasson, J.C. (1988) Scientific American, July, pp. 34–42.

THREE

STRUCTURE AND BIOSYNTHESIS OF FIRST MESSENGERS

Presumably there was a time, back in the days before the development of multicellular eukaryotes, when many of the proteins and metabolic pathways which are familiar today had evolved, but extracellular messengers did not exist. It is therefore not surprising that the first messengers should have evolved by simple modifications of common metabolites such as amino acids, fatty acids and cholesterol. This chapter will discuss how the structures of the first messengers are related to those common metabolites, and the metabolic pathways which interconvert them. The structure and synthesis of peptide messengers exhibits several features which are rather unique and sets them apart from the synthesis of other proteins, and they will be discussed separately at the end of the chapter.

3.1 *Messengers derived from amino acids and other water soluble metabolites*

Many first messengers are derived by simple modifications of water soluble metabolites. Amino acids are the most common precursors, but we start by considering two exceptions to this generalization.

3.1.1 *Acetylcholine*

Acetylcholine is a neurotransmitter acting at the synapse (**a neuromuscular junction**) between a motor neurone and its target cell, skeletal muscle, and at many synapses in the central and sympathetic nervous systems. It is synthesized from choline (a component of the major membrane phospholipid, phosphatidylcholine) and acetyl-CoA, with the formation of an ester linkage (Figure 3.1). This reaction occurs in the cytosol of the nerve ending, and the acetylcholine is then taken up into the synaptic vesicles.

At a neuromuscular junction, the concentration of released acetylcholine can reach high levels (up to 10^{-3} $mol.l^{-1}$). It is obviously essential that the acetylcholine is very rapidly broken down or removed, or the target muscle will go into a sustained contraction or **tetanus** which is very dangerous. Removal is

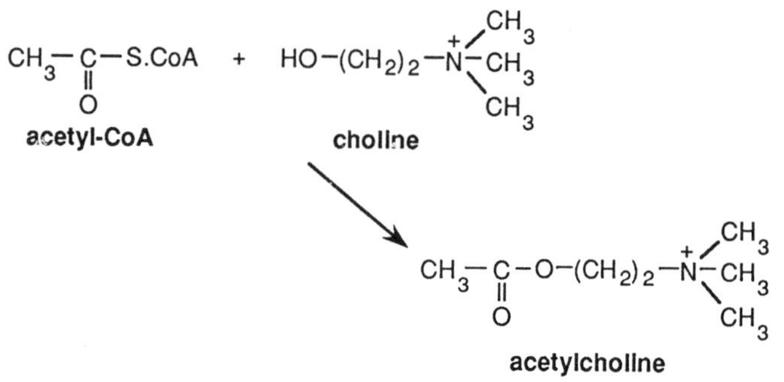

$$CH_3-\underset{\underset{O}{\|}}{C}-S.CoA \quad + \quad HO-(CH_2)_2-\overset{+}{N}(CH_3)_3 \longrightarrow CH_3-\underset{\underset{O}{\|}}{C}-O-(CH_2)_2-\overset{+}{N}(CH_3)_3$$

acetyl-CoA + **choline** → **acetylcholine**

Fig. 3.1 Synthesis of acetylcholine.

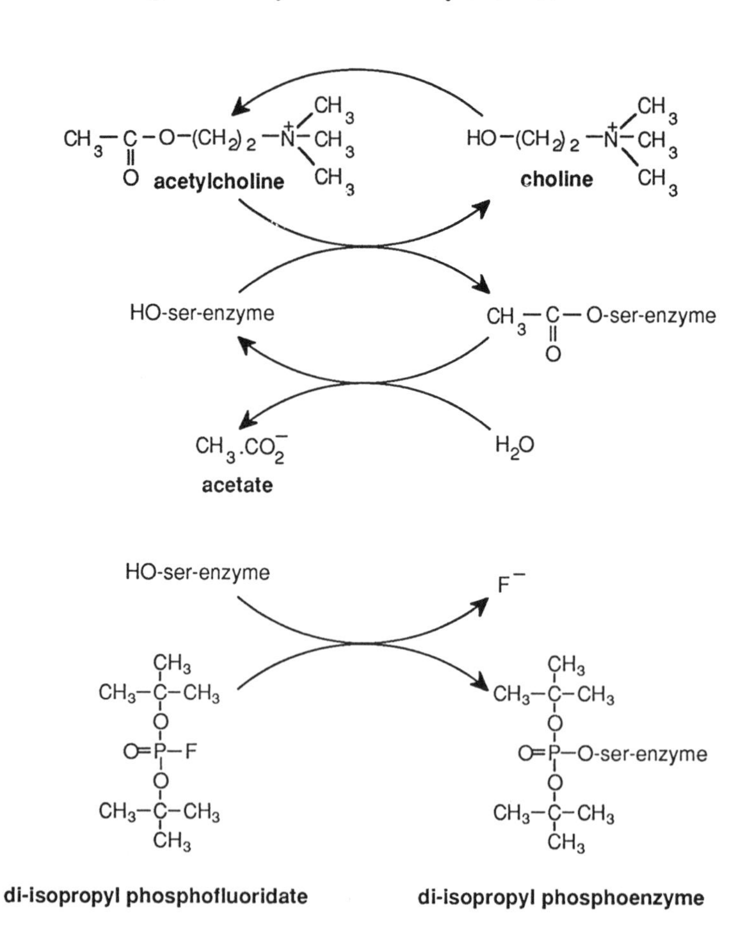

Fig. 3.2 Mechanism of breakdown of acetylcholine by acetylcholinesterase (top), and formation of a dead-end complex with the enzyme by the nerve gas, di-isopropyl phosphofluoridate (bottom).

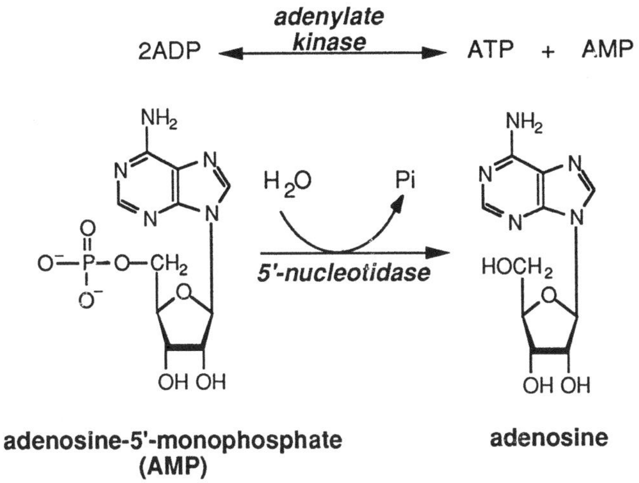

Fig. 3.3 Interconversion of ATP, ADP and AMP by adenylate kinase (top) and production of adenosine from AMP (bottom).

achieved by breakdown by the enzyme acetylcholinesterase, which is anchored in the synaptic cleft itself. The enzyme is a member of the class of enzymes known as **serine esterases**. The members of this group, which includes serine proteinases such as trypsin and chymotrypsin, have a reaction mechanism which involves the formation of a covalent acyl-enzyme intermediate, which is an ester between an acyl group and the hydroxyl of a reactive serine side chain. In the case of acetylcholinesterase, hydrolysis of the acetyl-enzyme intermediate results in the production of acetate (Figure 3.2).

Some of the original nerve gases, such as di-isopropyl phosphofluoridate, as well as many organic phosphates used as agricultural insecticides (e.g. malathion), exert their extreme toxicity by forming irreversible complexes with the reactive serine residue of serine esterases (Figure 3.2). Although there are many serine esterases and proteases in the body, the importance of acetylcholinesterase is emphasized by the fact that the dramatic signs and symptoms of organic phosphate poisoning are due almost entirely to inhibition of acetylcholine breakdown. Any synapse which uses acetylcholine as transmitter is effectively permanently switched on by the poison.

3.1.2 *The purines*

Adenosine and its related nucleotides, AMP, ADP and ATP (Figure 3.3), are all familiar as intracellular metabolites. As discussed in Section 2.3.5, it is now clear

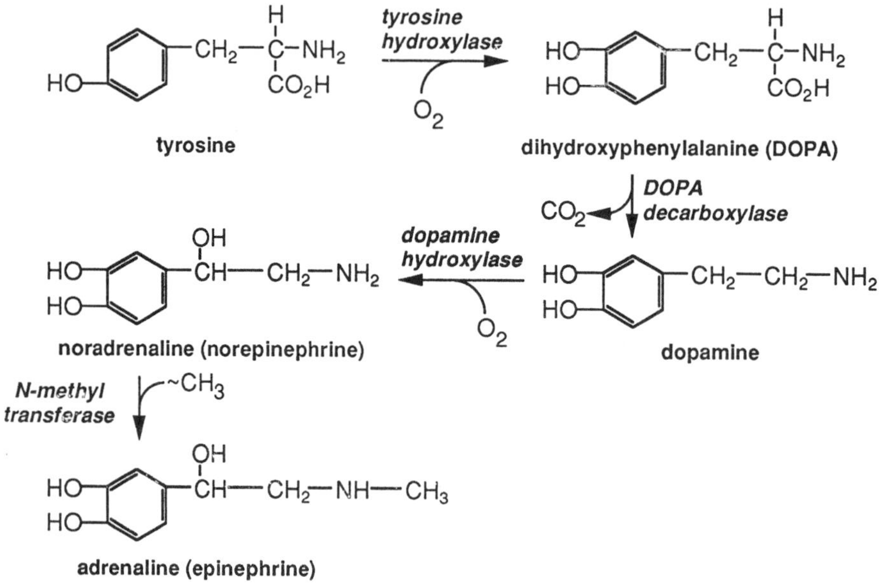

tyrosine

tyrosine hydroxylase (O_2)

dihydroxyphenylalanine (DOPA)

DOPA decarboxylase (CO_2)

dopamine

dopamine hydroxylase (O_2)

noradrenaline (norepinephrine)

N-methyl transferase ($\sim CH_3$)

adrenaline (epinephrine)

Fig. 3.4 Production of the catecholamines from the amino acid tyrosine.

that they are all also active as first messengers, via binding to a class of cell surface receptors known as **purinergic** receptors. ATP, ADP and AMP are interconverted by the enzyme **adenylate kinase**, which is very active in most cells, while adenosine is produced by a **5'-nucleotidase** from AMP (Figure 3.3).

3.1.3 *The catecholamines*

These important first messengers are synthesized by modification of the amino acid tyrosine, via oxidation to dihydroxyphenylalanine (DOPA) followed by decarboxylation to **dopamine** (Figure 3.4). Dopamine itself is an important neurotransmitter in the central nervous system, but in the sympathetic nervous system additional enzymes convert dopamine to the other catecholamines, i.e. **noradrenaline** (known as norepinephrine in the USA – released as a neurotransmitter by sympathetic nerves) and **adrenaline** (epinephrine – released, together with some noradrenaline, as a hormone by the chromaffin cells of the adrenal medulla).

Parkinsonism, a nervous disorder which occurs with fairly high frequency in elderly people, is caused by degeneration of the cells in the brain which secrete dopamine, and results in serious difficulties in controlling movement of voluntary muscles. The distressing symptoms can be relieved to some extent by giving the

glutamate γ-aminobutyrate histidine histamine

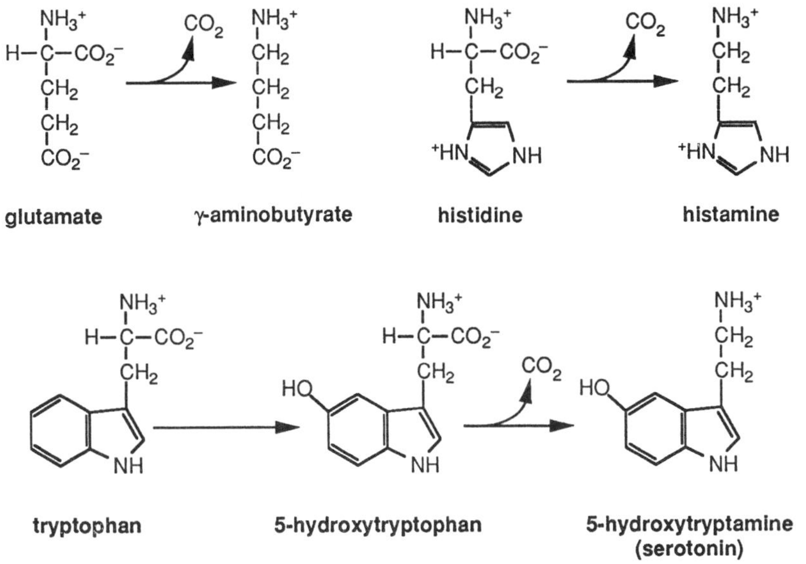

tryptophan 5-hydroxytryptophan 5-hydroxytryptamine (serotonin)

Fig. 3.5 Synthesis of γ-aminobutyrate (GABA), histamine and 5-hydroxytryptamine by decarboxylation of amino acids.

precursor, DOPA, although this is by no means a cure.

3.1.5 *Other messengers derived from amino acids*

Just as catecholamines are derived by decarboxylation of the modified amino acid dihydroxyphenylalanine, the messengers γ-aminobutyrate (GABA), histamine and serotonin (5-hydroxytryptamine) are produced by decarboxylation of the amino acids glutamate, histidine and 5-hydroxytryptophan respectively (Figure 3.5). Histamine is a local mediator (Section 2.3.3) while serotonin and GABA are neurotransmitters in the central nervous system. In addition some amino acids (e.g. glycine, glutamate) are used directly as neurotransmitters in the central nervous system without modification.

3.1.5 *Thyroid hormones*

The thyroid hormones are also made by modifications of amino acids, but unlike the other messengers discussed in this section they are not freely water-soluble, and are synthesized by a much more complex and remarkable mechanism. They are derived by the combination and modification of two iodinated tyrosine residues, but this does not occur on free tyrosine but on tyrosine residues incorporated into a

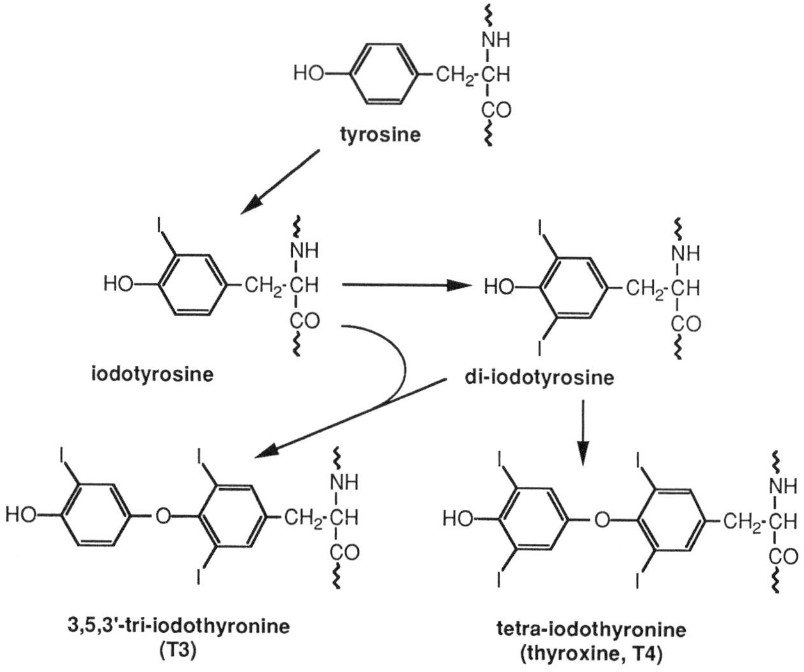

Fig. 3.6 Synthesis of thyroid hormones from tyrosine. All of the above reactions take place not on free tyrosine, but on specific tyrosine residues in the precursor protein, thyroglobulin. The last step (not shown) is the proteolytic cleavage of the hormones from the precursor.

very large precursor protein called **thyroglobulin**. Thyroglobulin is a dimer of two identical chains of 2800 amino acids: although each polypeptide chain contains nearly 70 tyrosine residues, only four molecules of thyroid hormone are formed, at sites close to the N- and C-termini. These tyrosine residues are iodinated by a very specific peroxidase enzyme to mono- or di-iodotyrosine. Two molecules then combine to form 3,5,3'-triiodothyronine (T3) or 3,5,3',5'-tetraiodothyronine (thyroxine or T4) (Figure 3.6). These are initially still incorporated into the polypeptide chain, but proteinases then hydrolyse the peptide bonds on either side, releasing T3 and T4. The thyroid gland secretes a mixture of the two hormones, but a large proportion of T4 is deiodinated to T3 in other tissues such as the liver. Since T3 acts with higher potency on target cells that T4, it is generally regarded as the active species of thyroid hormone. T3 and T4 are only sparingly soluble in water, and are carried in the blood bound to proteins, particularly **thyroxine binding globulin**.

Iodine is not known to be used in any other biological process, and is normally

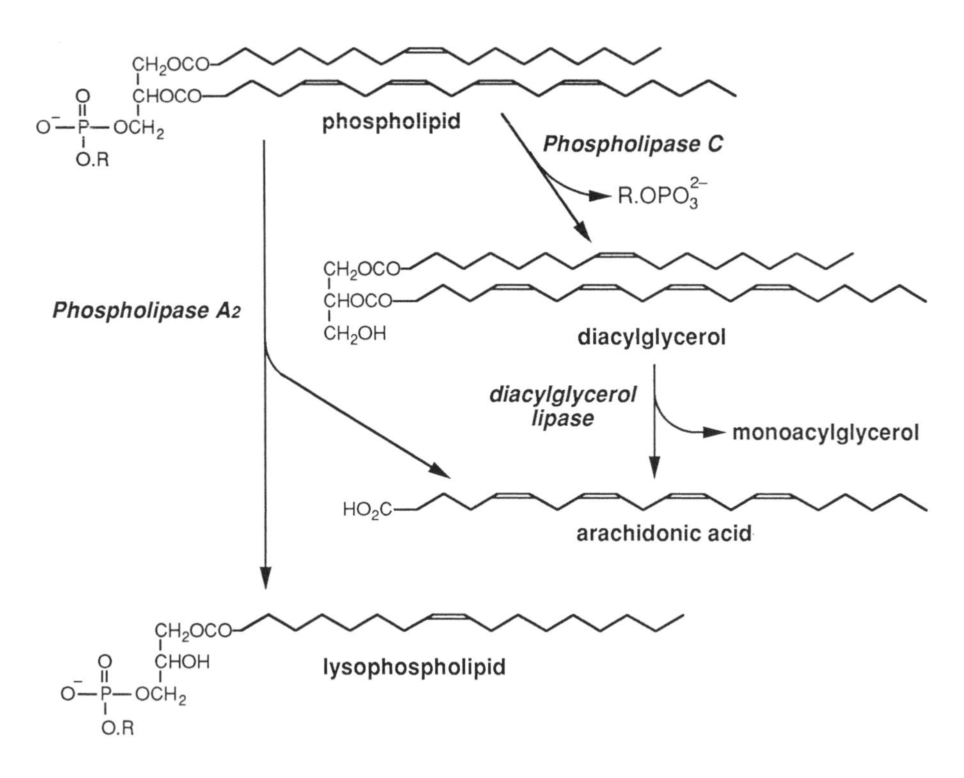

Fig. 3.7 Release of arachidonic acid from phospholipids via phospholipase A_2 (left), or via phospholipase C plus diacylglycerol lipase (right).

only a trace component of the diet. Lack of iodine leads to thyroid hormone deficiency and hypertrophy of the thyroid gland (**goitre**) because the lack of feedback by T3 leads to continuous thyrotropin release (Section 2.2.1). The thyroid gland has very active systems for uptake of iodine from plasma, and 90–95% of the iodine in the whole body is present in the gland. This explains why thyroid cancer resulting from uptake of γ-emitting isotopes of iodine is a particular problem after exposure to radioactive fallout.

3.2 *Messengers synthesized from lipids*

3.2.1 *Eicosanoids, including prostaglandins*

The eicosanoids are a diverse group of local mediators (Section 2.3.4) of low water solubility which are derived from polyunsaturated fatty acids, particularly arachidonic acid (Figure 3.7). Arachidonic acid cannot be synthesized from scratch

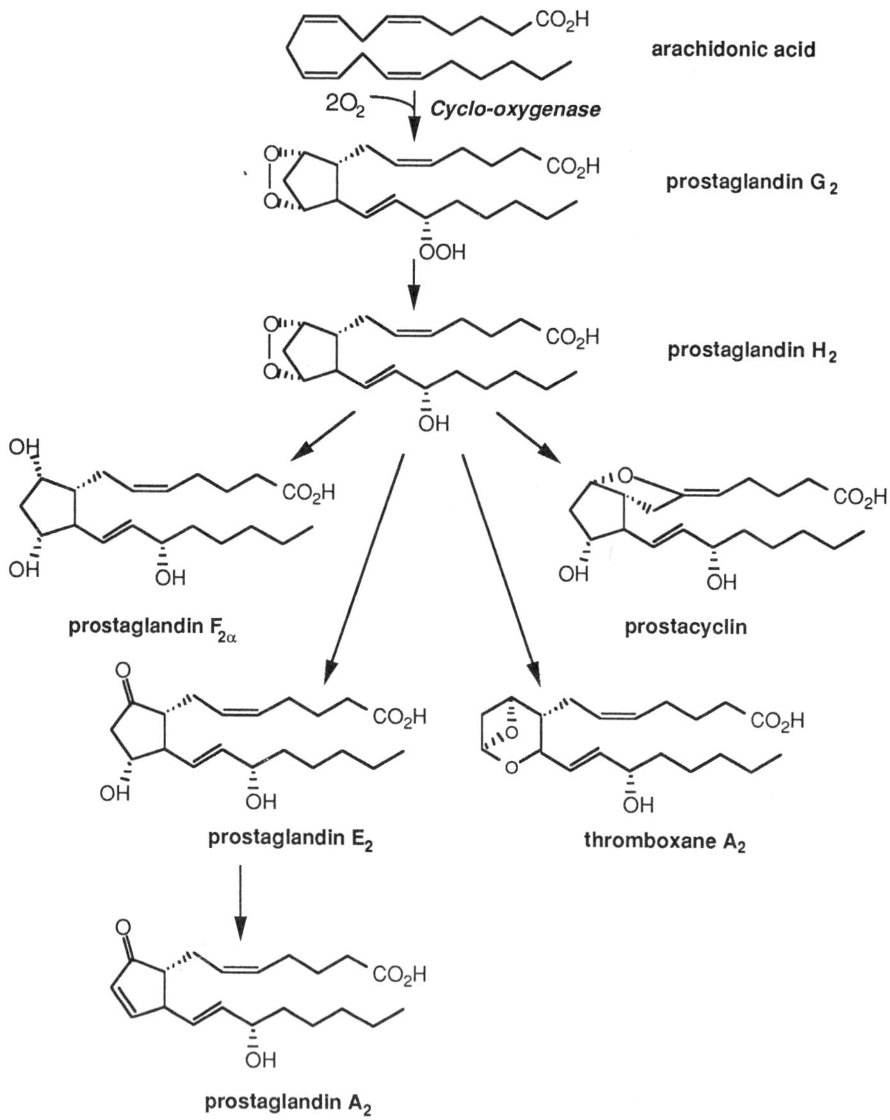

Fig. 3.8 Synthesis of the major prostaglandins, prostacyclin and thromboxane A_2 via the cyclooxygenase pathway.

in mammals, but can be produced by elongation and desaturation of **essential fatty acids** such as linoleic acid. The necessity for production of eicosanoids may at least partly explain the requirement for these fatty acids in the diet. Eicosanoids are not stored in cells in significant quantities, so their rate of release is determined by their rate of synthesis, which in turn is thought to be chiefly

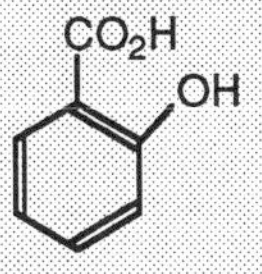

salicylic acid 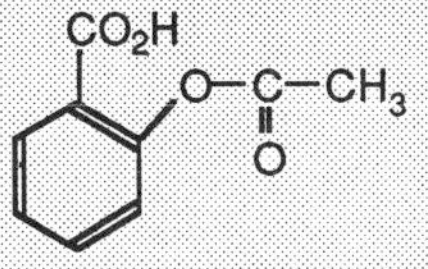 **acetyl salicylate (aspirin)**

One summer day in 1758, an English clergyman named Edward Stone was walking along a brook near Chipping Norton, past some willow trees (they still grow there today). For reasons that are not recorded, he put a strip of the bark to his lips and noticed that it tasted bitter. He knew that the bitter tasting bark of the South American cinchona tree (which contains quinine) was used as a treatment for malaria, and argued that willow bark might have similar curative properties. He found that it was very effective in the treatment of fevers and rheumatism, and published his results as a letter to the Royal Society in 1763.

The active ingredient of willow bark was later identified as salicylic acid. In fact it works by a completely different mechanism from quinine, so Edward Stone had made an important discovery by erroneous reasoning, not an uncommon event in Science! Unfortunately, salicylic acid itself is unpleasant to take and irritates the stomach. In 1899 German chemists working for Bayer, then a dye company, found that acetyl salicylic acid was much more convenient to take. They named it aspirin, *a* standing for acetyl and *spirin* after *Spiraea ulmaria,* the meadowsweet, an alternative source of salicylic acid.

Box 3.1 Discovery of aspirin

regulated by the availability of arachidonic acid. The fatty acid is mainly found in cells esterified at the 2-position in phospholipids, from where it can be released by the action of phospholipase A_2 enzymes (Figure 3.7). The regulation of phospholipases A_2 is not well understood, although there are inhibitor proteins (**lipomodulins**) whose synthesis is stimulated by glucocorticoids, and for which there is some evidence for regulation by protein phosphorylation. Arachidonic acid can also be generated by diacylglycerol lipases from diacylglycerol generated by phospholipase C (Figure 3.7). Phosphatidylinositol-specific phospholipases C are activated by a number of first messengers by mechanisms discussed in Chapter 7.

Three of the major classes of eicosanoid, i.e. prostaglandins, prostacyclin and thromboxanes, are produced from polyunsaturated fatty acid by the **cyclooxygenase** pathway (Figure 3.8). Cyclooxygenase is inhibited by the familiar drug, aspirin, which was originally extracted from willow bark (Box 3.1). This probably accounts for most, if not all, of the numerous effects of this drug. The cyclooxygenase enzyme complex catalyses oxidation reactions leading to a cyclic endoperoxide intermediate (prostaglandin H_2) with the characteristic 5-membered ring structure of the prostaglandins. From this intermediate, reduction

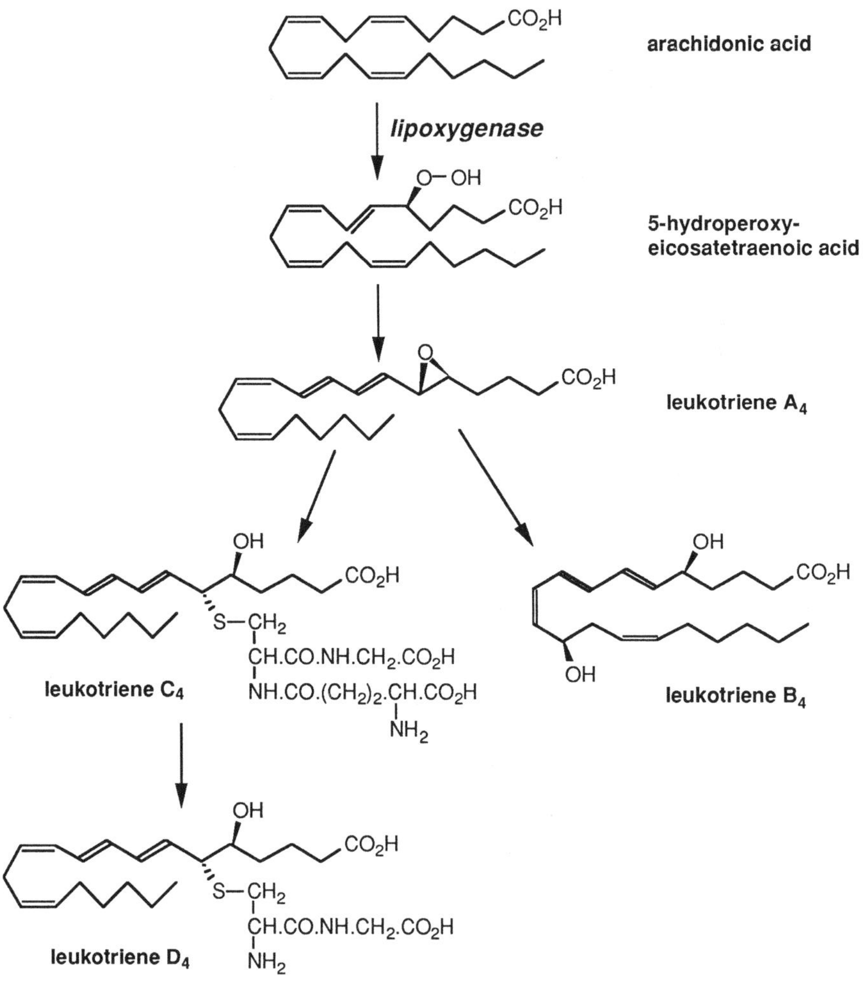

Fig. 3.9 Synthesis of the major leukotrienes via the lipoxygenase pathway.

or isomerization reactions on the endoperoxide lead to the three main series of prostaglandins, A, E and F, which differ in the number and nature of the substituents on the 5-membered ring. The prostaglandins are usually referred to by standard abbreviations, e.g. $PGF_{2\alpha}$. The "2" subscript indicates that the molecule has two double bonds (always the case for prostaglandins derived from arachidonic acid) while the "α" subscript refers to the configuration of the substituent on the 9-position in the 5-membered ring. Other, less abundant, series of prostaglandins (PGE_1, $PGF_{1\alpha}$, PGA_1; PGE_3, $PGF_{3\alpha}$, PGA_3) are synthesized from 20-carbon fatty acids which have one less or one more double bond than arachidonic acid.

Prostacyclin (PGI_2) and thromboxane A_2 are synthesized from the cyclic endoperoxide intermediate via isomerization reactions (Figure 3.8). In thromboxane A_2 the 5-carbon ring is converted to a 6-membered ring containing one oxygen atom, while in prostacyclin an adjacent 5-membered ring is formed, also containing an oxygen atom. Prostacyclin and thromboxane A_2 are antagonistic mediators which control the aggregation of platelets and, hence, the formation of blood clots.

Another class of eicosanoid, the leukotrienes, are not synthesized via the cyclooxygenase pathway but via a **lipoxygenase** which catalyses direct oxidation at the 5-carbon of arachidonic acid to form a hydroperoxy fatty acid intermediate (Figure 3.9). There are several leukotrienes, some of which are conjugated with more polar substituents such as glutathione (leukotriene C_4) or cysteinylglycine (leukotriene D_4). Leukotrienes cause contraction of smooth muscle in the airways leading to the lung, and play an important role in the disease asthma.

3.2.2 *Steroid hormones*

The steroids are amongst the most familiar group of hormones to the general public due to their widespread use and abuse for such purposes as contraception and bodybuilding. Their medicinal use has a long history, since reasonably pure preparations of steroids were apparently prepared by Chinese alchemists in the 10th to 15th centuries. There are over 200 naturally occurring steroids, of which a few important representatives are shown in Figure 3.10. The naturally occurring steroids are all synthesized from cholesterol, and have the characteristic **sterane** nucleus of five hexane and one pentane rings (Figure 3.10). There are also very large numbers of biologically active synthetic steroids which are of great importance in the pharmaceutical industry (Figure 3.10). The female contraceptive pill is made of a combination of synthetic analogues of progesterone (e.g. norethindrone) and of oestrogens such as oestradiol (e.g. ethinyl oestradiol). The ethinyl groups on these analogues enhance their activity when taken orally. Anabolic steroids like stanozolol are analogues of the male sex steroid, testosterone, and have become notorious because of their abuse by athletes and bodybuilders in order to increase muscle bulk. Dexamethasone is a potent analogue of the glucocorticoid, cortisol, and is widely used as an anti-inflammatory agent in diseases such as arthritis.

The common intermediate in the biosynthesis of most steroid hormones is **pregnenolone**, which is produced from cholesterol by two successive hydroxylations, followed by a cleavage of the side chain (Figure 3.11), the cleavage step generally being thought to be the regulatory step which is stimulated by the trophic hormones (adrenocorticotropin in adrenal cortex, and lutropin and choriogonadotropin in ovary and testis). In adrenal cortex, many steroids are produced, but these fall into two main classes, which are made in different regions of the cortex. The glucocorticoids (mainly cortisol in humans; corticosterone in rat)

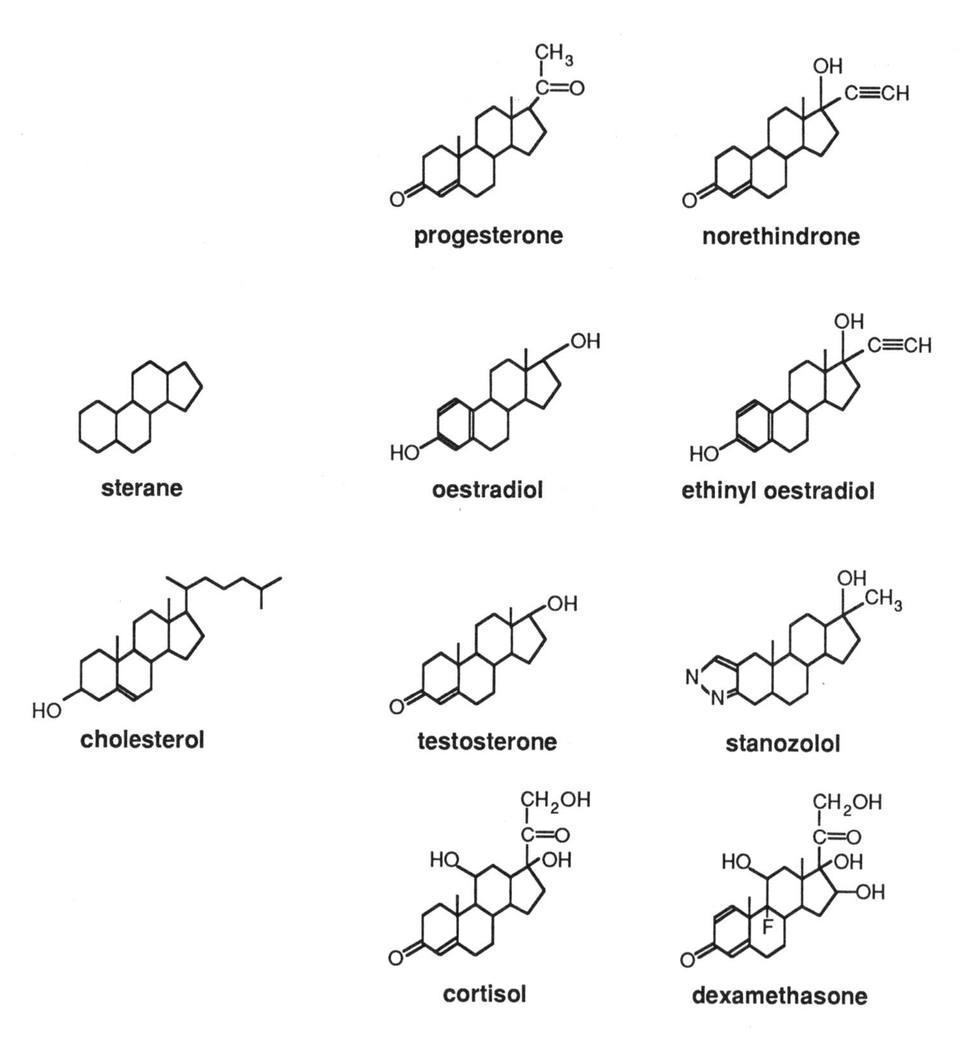

Fig. 3.10 Selected examples of naturally occurring (centre) and synthetic (right) steroid hormones. All are based on the sterane nucleus and synthesized from cholesterol (left).

have many effects on metabolism but are particularly associated with long-term stress, while the mineralocorticoids (mainly aldosterone in humans) are principally concerned with controlling water and salt balances in the body. The sex steroids are produced in specialized cells in the testes (Leydig cells) in males, and in the ovary in females. The principal sex steroid hormones in males and females are testosterone and the oestrogens (e.g. oestradiol) respectively. These are the hormones responsible for the secondary sex characteristics such as the distribution of body hair, muscle and fat. Females also produce the hormone progesterone in

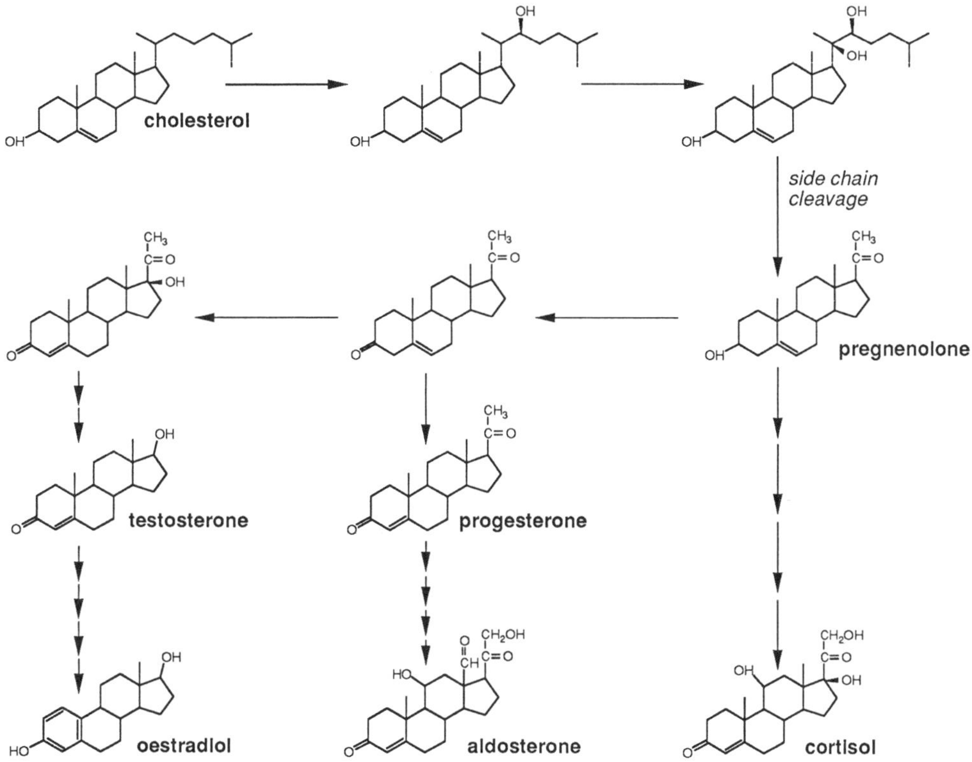

Fig. 3.11 Routes of biosynthesis of some of the major steroid hormones from cholesterol.

the corpus luteum. This hormone plays a particularly important role in maintaining the pregnant state.

A hormone related to the steroids is vitamin D_3. It was originally defined as a vitamin because children developed **rickets**, a defect in bone development, due to dietary deficiency of Vitamin D. However, it can be synthesized by a photochemical reaction from 7-dehydrocholesterol in human skin (Figure 3.12), and rickets is only a problem in malnourished children whose skin is not exposed to sunlight, e.g. in overcrowded city slums. Vitamin D_3 is converted in the kidney into the active hormonal form, which is hydroxylated at two positions (Figure 3.12). This active form interacts with calcitonin and parathyroid hormone to regulate whole body Ca^{2+} and phosphate metabolism. It promotes uptake of Ca^{2+} and phosphate in the intestine, reduces urinary excretion of these ions, and promotes remodelling and mineralization of bone.

Like the thyroid hormones and eicosanoids, steroid hormones and vitamin D_3

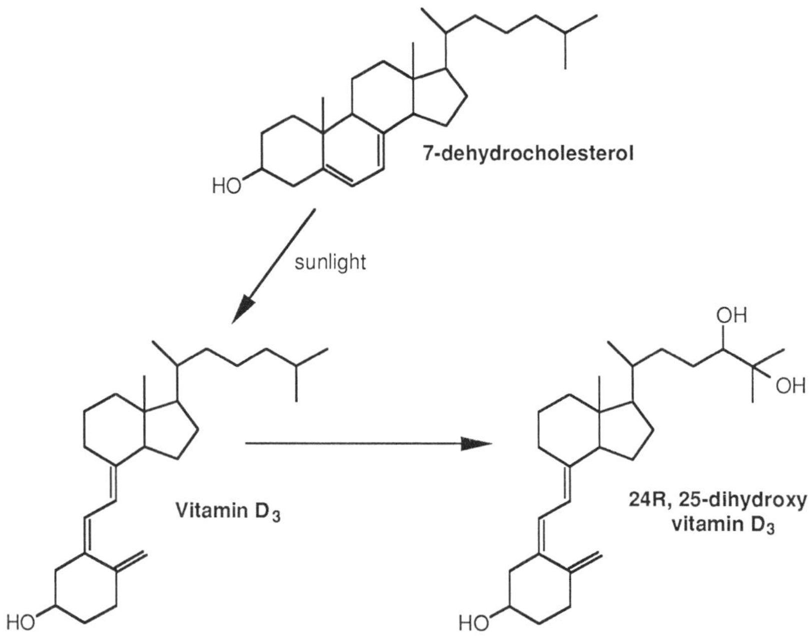

Fig. 3.12 Photochemical conversion of 7-dehydrocholesterol to vitamin D_3, and its conversion to the active, hydroxylated hormone.

have very low solubility in water, and are carried in the blood bound to specific carrier proteins.

3.3 *Peptide messengers*

Peptide messengers are constructed of amino acids linked by peptide bonds. They are a very diverse group with examples being found in all three functional classes of first messenger, i.e. neurotransmitters, local mediators and hormones. Unlike the other messengers derived from amino acids which were discussed in Section 3.1, they are of course encoded in the genome, and synthesized via successive transcription and translation. However the synthesis of peptide messengers shows some unique features compared with the synthesis of other proteins, and these are discussed in detail in Section 3.3.2.

3.3.1 *Structure of peptide messengers*

The amino acid sequence determination of peptide messengers has brought

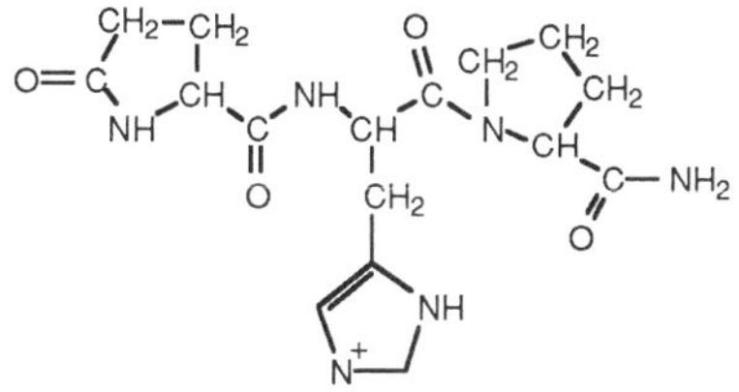

(pyro-glutamyl-histidyl-prolyl-amide)

Fig. 3.13 Structure of the tripeptide messenger, thyrotropin releasing hormone. The N-terminal glutamic acid is cyclized to pyro-glutamic acid, while the C-terminus is amidated.

considerable rewards, both scientific and medical. Firstly, it has resolved a confusing morass of biological activities into a series of defined structures which can be classified into "families", as discussed below. Secondly, although few peptide messengers can be purified from their natural sources in large amounts, sequence determination has allowed the chemical synthesis of many of the smaller peptides for therapeutic use, as well as aiding the design of analogues which may be useful agonists or antagonists. An example of this is the use of synthetic oxytocin to induce childbirth.

Peptide messengers show great variation in length, from the three amino acids of thyrotropin releasing hormone (TRH, Figure 3.13), to the 231 amino acids of human chorionic gonadotropin. The smaller peptides are often modified at the N- or C-termini. TRH is a typical example, in which the N-terminus is a cyclized glutamic acid (pyroglutamic acid), while there is an amide on the C-terminus (Figure 3.13). These modifications may protect the peptides against degradation by exopeptidases.

The number of peptide messengers which have been identified in eukaryotes is probably now in three figures. However, the sequence data have revealed that this complexity can be partially reduced by grouping the peptides into families of related sequence, which have probably arisen by gene duplication and evolutionary divergence. These sequence similarities can sometimes, but not always, be correlated with functional similarities. This will be illustrated by considering a few examples:

Oxytocin/vasopressin

These peptide hormones are both produced in the posterior pituitary, and are obviously related in sequence (Figure 3.14), including the position of the single disulphide bridge and the C-terminal glycine amide. Both cause contraction of smooth muscle, but at different sites in the body. The major effect of vasopressin is to increase blood pressure by causing contraction of the smooth muscle lining

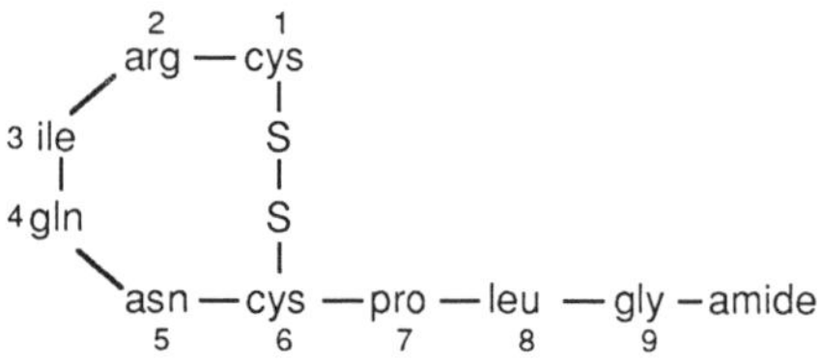

```
          2       1
         arg — cys
        /         |
   3 ile          S
      |           |
   4 gln          S
        \         |
         asn — cys — pro — leu — gly — amide
          5     6     7     8     9
```

oxytocin

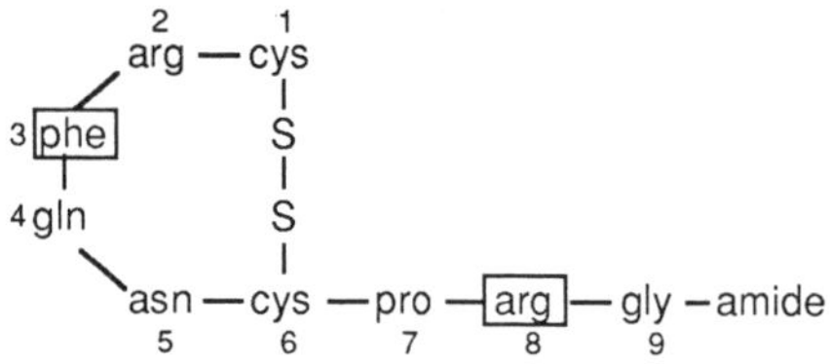

```
          2       1
         arg — cys
        /         |
  3 [phe]         S
      |           |
   4 gln          S
        \         |
         asn — cys — pro — [arg] — gly — amide
          5     6     7      8      9
```

arginine vasopressin

Fig. 3.14 Structure of the peptide hormones, oxytocin and arginine vasopressin. The residues are numbered from the N-terminus, and residues which differ in arginine vasopressin are boxed.

blood vessels (and by promoting water retention in the kidneys). Oxytocin, by contrast, causes contraction of smooth muscle in the uterus and mammary gland and is involved in childbirth and the ejection of milk during suckling. However, the similar evolutionary ancestry of oxytocin and vasopressin can be seen by the finding that oxytocin also has vasopressin-like effects on blood pressure. This can lead to problems when oxytocin is used to induce labour during childbirth.

Somatotropin/prolactin/placental lactogen

Somatotropin (growth hormone) and prolactin are products of the anterior pituitary, while placental lactogen is produced by the placenta during pregnancy. All three are related in sequence, with somatotropin and placental lactogen being particularly closely related, and prolactin being somewhat more distantly related (Figure 3.15). This is at first sight surprising when one considers that the main effects of prolactin and placental lactogen are very similar, i.e. to stimulate differentiation of the milk-producing systems in the mammary gland, and their synthesis of milk proteins, whereas somatotropin is a growth hormone. However, in one sense prolactin and placental lactogen are also growth hormones, although their effects are mainly on the mammary gland rather than the skeleton. This is reflected in an alternative name for placental lactogen (**chorionic somatomammotropin**).

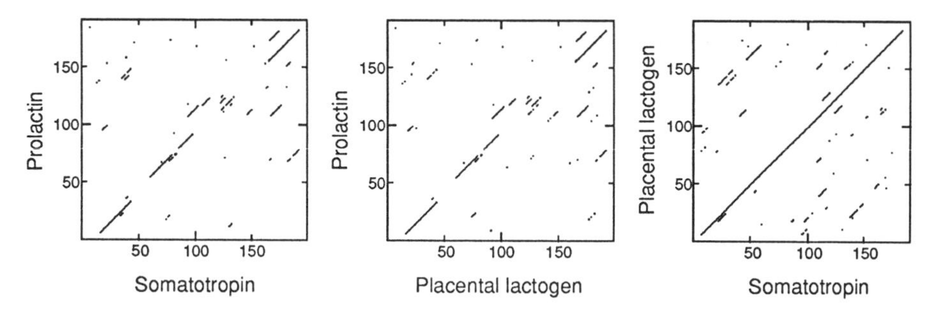

Fig. 3.15 Dot matrix comparisons of the sequences of somatotropin, prolactin and placental lactogen. In this method of comparison, every stretch of n residues (in this case $n = 15$) in one sequence is compared by computer with every stretch of n residues in the other sequence. If the comparison indicates a certain degree of similarity (using in this example the scoring system of Dayhoff) a dot is plotted at that position. Although many isolated dots may be plotted due to chance similarities, if the two proteins are truly related this will be evident as distinct diagonal lines. Deletions or insertions in one sequence appear as offsets in the diagonals. In the examples above, it is clear that somatotropin and placental lactogen are closely related throughout their lengths, whereas prolactin contains at least one region (residues 100–150 approx.) which is somewhat different from the other two.

The glycoprotein hormones

Three trophic hormones of the anterior pituitary, and one of the placenta, i.e. thyrotropin, follicle stimulating hormone, lutropin and chorionic gonadotropin, are all very closely related in structure. All are heterodimers of α and β subunits, with the α subunits of 92 amino acids being identical. The β chains vary from 110 to 145 amino acids and, although distinct gene products, are clearly related (Figure 3.16). This has led to some technical difficulties in developing specific antibody probes to measure these hormones, the levels of which are important diagnostic aids in monitoring ovarian and thyroid function. Both α and β subunits of all four hormones contain numerous N-linked, branched carbohydrate chains, hence their designation as the **glycoprotein hormone family**.

Insulin/relaxin/insulin-like growth factors-1 and -2

All of these messengers were met with in Chapter 2 except **relaxin**. The latter has been isolated and purified from the corpus luteum in the ovary, and receives its name because it is thought to cause relaxation of the birth canal, via softening of connective tissue, in preparation for childbirth. At first sight these four peptide messengers would appear to have disparate actions, yet sequence analysis indicates that they are all related and have presumably derived from gene duplication and evolutionary divergence. The conservation of the disulphide bridges is particularly

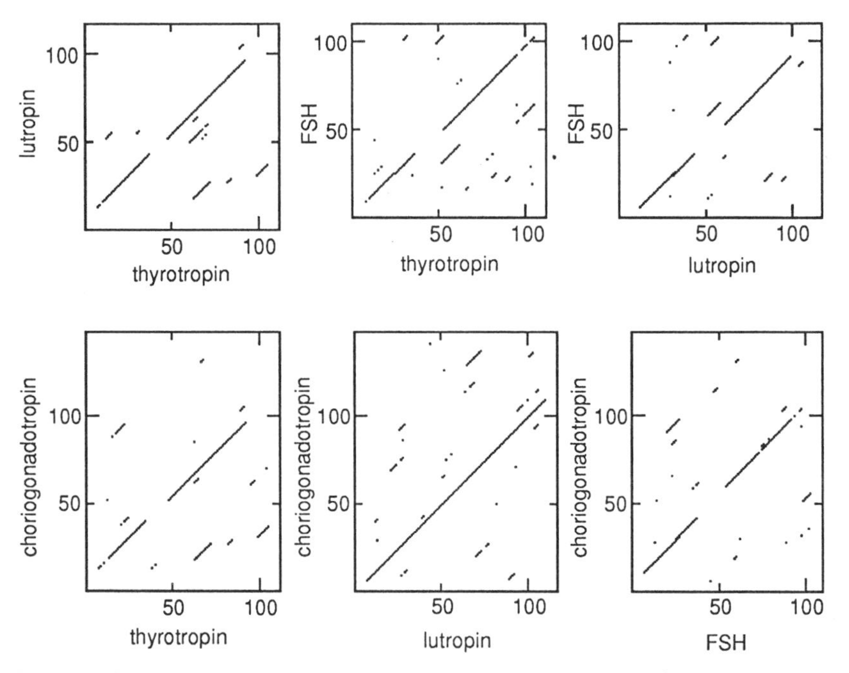

Fig. 3.16 Dot matrix comparisons (see Fig. 3.15) of the β-subunits of the glycoprotein hormones. The plots involving choriogonadotropin (bottom panels) indicate that it is related to the other three, particularly lutropin, but has an extension of about 30 residues at the C-terminus.

striking (Figure 3.17). There is one intriguing difference between insulin/relaxin and the insulin-like growth factors (IGF-1/IGF-2). The former pair consist of separate A and B chains, whereas the latter pair consist of a single, longer polypeptide. The N-terminal sections of IGF-1/IGF-2 are related to the B chains of insulin/relaxin, while the C-terminal sections are related to the A chains. At first sight the central sections of IGF-1 and IGF-2 have no equivalent in insulin/relaxin: the problem of what has happened to these "missing pieces" of the latter two hormones will be addressed in the next section.

3.3.2 *Synthesis of peptide messengers*

Given a basic knowledge of the normal mechanism of protein synthesis, several questions can be raised regarding the special problems of synthesis of peptide messengers:

1) How is a two chain structure such as insulin synthesized?

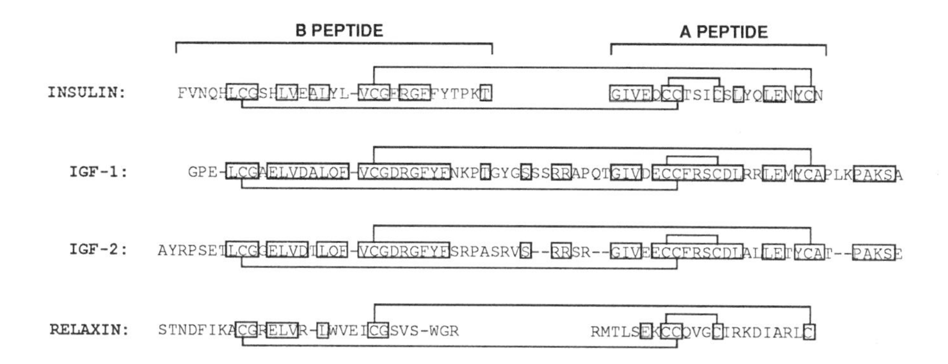

Fig. 3.17 Aligned sequences of the insulin/insulin-like growth factor/relaxin family, shown in the single letter amino acid code (see Appendix). Cysteine (C) residues joined by lines are linked by disulphide bridges. Identical residues are boxed. A few gaps have been introduced to optimize the alignment.

2) Are the small peptides, such as thyrotropin releasing hormone, synthesized by the normal protein synthetic machinery, and if so, how?
3) How are the N- and C-terminal modifications found in many small peptide messengers produced?

Due to the high frequency in Western society of insulin-dependent diabetes, in which there is a failure to secrete sufficient insulin, synthesis of that hormone has been a much studied system which has often led the way in our understanding of these questions. Insulin represents >70% of the protein synthesized by the β cells of the pancreas. However, when insulin-secreting cells were incubated briefly (**pulsed**) with radioactive amino acids, the major radioactive species was found to be a protein larger than insulin, but which was recognized by anti-insulin antibody. Sequencing of this protein showed that it was indeed a precursor of insulin, which was named proinsulin (Figure 3.18). The N-terminal section of proinsulin corresponds to the B chain of insulin, and the C-terminal section to the A chain. The central section of proinsulin (C peptide) is removed proteolytically during maturation of insulin by a proteinase which appears to recognize two adjacent basic residues.

The C peptide of proinsulin corresponds, of course, to the "missing piece" when the sequences of insulin and relaxin are compared with those of the insulin-like growth factors (Section 3.3.1). It also explained another problem, which was that if insulin was denatured using urea in the presence of a reducing agent to cleave the disulphide bonds, it could not be renatured into an active form by removal of the denaturants. This had suggested that the three-dimensional structure of insulin was **not** the structure of minimum free energy, and that information

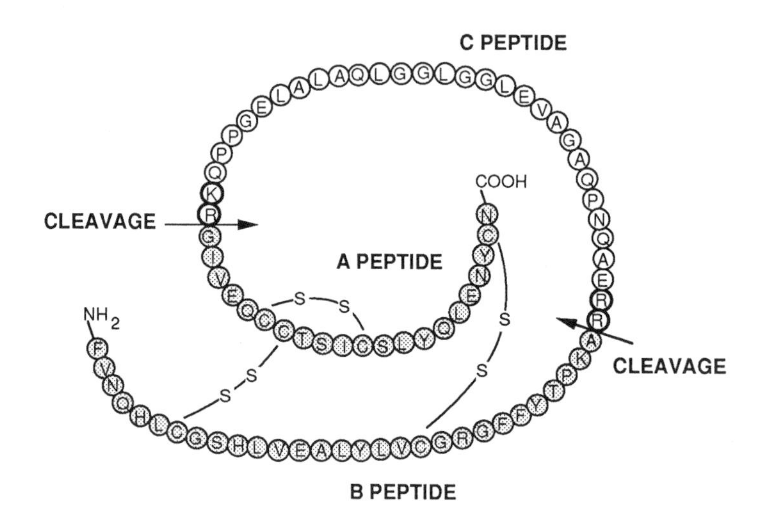

Fig. 3.18 Sequence of proinsulin, showing adjacent basic residues (RR, KR) at which cleavage to insulin occurs.

additional to its amino acid sequence must have been used to direct folding. However this paradox was solved by the finding that proinsulin will renature spontaneously, showing that the "additional information" was the C peptide. Once insulin is cleaved from proinsulin, the "correct" folding is of course maintained by the three disulphide bridges. These observations are not just an abstruse academic point, because a large proportion of the insulin used for treatment of diabetics is now made by expression of recombinant DNA in bacteria, and obtaining correct folding of the product was one of the technical problems that had to be overcome.

This was not quite the end of the story of insulin precursors, because it was subsequently found that proinsulin, like nearly all secreted proteins, is itself initially synthesized as a larger precursor (preproinsulin) containing an N-terminal **signal sequence** which is responsible for targetting of the nascent polypeptide to the endoplasmic reticulum and, hence, for secretion. The signal peptide is removed by proteolysis as the proinsulin is extruded into the lumen of the endoplasmic reticulum. This topic will be discussed in more detail in the next chapter.

Subsequent to the discovery of proinsulin, it has been shown that almost all peptide messengers are derived from larger precursors (promessengers) which are themselves derived by loss of signal sequences from yet larger precursors (prepromessengers). In some cases one can get more than one messenger, which can even be overlapping, from a single precursor. In these cases the precursors can be referred to as **polyproteins**. The most remarkable examples of this are furnished by the precursors which give rise to the opioids, the peptide messengers whose actions are mimicked by the opiate drugs, morphine and heroin. The

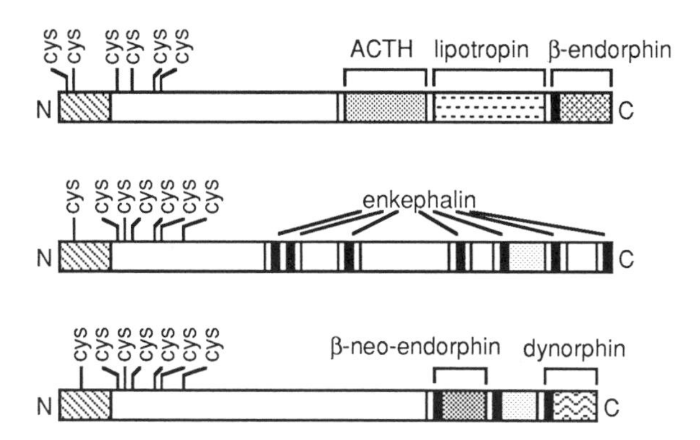

Fig. 3.19 Alignment of sequences of prepro-opiomelanocortin (top), prepro-enkephalin (centre) and prepro-dynorphin (bottom), showing the cysteine-rich N-terminal regions, and the locations of peptide messenger sequences. Enkephalin sequences are shown by black boxes: these are found in all three precursors. Double basic regions at which cleavage occurs are shown by double lines. The hatched regions at the N-termini are the signal sequences which are cleaved during synthesis of the precursor on the rough endoplasmic reticulum (Chapter 5). After Herbert *et al.* (1985) in Biochemical Actions of Hormones (Litwack, G., ed.) Academic Press, Orlando, vol. 12, pp. 1-36.

smallest opioid peptides are Met-enkephalin (tyr-gly-gly-phe-met) and Leu-enkephalin (tyr-gly-gly-phe-leu), and the others are C-terminal extensions of either Met-enkephalin (e.g. β-endorphin) or Leu-enkephalin (e.g. dynorphin, β-neo-endorphin). All are produced by cleavage of only three promessenger precursors (Figure 3.19):

1) Pro-opiomelanocortin, synthesized in the anterior pituitary, gives rise to the opioid β-endorphin, as well as several other peptide hormones including adrenocorticotropin and α-, β- and γ-melanotropins.
2) Pro-enkephalin, isolated originally from adrenal medulla, contains six copies of Met-enkephalin and one of Leu-enkephalin.
3) Pro-dynorphin contains three copies of Leu-enkephalin and also gives rise to dynorphin and β-neo-endorphin.

These three promessengers are in fact remarkably similar, all having ~260 residues, an N-terminal region containing a cluster of conserved cysteine residues, and a C-terminal region containing the messenger sequences. This similarity also extends to the location of the intron–exon boundaries in the corresponding genes.

As is the case for insulin, the messenger sequences within the opioid precursors are almost always bounded by pairs of adjacent basic residues (lys-arg,

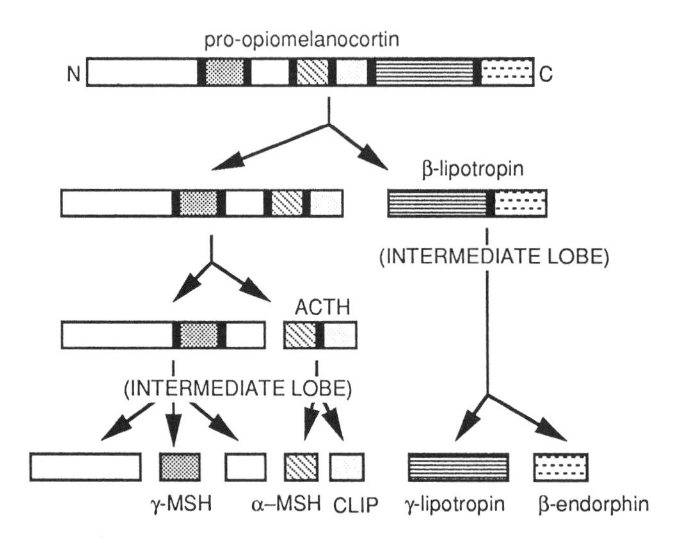

Fig. 3.20 Cell-specific cleavage of pro-opiomelanocortin in the anterior and intermediate lobe of the pituitary gland. Sites of cleavage are shown by thick black lines. Certain cleavages occur in the intermediate lobe only. After Herbert *et al.* (1985) in Biochemical Actions of Hormones (Litwack, G., ed.) Academic Press, Orlando, vol. 12, pp. 1-36.

arg-arg, or lys-lys) which must represent recognition sites for trypsin-like processing enzymes. Intriguingly, the processing events can be tissue-specific (Figure 3.20). Thus in the anterior pituitary, pro-opiomelanocortin is processed largely to adrenocorticotropin (ACTH) and β-lipotropin. However, the precursor is also expressed in the intermediate lobe of the pituitary, in which case more extensive processing occurs to yield α- and γ-melanotropins (α-, γ-MSH), corticotropin-like intermediate lobe peptide (CLIP), γ-lipotropin and β-endorphin. This tissue specificity is thought to be due to the expression of different processing enzymes.

Sequencing of promessengers also helped to solve part of the third question posed at the beginning of this section, i.e. how peptides like thyrotropin releasing hormone, vasopressin and oxytocin are amidated at their C-terminus. It was found that these amidated peptides always had a glycine residue at the C-terminus within the promessenger, which was lost in the mature peptide. Subsequently, enzymes have been found which remove the glycine, apparently via a dehydrogenation to an imino linkage, followed by hydrolysis. This leaves the amino group that was part of the glycine residue at the C-terminus of the peptide (Figure 3.21).

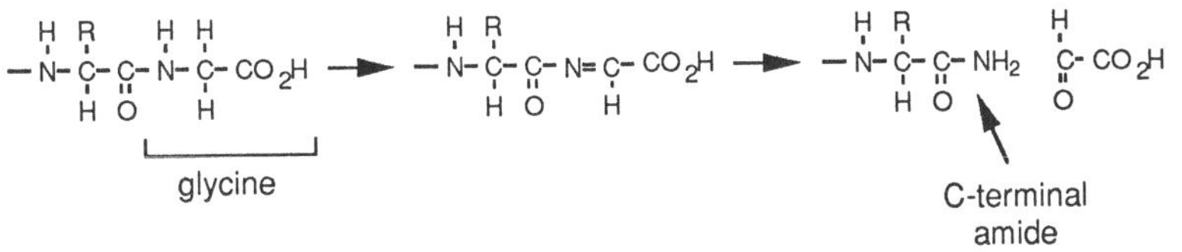

glycine

C-terminal amide

Fig. 3.21 Mechanism of formation of a C-terminal amide on a peptide messenger via reduction to an imino-linkage, followed by hydrolysis.

3.3.3 *Why are peptide messengers made as larger precursors?*

This is a teleological question which at present we can only answer by speculation rather than experiment, but several possibilities can be discussed. Probably there is some truth in all of them:

1) *A minimum size of peptide may be necessary for protein synthesis*

   Proinsulin (~80 residues) is among the smallest polypeptides known to be synthesized on ribosomes, and it seems possible that there is a minimum size for protein synthesis to occur. This would particularly be true for a secreted protein (i.e. all peptide messengers) since the peptide has to be long enough to extrude from the ribosome and bind to the signal recognition particle (next chapter) while still undergoing synthesis. It certainly seems likely that direct synthesis of a peptide as short as thyrotropin releasing hormone (3 residues) may not be possible on ribosomes.

2) *The mechanism protects the secreting cell against active messenger*

   This is a possibility because most promessengers are much less potent biologically than the mature messengers. If the secreting cell also had receptors for the messenger, a potentially lethal self-stimulation (**autocrine stimulation**) might result. An analogous argument can be used to explain why proteinases such as trypsin are stored in the pancreas as inactive precursors or **zymogens**. However, unlike the latter case where processing of the zymogen occurs outside the cell, in most cases conversion to the mature messenger occurs **within** the cell (Section 5.1.2).

3) *Ensures correct folding of the mature messenger*

   This argument could be used in the case of insulin, since the presence of the C peptide appears to be necessary to obtain the correct folding of proinsulin, and hence of insulin (Section 3.2.2). The same argument can hardly be used, however, for peptides like thyrotropin releasing hormone (3 residues), which are too small to have a stable secondary or tertiary structure in solution.

4) *Increases solubility of the messenger during synthesis*

   Many peptide messengers are only sparingly soluble under neutral conditions, a fact that does not cause a problem in the extracellular environment since they are usually active at very low concentrations anyway. This argument could be applied to insulin, which is much less soluble than proinsulin, and does indeed precipitate out in the storage granules as zinc–insulin crystals. Conceivably this could cause problems if it happened at an earlier stage during synthesis.

5) *It is accidental – reflects evolutionary origin*

   A final possibility is that there is no functional reason for synthesis of messengers as promessengers, but that it merely reflects their evolution from larger proteins. This argument can be illustrated once again by considering the example of insulin, which is processed from proinsulin by loss of the central C peptide. However, in the very closely related messenger, insulin-like growth factor-1, this does not happen, and the mature messenger remains as a single chain structure. The structural requirements for a peptide messenger are in many ways less stringent than those for an enzymic or structural protein: the ability to be recognized by a receptor, which can evolve along with the messenger, is the only real constraining feature. Possibly, the development of different cleavage patterns of peptides from a small number of ancestral precursors has merely been a rapid way of evolving a very diverse collection of messengers.

SUMMARY

Many first messengers are synthesized by simple modifications of common intermediary metabolites. Acetylcholine is derived from acetyl-CoA and choline, the latter a normal constituent of phospholipids. Extracellular purines like adenosine and ATP are normal intracellular metabolites. Many messengers are produced by decarboxylation of amino acids: examples include the catecholamines, histamine, serotonin and γ-aminobutyrate. Some amino acids, e.g. glycine and glutamate, are used as neurotransmitters without modification. Thyroid hormones are synthesized by the iodination and condensation of tyrosine residues. This does not take place with free tyrosines but with specific tyrosine residues located in a large precursor protein, thyroglobulin.

Eicosanoids are synthesized from arachidonic acid and other related fatty acids, released from phospholipids by phospholipases. The prostaglandins, prostacyclin and thromboxane contain ring structures and are synthesized by the cyclooxygenase pathway. The leukotrienes are synthesized by oxidation at carbon-5 catalysed by lipoxygenase.

The steroid hormones, of which there are now over 200 natural and synthetic examples, are lipophilic molecules based on the four ring sterane nucleus. They are synthesized from cholesterol via cleavage of the aliphatic side chain.

The peptide messengers are a large and important class which are encoded in the genome. Many have been derived by duplication and divergence of ancestral genes and they can therefore be classed into families of related structure. Most peptide messengers are produced by proteolysis of larger precursors, and in some cases more than one messenger can be produced from a single precursor.

FURTHER READING

1. "Hormones" – general text on the structure, metabolism and physiology of hormones. Norman, A.W. and Litwack, G. (1987) Academic Press, New York.
2. "Mechanism of C-terminal amide formation by pituitary enzymes". Bradbury, A.F., Finnie, M.D.A. and Smyth, D.G. (1982) Nature **298**, 686–688.
3. "Generation of diversity of opioid peptides" – review. Herbert, E., Civelli, O., Douglas, J., Martens, G. and Rosen, H. (1985) in Biochemical Actions of Hormones (Litwack, G., ed.) Academic Press, Orlando, vol. 12, pp. 1–36.

FOUR

IONIC HOMEOSTASIS AND NERVOUS CONDUCTION

Since many chemical messengers exert their effects by producing transient, yet crucially important, variations in the ionic composition of the target cell, it is necessary to understand how cells maintain a constant ionic composition (**ionic homeostasis**) in the absence of stimulation. In this chapter we will also consider the special ionic properties of excitable cells, including nerve cells, that allow them to transmit information along their surfaces in the form of electrical impulses.

4.1 *Ionic homeostasis*

As anyone who has experimented with enzymes or other biological systems will know, they are notoriously labile and sensitive to mishandling when extracted from the cell. It is crucial that the conditions of pH, temperature, ionic composition and redox state are carefully controlled, in order that the structures and activities of the fragile enzyme systems may be preserved. An important function of the plasma membrane is to maintain a constant ionic composition inside the cell so that the activities vital to life can function.

The lipid bilayer component of the plasma membrane is freely permeable to small, uncharged molecules such as H_2O, O_2 or CO_2. However ions like Na^+, K^+ or Ca^{2+}, while small in themselves, are surrounded by a large shell of water molecules (**solvated**) in solution, and this means that they do not permeate through lipid bilayers at significant rates unless this is catalysed by a protein. Facilitation of ionic flux across membranes can be via transport proteins (**porters**), or via **channel proteins** (Figure 4.1).

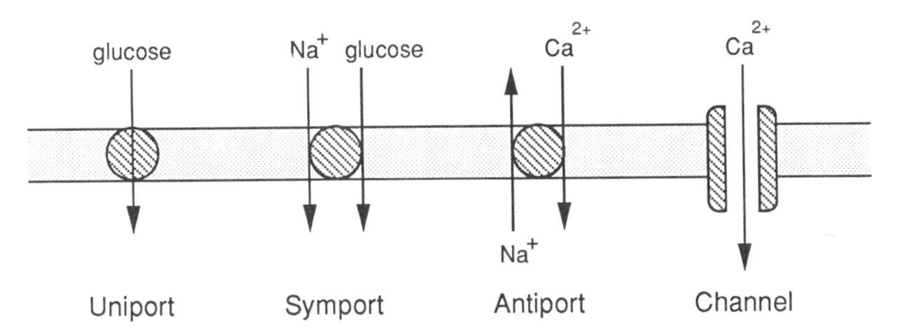

Fig. 4.1 Examples of four classes of membrane transport system.

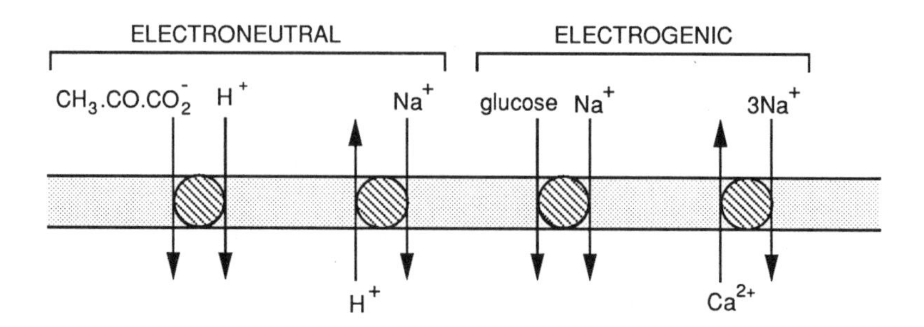

Fig. 4.2 Examples of electroneutral and electrogenic transport processes.

Porters can transport a single molecule across the membrane (**uniport**), but more commonly there is obligatory co-transport of two different molecules: this can be in the same direction (**symport**) or in opposite directions (**antiport**). Porters are similar to enzymes in many respects: they contain specific binding sites, their turnover numbers are typically of the order of 1000 molecules transported per second, and co-transport occurs with **fixed stoichiometry**, e.g. $1Na^+/1$glucose or $1Ca^{2+}/2Na^+$. Although we do not yet know the three-dimensional structure of any transporter, their function is thought to involve conformational changes which expose the binding site for the transported molecule on alternate sides of the membrane.

Like all catalysts, porters do not affect the position of the equilibrium for the reaction they catalyse, so net transport of ions can only occur if there is a driving force (i.e. a negative ΔG). If the transport is **electroneutral**, i.e. involves no net movement of charge (Figure 4.2), the molecules will merely tend to flow down concentration gradients. It is easy to calculate the ΔG from the ratio of concentrations of the transported molecules on each side of the membrane (Box 4.1). If, however, transport is **electrogenic**, i.e. involves a net movement of charge (Figure 4.2), any electrical potential difference across the membrane (**membrane potential**) must also be taken into account. Ions will tend to flow in the direction which abolishes the membrane potential. The ΔG for transport in these cases, taking into account both electrical (charge) and chemical (concentration) gradients, can be calculated as shown in Box 4.1. In a limited number of special cases, transport can occur against an electrochemical gradient because there is an obligatory coupling to the release of free energy from a chemical reaction, usually the hydrolysis of ATP (Figure 4.3). This type of transport protein is referred to as a **pump** or **transporting ATPase**.

Channel proteins are integral membrane proteins which form a pore in the membrane (Figure 4.1). There is no fixed stoichiometry for the ion flux produced, which depends on the size of the pore, its surface chemical properties, the length of time it is open, and the driving force (electrochemical gradients for the ions). However, due to the size of the pores and the nature of the amino acids lining

For an uncharged molecule X which is distributed across a membrane, the free energy change for the movement of the molecule into the cell is given by:

$$\Delta G = -RT \ln \frac{[X]_{out}}{[X]_{in}} \quad (kJ.mol^{-1})$$

If a charged molecule Xm^+ is present at equal concentration on both sides of a membrane but there is a membrane potential of $\Delta\Psi$ volts across the membrane, the free energy change ($kJ.mole^{-1}$) is given by:

$$\Delta G = -mF\Delta\Psi \quad \text{(F is Faraday's constant, 0.0965 } kJ.mol^{-1}.mV^{-1}\text{)}$$

In the presence of both a concentration gradient and a membrane potential, the two terms are combined:

$$\Delta G = -mF\Delta\Psi - RT \ln \frac{[Xm^+]_{out}}{[Xm^+]_{in}} \quad (kJ.mol^{-1})$$

If the ion is at equilibrium across the membrane, $\Delta G = 0$, then the above equation reduces to the **Nernst equation:**

$$\Delta\Psi = \frac{-RT}{mF} \ln \frac{[Xm^+]_{out}}{[Xm^+]_{in}} \quad (mV)$$

Box 4.1 The Nernst equation

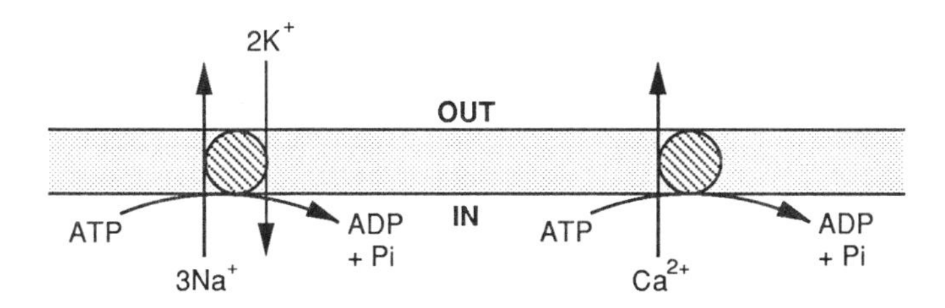

Fig. 4.3 Examples of ion-transporting ATPases (ion pumps). The plasma membrane Ca^{2+} pump may also transport protons in the opposite direction.

them, channels are generally selective for the charge, and the maximum size, of ion passed. Rates of flux via channel proteins are much higher than those of porters, typically 10^6–10^7 molecules per second. A channel clearly can only catalyse a net ion flux down an electrochemical gradient, and if active ion channels were open permanently, they would rapidly produce the same ionic environment on both sides of the membrane. This does not happen because the opening of the ion channel is usually transient, and is controlled by some external factor, i.e. it is **gated**.

The typical free concentrations of the most important ions on the inside (cytosol) and outside of a living mammalian cell are shown in Table 4.1:

Table 4.1 Approximate basal ionic concentrations inside and outside of a mammalian cell (mM = mmol.l^{-1}).

	Na^+ (mM)	K^+ (mM)	Ca^{2+} (mM)	Mg^{2+} (mM)	H^+ (mM/pH)	Cl^- (mM)
Intracellular:	10	140	10^{-4}	0.5	8×10^{-5}/7.1	10
Extracellular:	150	5	1	1	4×10^{-5}/7.4	110

It is clear that the concentrations inside the cell are very different from those outside. There are 10- to 20-fold differences in the concentrations of Na^+ and K^+ inside and out, in opposite directions. Even more dramatic is the 10,000-fold difference in free Ca^{2+} concentration. The maintenance of a low internal Ca^{2+} concentration is one of the fundamental properties of a living cell, and is also crucial in the action of many chemical messengers (Chapter 7). If a cell failed to maintain this gradient, phosphate and phosphate-containing compounds would tend to precipitate out, Ca^{2+}-activated proteinases would degrade the cell proteins, and the cell would die.

I will now discuss in detail the manner in which the various transport proteins and channels in the cell maintain this vital constant internal ionic environment.

4.1.1 *Regulation of Na^+ and K^+*

The most important activity in maintaining cellular ionic homeostasis is the Na^+/K^+ pump, which is present in at least certain areas of the plasma membrane of all animal cells (Figure 4.3), but has been most studied in red blood cells (**erythrocytes**). If erythrocytes are suspended in hypotonic medium, they swell and burst, releasing their contents. Under the right conditions the membrane will reseal to form a functional isolated membrane: a so-called **erythrocyte ghost**. One can also prepare **inside-out** membrane vesicles in which the intracellular face of the membrane is exposed on the outside.

In the 1950s it was already known that cells actively pump Na^+ out and K^+ in,

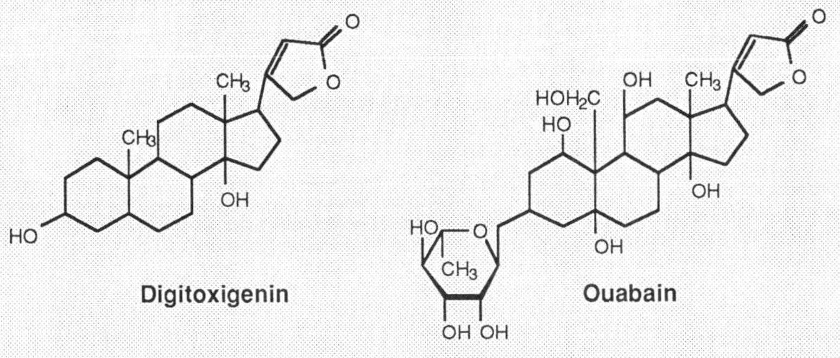

Digitoxigenin

Ouabain

In 1785, Dr. William Withering of Birmingham, England, published a monograph on the use of extracts of the Foxglove (*Digitalis purpurea*) in the treatment of dropsy (now known as congestive heart failure). The foxglove and certain other plants contain *cardiac glycosides* such as digitoxigenin and ouabain which are in fact steroids (see Section 3.2.2). Among other actions, they powerfully stimulate the contraction of the heart. Their molecular effect is to inhibit the plasma membrane Na^+/K^+-transporting ATPase. This may increase the amount of Ca^{2+} in heart muscle cells by reducing the driving force for the Na^+/Ca^{2+}exchanger in the plasma membrane which normally removes Ca^{2+} from the cells (Section 4.1.2).

Box 4.2 Cardiac glycosides

processes dependent on active ATP synthesis. Then it was found that there was an ATPase activity associated with the plasma membrane that required the simultaneous presence of both Na^+ and K^+. The crucial observation linking the two was the discovery that cardiac glycosides like ouabain (Box 4.2) inhibited both processes. Using erythrocyte ghosts and inside-out vesicles, with ouabain as an inhibitor, it was possible to show that the pump had a fixed stoichiometry of three Na^+ pumped out to two K^+ pumped in per molecule of ATP hydrolysed (Figure 4.3). Since the net effect is one positive charge extruded, the pump is electrogenic, i.e. it creates an electrical gradient or membrane potential, as well as concentration gradients of the two ions.

Although the detailed molecular mechanism of the pump is not known, it has been purified, cloned and sequenced, and consists of a catalytic α subunit of ~1000 amino acids, plus a smaller glycosylated β subunit which appears to be essential, but whose function is not known. ATP causes phosphorylation of the α subunit on a specific aspartate side chain, and a current model for the mechanism of action of the pump is shown in Figure 4.4.

The importance of the Na^+/K^+ pump is made clear from estimates that about one third of the energy expended by a human at rest is required for this one reaction. It maintains the 10- to 20-fold gradient of Na^+ concentration across the

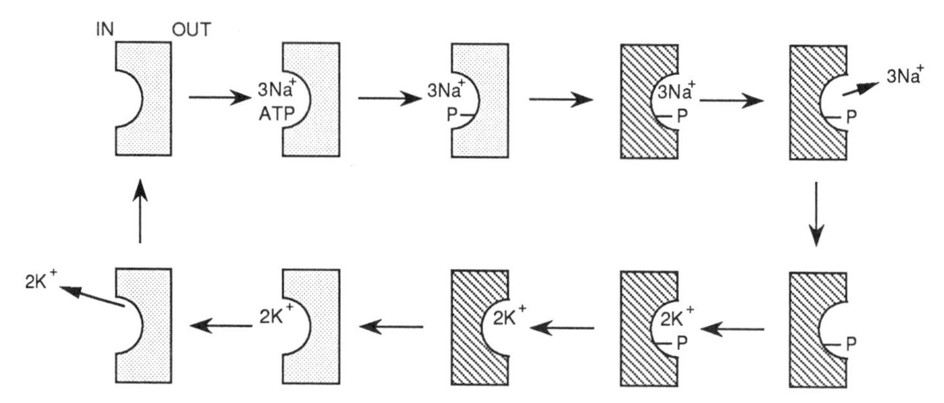

Fig. 4.4 Proposed mechanism for the action of the Na^+/K^+-transporting ATPase. Phosphorylation of the aspartate residue promotes a conformational change such that the cation-binding site is exposed on the inside: dephosphorylation has the reverse effect.

cell membrane shown in Table 4.1. The lipid bilayer is extremely impermeable to Na^+, and Na^+ can only re-enter via co-transport processes such as the Na^+/glucose symport in intestine shown in Figure 4.1, or the Na^+/Ca^{2+} or Na^+/H^+ antiports discussed in Sections 4.1.2 and 4.1.3. Because of the large electrical and concentration gradient favouring Na^+ re-entry, the pump can indirectly "drive" the transport of these co-transported molecules against their own concentration gradients.

Due to its electrogenic nature, the Na^+/K^+ pump plays a part in setting up the membrane potential, positive outside, that exists across the plasma membrane of all living cells. However it does not play a crucial part in determining the value of this potential. Unlike the case for Na^+, K^+ can cross the membrane at a slow but significant rate via **K^+ leak channels,** which are not yet well characterized in biochemical terms. Even in the absence of the Na^+/K^+ pump, K^+ tends to be higher inside the cell than out in order to balance the charges on the non-diffusible anions in the cell (particularly the phosphates on phospholipids, nucleic acids, nucleotides and other metabolites). Since this creates a concentration gradient of K^+, the ion tends to diffuse out through the "leak channels" until the system reaches equilibrium where the concentration gradient of K^+ is balanced by the electrical gradient created by the outward flow. This is an example of a **Donnan potential,** obtained in any system where ions carrying the same charge are constrained by a membrane which is permeable to the counter-ions. The action of the Na^+/K^+ pump increases the membrane potential still further, so that in a typical energized cell it is of the order of –70 mV (negative inside).

Assuming a ratio of concentrations of K^+ concentration (inside/outside) of 30 (Table 4.1) it is easy to calculate from the Nernst equation (Box 4.1) that at 37°C

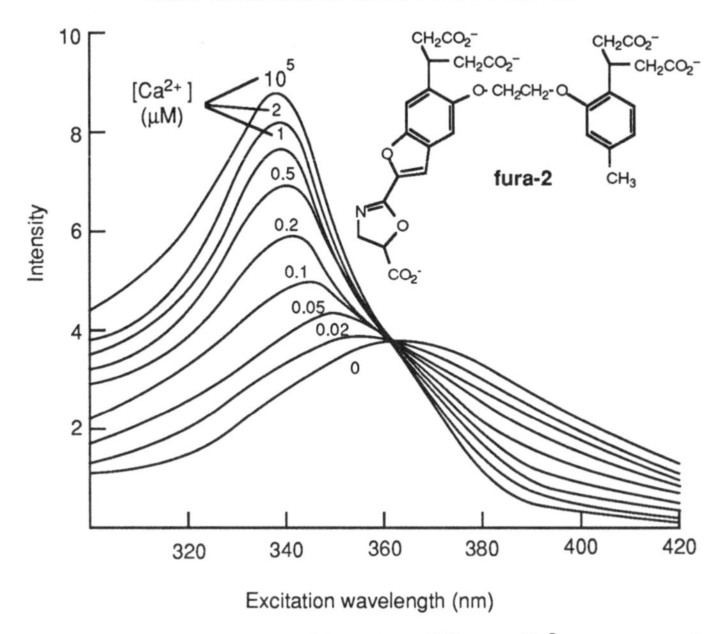

Fig. 4.5 Excitation spectra of fura-2 at different Ca^{2+} concentrations and (inset) structure of the unesterified form of fura-2.

the ion would be at equilibrium if the membrane potential was –90 mV (negative inside). At this value the efflux due to the concentration gradient would be exactly balanced by the influx due to the membrane potential. K^+ is therefore close to equilibrium across the membrane of a resting cell, which is not surprising given the existence of the K^+ leak channels. On the other hand, both the concentration gradient of Na^+ (15:1, outside:in) and the membrane potential yield free energies for Na^+ re-entry which are roughly equal and both negative, and so Na^+ is far from electrochemical equilibrium (ΔG for entry = –13.7 kJ/mole). This means that it can be made to do useful work such as driving co-transport (see above) or the passage of an action potential (Section 4.2).

4.1.2 *Regulation of cytosolic Ca^{2+} concentration*

As already mentioned, any serious elevation of intracellular Ca^{2+} is not tolerated and leads to autolysis and cell death. Since much of the Ca^{2+} inside the cell is bound to proteins or phospholipids or sequestered inside organelles, until recently it was not possible to accurately estimate the free concentration in the cytosol. However, in recent years fluorescent "reporters" of free Ca^{2+} concentration, such as quin-2 and fura-2 (Figure 4.5), have been developed. These are chelating agents for Ca^{2+} containing four negatively charged carboxyl groups which bind the

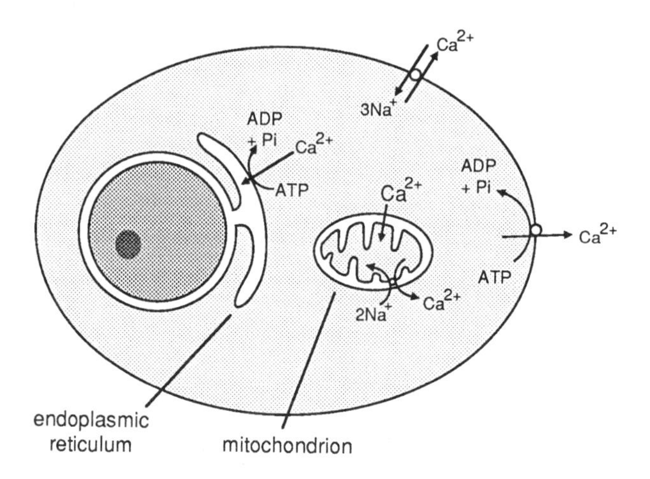

Fig. 4.6 Summary of cellular Ca^{2+} transport mechanisms.

cation, analogous to the well-known Ca^{2+} chelator, EGTA. However, unlike the latter they also contain fluorescent groups whose spectra are affected by Ca^{2+} binding. In the case of fura-2, changes in Ca^{2+} concentration within the physiological range dramatically shift the ratio of excitation efficiency obtained at 340 to 380 nm (Figure 4.5), without significantly affecting the emission spectra. A particularly clever feature of this approach is that the reporter molecules are given in the form of uncharged esters, which are permeable to the plasma membrane, but that once inside they are cleaved to the impermeant and active carboxylates by cellular esterases. Using suitable microscope equipment in which the fluorescence is excited alternately at 340 and 380 nm, and emission is recorded by a video camera, it is even possible to measure the cytosolic Ca^{2+} concentration in real time at the subcellular level. An example is shown on the front cover of this book. Using this approach, it is now generally found that the Ca^{2+} concentration in the cytosol of energized but unstimulated cells is ~10^{-7} mol.l^{-1}, i.e. 10,000 times lower than the concentration outside the cell. This extraordinary gradient is maintained by several Ca^{2+} transport systems which are summarized in Figure 4.6.

Among the most important of these systems are Ca^{2+} pumps which transport the ion across the membrane, coupled to hydrolysis of ATP on the cytoplasmic surface. These exist in the plasma membrane of all cells, but are best characterized from erythrocyte membranes. Reconstitution of the erythrocyte pump suggests that it transports one molecule of Ca^{2+} per ATP hydrolysed. It has a high affinity for Ca^{2+}, and its activity increases markedly if the Ca^{2+} concentration rises above the resting level of 10^{-7} mol.l^{-1}, by mechanisms discussed in Section 8.2.1. It will therefore tend to maintain cytosol Ca^{2+} at this level.

Similar Ca^{2+} pumps cause sequestration of Ca^{2+} at intracellular sites, now believed to be specialized regions of the endoplasmic reticulum. As discussed in

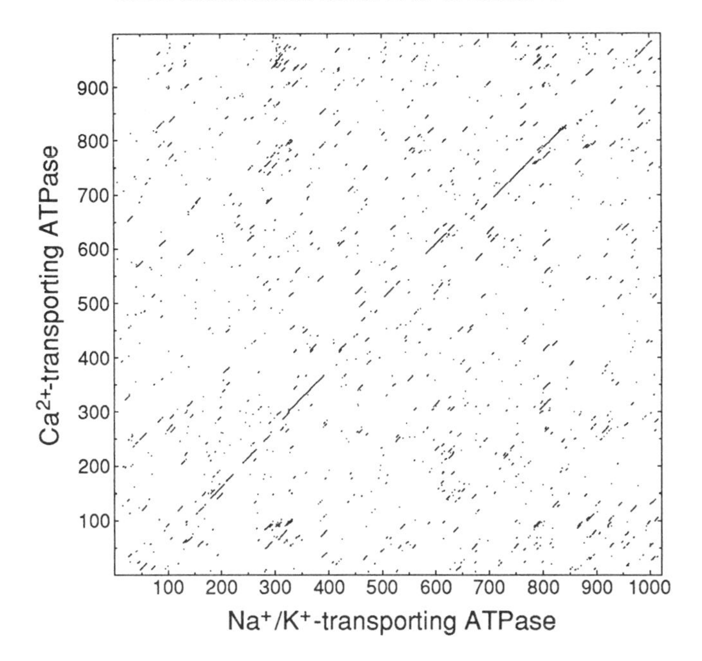

Fig. 4.7 Dot matrix plot comparing the sequences of the α subunit of the plasma membrane Na^+/K^+-transporting ATPase and the Ca^{2+}-transporting ATPase of sarcoplasmic reticulum. The distinct diagonal line indicates that the two sequences are related (Figure 3.15).

Chapter 7, these pools of internal Ca^{2+} have particularly important roles in the response to first messengers. A special case is the function of Ca^{2+} in initiating contraction of muscle (Section 8.1.3), in which there is a specialized type of endoplasmic reticulum known as **sarcoplasmic reticulum** that is very active in uptake and storage of the ion. The very abundant Ca^{2+} pumps of skeletal and heart muscle sarcoplasmic reticulum are particularly well characterized: unlike the plasma membrane pump they appear to transport two Ca^{2+} ions per ATP. Both have been cloned and sequenced, and their sequences are related to those of the α subunit of the Na^+/K^+ pump (Figure 4.7), suggesting that both types of pump arose by duplication and divergence of an ancestral, perhaps non-specific, ion pump. Both the Ca^{2+} and Na^+/K^+ pumps form catalytic intermediates in which a single aspartate residue is phosphorylated by ATP, and the amino acid sequences around these aspartate residues are particularly well conserved (Figure 4.8).

Although the plasma membrane Ca^{2+} pump has a high affinity, it does not appear to have the capacity to rapidly remove Ca^{2+} in cells in which there are regular and large rises in Ca^{2+} such as muscle cells and secretory cells. These and many other cells also have an active Ca^{2+}/Na^+ antiport system which is thought to have the stoichiometry $1Ca^{2+}:3Na^+$. This of course would be electrogenic, so that in a normally polarized cell the efflux of Ca^{2+} would be driven by the membrane

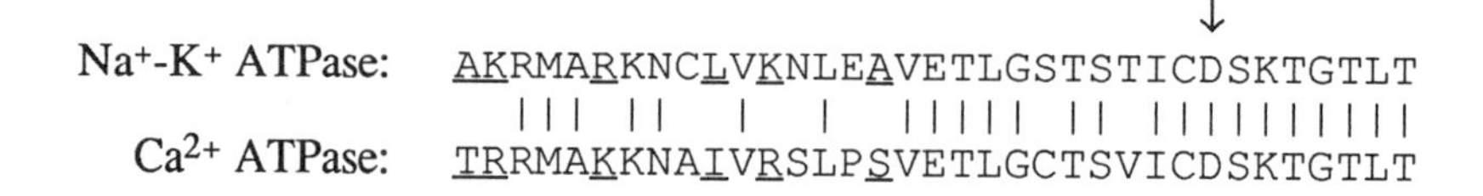

Fig. 4.8 Amino acid sequences around the reactive aspartate residue (D – arrowed) of the Na^+/K^+- and Ca^{2+}-transporting ATPases. The sequences are represented using the one-letter code (see Appendix). Identities are shown by vertical lines and conservative replacements are underlined.

potential as well as the concentration gradient of Na^+. This system appears to have a lower affinity for Ca^{2+} than the plasma membrane pump, and is probably a high capacity system for rapidly removing Ca^{2+}, with the fine-tuning of the concentration being the responsibility of the pump. The Ca^{2+}/Na^+ antiporter may help to explain why cardiac glycosides stimulate the force of contraction of the heart (Box 4.2), since inhibition of the Na^+/K^+ pump would reduce the ability of the Ca^{2+}/Na^+ antiporter to remove Ca^{2+} from the cell.

Transport of Ca^{2+} to the outside of the cell and into endoplasmic/sarcoplasmic reticulum are not the only mechanisms which maintain the low cytosolic Ca^{2+} concentration. Mitochondria also take up Ca^{2+} via uniport carriers. Like the plasma membrane, the inner mitochondrial membrane also carries a membrane potential generated by the vectorial uptake and release of H^+ by components of the electron transport chain: it is of course this electrochemical gradient of H^+ that "drives" synthesis of ATP in the mitochondria. However, the membrane potential component (typically ~160 mV, positive outside) also provides a driving force for Ca^{2+} entry which is particularly large because of the two charges on the ion. If this membrane potential was maintained, the equilibrium distribution of Ca^{2+} across the inner mitochondrial inner membrane (in:out) would be no less than 10^5:1! However, the system does not reach equilibrium because there are also electroneutral efflux pathways for Ca^{2+} in the mitochondrial membrane which reverse the uptake. The predominant pathway in most cells appears to be a $Ca^{2+}/2Na^+$ antiport. The kinetics of the mitochondrial uptake and efflux system are such that at extramitochondrial Ca^{2+} concentrations above 1 $\mu mol.l^{-1}$, the efflux pathway is saturated, and mitochondria will carry out a net uptake of Ca^{2+} until the extramitochondrial concentration falls to 1 $\mu mol.l^{-1}$ and the influx and efflux pathways balance. Mitochondria therefore act to buffer cytosolic Ca^{2+} concentration if it rises above 1 $\mu mol.l^{-1}$, a system which may prevent deleterious rises in Ca^{2+} under conditions where the rate of ATP generation cannot keep pace with the demands of the plasma membrane transporters. For more physiological changes in cytosol Ca^{2+} in the range 0.1 to 1 $\mu mol.l^{-1}$, there is now good evidence

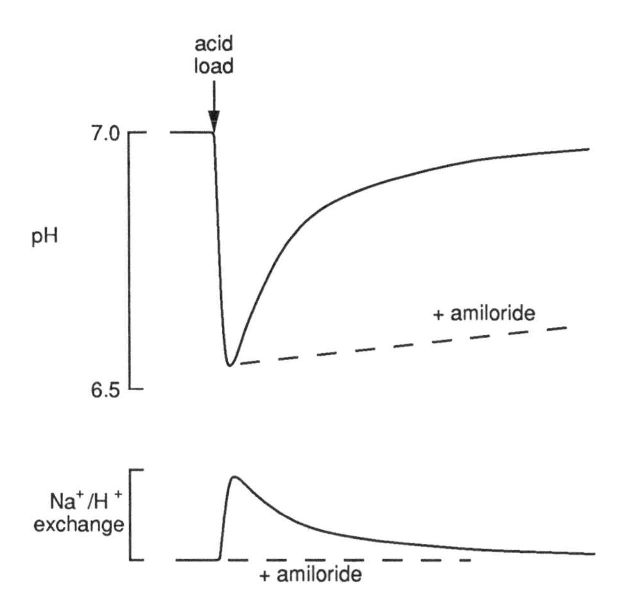

Fig. 4.9 Response of intracellular pH to an acid load (top), and the rate of Na^+/H^+ exchange measured in the same experiment (bottom). The effect of amiloride on both parameters is also shown.

that these mitochondrial transport systems have the rather different function of transmitting Ca^{2+} signals produced by chemical messengers into the mitochondrial matrix (Section 8.2.3).

4.1.3 *Regulation of cytosolic H^+ concentration (pH)*

The pH of the extracellular medium is generally ~7.4. For a normally polarized cell with a plasma membrane potential of 70 mV (positive outside), the Nernst equation predicts that H^+ would equilibrate across the membrane with a ratio of 10:1 (in:out) giving an internal pH of 6.4. In addition to the effects of membrane potential, metabolism produces acidic components such as lactic acid, CO_2 and ketone bodies which create an additional acid load. However pH values below 7.0 would be toxic to most enzymic processes, and in fact healthy cells maintain their cytosol pH between 7.0 and 7.4. Clearly there must be mechanisms for exporting H^+ from the cell, and the major mechanism in mammalian cells appears to be an electroneutral Na^+/H^+ antiport. Although not yet well characterized in structural terms, its activity can be studied by virtue of its inhibition by the diuretic drug, amiloride. Intracellular pH can be monitored via fluorescent reporters which bind protons at physiological concentrations, and which can be introduced into cells as permeant

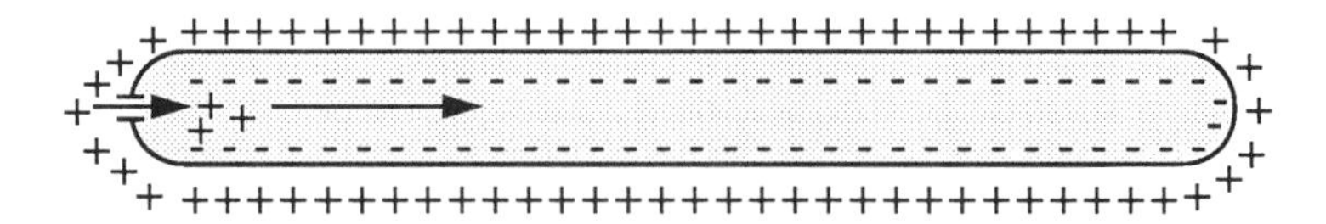

Fig. 4.10 Diagram showing how depolarization at one end of a cell would cause a current to flow towards the other end.

esters in an analogous manner to the Ca^{2+} reporters like fura-2. Using these methods, it can be shown that in response to an acid load, cells rapidly return cytosol pH to ~7, a process that is blocked by amiloride, and is dependent on extracellular Na^+ (Figure 4.9).

Since the ratio of $[Na^+]_{out}$:$[Na^+]_{in}$ is maintained at ~10:1 by the Na^+/K^+ pump, one would expect $[H^+]_{out}$:$[H^+]_{in}$ to reach ~1:10 if this antiport system came to equilibrium, i.e. an internal pH of ~8.4. That this does not happen indicates that the antiporter is being restricted kinetically. The exchanger appears to be dramatically activated by falls in pH (increases of $[H^+]$) below the physiological range, but to be almost inactive at physiological pH. This is most easily explained if H^+ is an allosteric activator as well as a substrate. The possibility that the antiporter is regulated by protein phosphorylation is discussed in Section 8.2.5.

As well as the well-known Na^+/H^+ antiport, at least three types of HCO_3^- porters have been described in different cell types. Their contribution to regulation of intracellular pH is often underestimated, partly because no specific inhibitors are available, but mainly because experiments are often carried out in the absence of HCO_3^- in the medium. One system is a Cl^-/HCO_3^- antiport, which is activated by a **rise** in pH, transporting out the base HCO_3^- to reduce pH. A specialized and abundant form of this system (**band 3 protein**) is found in red blood cells, which transport CO_2 from the tissues back to the lungs.

4.2 *Excitable cells and nervous conduction*

As already discussed in Chapter 1, neurones (nerve cells) are rather like endocrine cells which have grown out a long process (the axon) which means that the chemical messenger is released in very close proximity to the target cell. In order to fulfil its role as a means of very rapid cell–cell communication, however, a neurone must have a means of transmitting a signal from one end of the cell to the other. This is achieved via an electrical impulse known as the **action potential**.

4.2.1 *The action potential*

Neurones have all the systems for ionic homeostasis described in Section 4.1. As

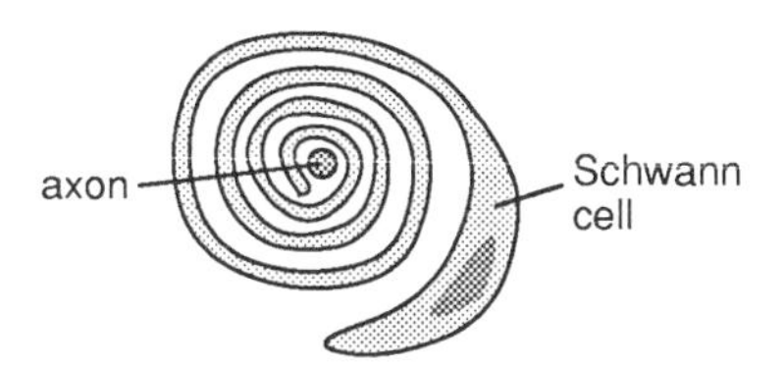

Fig. 4.11 Schematic diagram of myelination of a peripheral nerve by a Schwann cell. In reality the layers of membrane are much more tightly packed than shown.

in other cells, they have a resting membrane potential of about –70 mV (negative outside), and for K^+ the electrical gradient (favouring influx) is roughly balanced by the concentration gradient (favouring efflux), so that the distribution of K^+ across the membrane is close to equilibrium. However, for Na^+ both the membrane potential and the concentration gradient favour influx, so that there is a large driving force promoting entry of Na^+ into the cell.

Now imagine what would happen if, at one point on the membrane, a channel which allowed cations like Na^+ to pass through was to open transiently. This is just what happens when one nerve cell excites another (do not worry for now about the mechanism, which is discussed in Section 7.1). Na^+ ions would flood in down the concentration gradient and electrical gradient. It might be expected that this would collapse both gradients, but in practice it can be calculated that a very small flux of Na^+ ions is sufficient to collapse or even reverse the membrane potential, and that this will hardly affect the concentration gradient of Na^+. Because the membrane potential is reduced, the membrane at that point is said to be **depolarized**. The inside of the cell would at that point have an excess of positive ions: these would diffuse away rapidly, so that if the depolarization was at one end of the nerve cell an electrical current would flow towards the other end (Figure 4.10). However, because of the low voltages involved, the relatively low conductance of the cell interior, and the rather leaky nature of the plasma membrane, this current would not travel very far (a few millimetres at most). Such losses also occur in transmission of electricity down copper wire, but are minimized by the high conductivity of copper and the use of effective insulators around the wire. Vertebrate nerve cells have developed one method which partially overcomes this problem: myelination. Accessory cells called **oligodendrocytes** (central nervous system) or **Schwann cells** (peripheral nervous system) wrap numerous layers of membrane around the axon of the nerve cell to form a **myelin sheath** (Figure 4.11). This provides much better insulation than a single membrane, and leakage out of the side of the axon is much reduced. However, the strength of the signal would still diminish as it passed down the axon. The final solution to this problem which neurones exhibit is to have a system whereby the

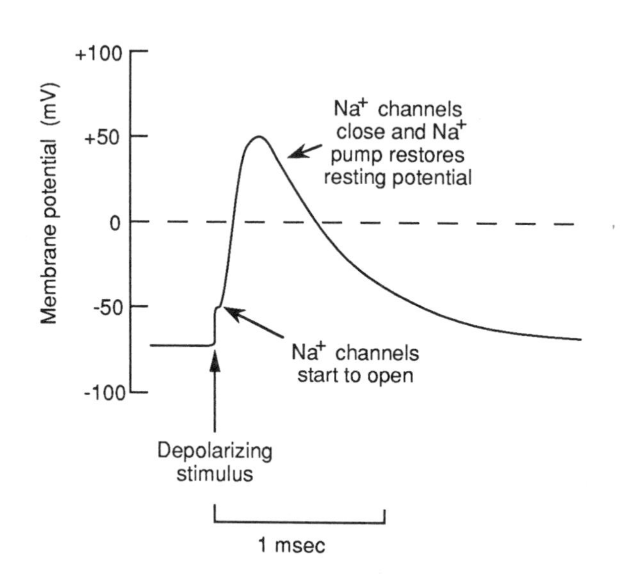

Fig. 4.12 Time course of voltage changes after a small depolarization at a particular point on excitable membrane.

depolarization is amplified as it passes down the axon.

This amplification is achieved by means of **voltage-gated ion channels**. It is the existence of these channels which makes a nerve cell membrane **excitable**, and which is one of the essential differences between a nerve cell and most other cell types. Although several different types of these channels exist, the most important for the action potential is the voltage-gated Na^+ channel. As their name suggests, these channels respond to changes in voltage. In particular a large number of these channels open transiently if the membrane is depolarized over a certain threshold, causing even more Na^+ ions to flood through the membrane – a kind of positive feedback system. Voltage recordings made using microelectrodes (which is relatively easy in the giant unmyelinated axons of molluscs like the squid) show that the polarity of the membrane potential actually reverses, to about +50 mV (Figure 4.12), which is close to the equilibrium potential for Na^+. Since the channels only open transiently, the influx of ions ceases after a few milliseconds, and the Na^+/K^+ pump and the K^+ leak channels more slowly re-establish the normal resting membrane potential (Figure 4.12). During this phase the Na^+ channels are in a refractory closed state, and will not respond to any further voltage changes until the normal resting membrane potential has been restored.

In myelinated nerves the voltage-gated Na^+ channels are clustered in regions where there is a break in the myelination called the **nodes of Ranvier**. The signal passes via simple ion conduction from one node to the next, where the depolarization is amplified back to the original strength. In non-myelinated nerves

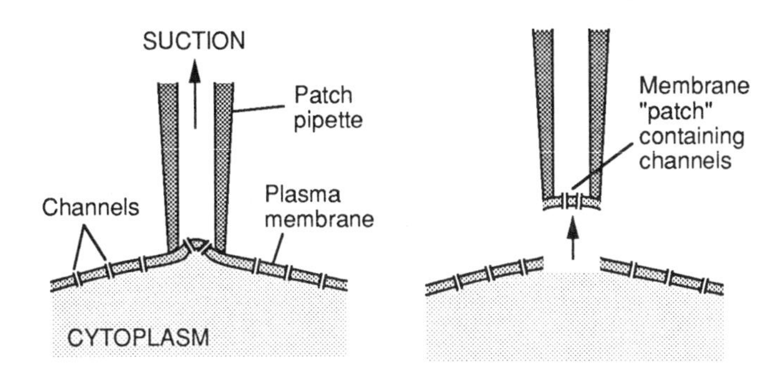

Fig. 4.13 The principle of the **patch-clamp** technique. The current flowing through the isolated patch of membrane can be monitored using sensitive electrical equipment. An alternative approach, to study the current flowing across the whole cell, is to suck away the membrane patch, leaving the remainder of the cell membrane still attached to the pipette (Figure 7.23).

the channels are more evenly distributed, and a self-sustaining wave of depolarization, the action potential, passes down the axon at speeds up to 100 $m.s^{-1}$. At any one point along the axon, the changes in membrane potential have the time course shown in Figure 4.12. When this particular point on the axon reaches maximum depolarization, the region just upstream will be in its recovery phase, while the region just downstream would be experiencing an initial depolarization and its voltage-gated channels would be opening. An important point to note about action potentials is that because of their self-sustaining nature by positive feedback they are always of the same magnitude: their frequency can be increased (up to a limit) or decreased, but not their amplitude.

There are now several ways in which the voltage-gated Na^+ channels can be studied in more detail. The first is patch clamping (Figure 4.13), in which a micropipette is pressed onto the membrane and a gentle suction applied to detach a tiny portion of membrane less than 1 μm in diameter. With luck this "patch" will contain a few channels, and it is possible to maintain (or "clamp") a voltage across this patch, and measure the current that flows using very sensitive recording equipment. With great care and some luck it is possible to record the activity of a single channel using this technique, a remarkable opportunity to study the function of an individual protein molecule. The results show that individual channels open and close abruptly, and their open conductance is always the same (Figure 4.14): in other words, there is no such thing as a half-open channel. At a potential of 100 mV, about 6000 Na^+ ions flow through in one millisecond. Individual channels open at random intervals: however, the probability that they will open depends on the membrane potential.

The second technique to study the channels is the biochemical approach. A particularly useful finding made by the electrophysiologists was that certain marine

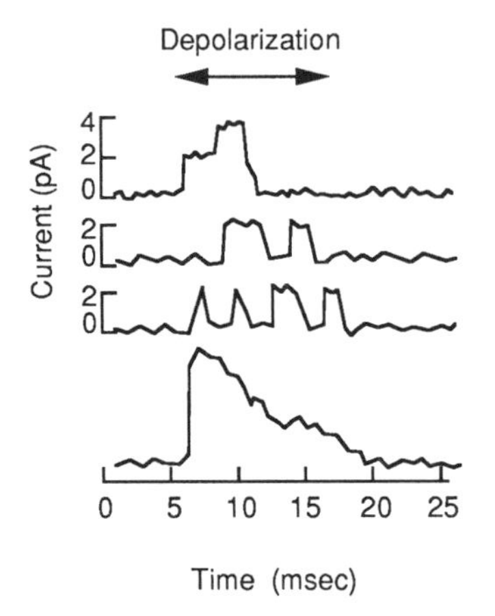

Fig. 4.14 Idealized voltage recordings from patch clamp experiment on voltage-gated Na^+ channels subjected to a transient depolarization. Due to the high amplification involved, there is a certain amount of background noise. Individual channels open and close at different times, but when open the current that flows is always ~2 pA. In the top trace, two channels in the patch are briefly open together, and the current is ~4 pA. The bottom trace shows a summation of a large number of individual recordings.

Tetrodotoxin **Saxitoxin**

Tetrodotoxin and saxitoxin potently block the action of the voltage-gated sodium channel. Tetrodotoxin is produced by the *fugu* or puffer fish, considered a great delicacy in Japan. The liver and other parts of the fish contain lethal concentrations of the toxin, so *fugu* can now only be prepared legally in Japan by specially licenced chefs. Truly a dish to tickle the palate! Saxitoxin is produced by dinoflagellates, blooms of which form the so-called "red tide". It accumulates in filter feeders such as mussels, in which it too can occur in concentrations lethal to humans.

Box 4.3 Marine toxins which block the voltage-gated sodium channel

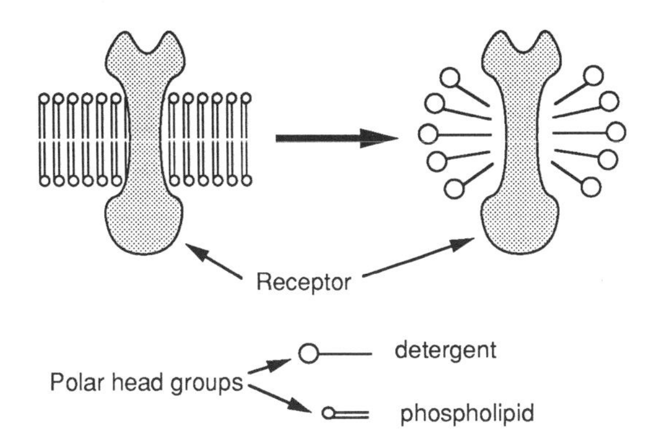

Fig. 4.15 Solubilization of an integral membrane protein using detergent. Detergents are amphipathic molecules which tend to form micelles (as opposed to the bilayers formed by phospholipids) and which can shield the hydrophobic portions of proteins to prevent aggregation.

toxins such tetrodotoxin and saxitoxin (Box 4.3) inhibit the channels. This was crucial because it allowed purification of the channel protein. The problem had been that in order to purify such a membrane protein, it has to be solubilized first in detergent (Figure 4.15), but that when this is done with an ion channel, it becomes of course completely non-functional. Toxin binding provided a function of the channel that could still be monitored in a solubilized preparation. Eventually a single large polypeptide of about 2000 residues was purified: this was cloned and sequenced, and complementary RNA was synthesized and injected into oocytes (egg cells) of the toad *Xenopus,* which do not normally express the channel (Figure 4.16). Functional, tetrodotoxin-sensitive voltage-gated Na^+ channels were obtained. This proved beyond doubt that the single polypeptide alone was responsible for channel activity. The sequence of the polypeptide showed that it contained four internally homologous repeats, which are clearly seen in the hydropathy profile (Figure 4.17). The hydropathy profile suggests that each of these repeats has six stretches of hydrophobic amino acids that might form transmembrane helices. It is proposed that these four repeats form a pseudosymmetrical structure with a four-fold axis at right angles to the membrane (Figure 4.18), lining the central pore which represents the ion channel. This structure is analogous to the six subunit structure of the gap junction (Section 1.3.1), except that the pore is much larger in the latter case. The sequences of the fourth putative helix in each repeat of the Na^+ channel sequence are particularly interesting in that they contain conserved, regularly spaced positively charged residues between the hydrophobic residues (Figure 4.19). It is speculated that this helix moves in the membrane in response to depolarization, somehow opening the ion channel. Although the voltage difference across the plasma membrane is small,

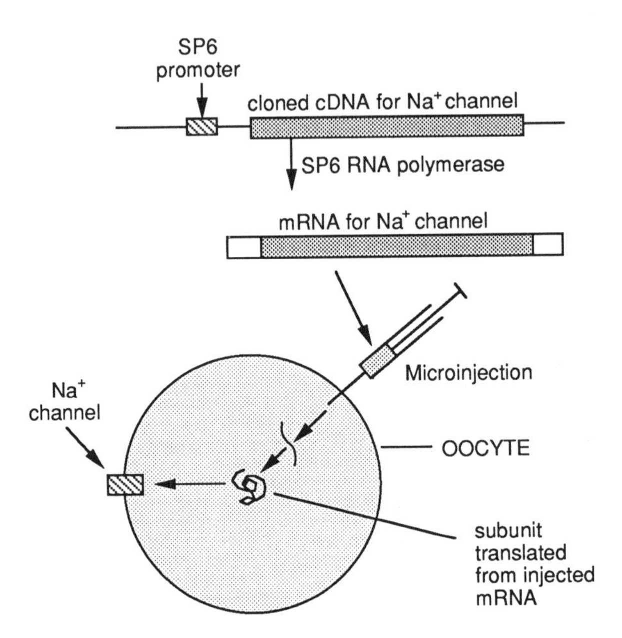

Fig. 4.16 Principle of the method by which cloned Na^+ channels can be expressed in *Xenopus* eggs.

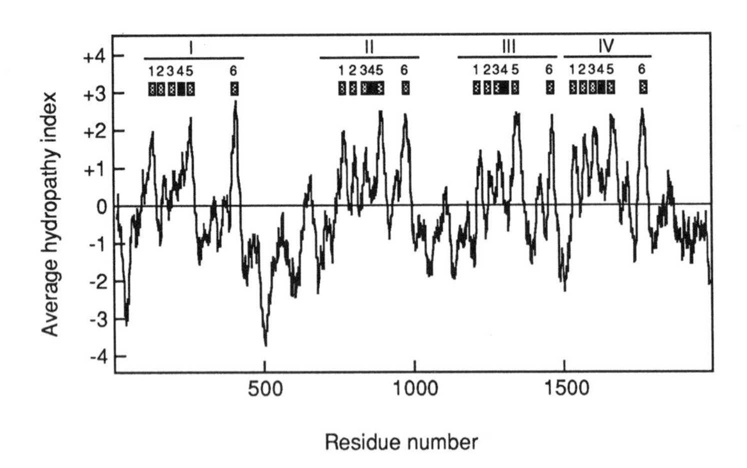

Fig. 4.17 Hydropathy profile of the voltage-gated Na^+ channel, showing the four sequence repeats, each of which contains six hydrophobic segments. The 4th segment in each repeat (marked by black boxes) has a lower hydropathy index than the others due to the presence of charged residues. See Figure 1.12 for an explanation of hydropathy profiles.

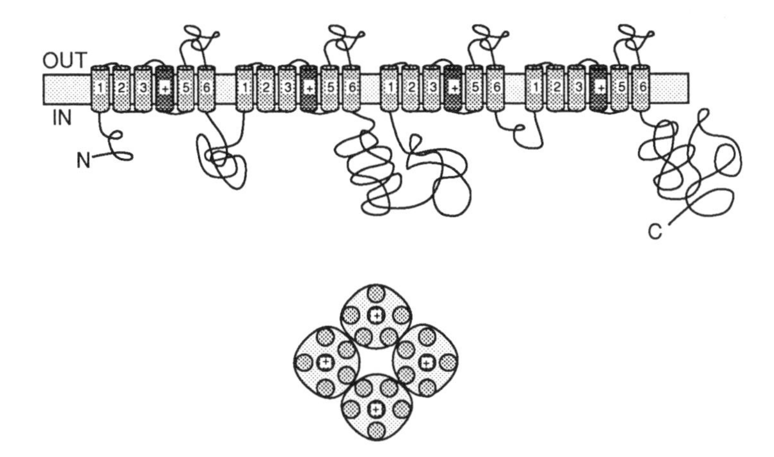

Fig. 4.18 Speculative model for the membrane topology of the voltage-gated Na^+ channel viewed from within the plane of the membrane (top), and a model for the manner in which the 24 helices might pack together, viewed at right angles to the plane of the membrane (bottom). It is assumed that all 6 hydrophobic segments have a transmembrane orientation: the very basic 4th segment is marked "+".

Repeat 1: V-S-A-L-**R**-T-F-**R**-V-L-**R**-A-L-**K**-T-I-S-V-I

Repeat 2: L-S-V-L-**R**-S-F-**R**-L-L-**R**-V-F-**K**-L-A-**K**-S-W

Repeat 3: I-**K**-S-L-**R**-T-L-**R**-A-L-**R**-P-L-**R**-A-L-S-**R**-F

Repeat 4: I-**R**-L-A-**R**-I-G-**R**-I-L-**R**-L-I-**K**-G-A-**K**-G-I

Fig. 4.19 Alignment of sequences of hydrophobic segment 4 in each repeat of the voltage-gated Na^+ channel. The single letter amino acid code is used (see Appendix). Basic residues (K = lysine, R = arginine) are shown in bold type.

it exists across a very narrow gap, so the actual voltage gradient is about 150,000 $V.cm^{-1}$! It would not be surprising if a lessening in this voltage gradient could trigger a conformational change. A second conformational change must be responsible for closing of the channel when depolarization is complete.

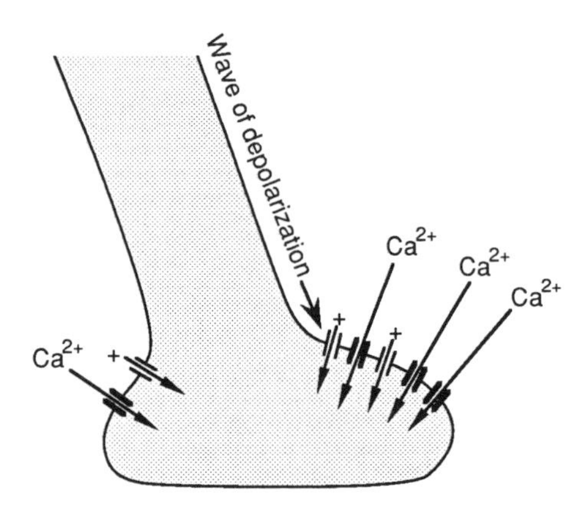

Fig. 4.20 Diagram showing how voltage-gated Ca^{2+} channels in the nerve ending turn an electrical signal (the action potential) into a chemical signal (an increase in Ca^{2+} concentration).

4.2.2 *Release of neurotransmitters at the synapse*

Although the passage of an action potential involves movement of Na^+ and K^+ ions across the membrane, the flux of these ions is so small compared with their concentrations in the cytosol and extracellular fluid that there is no significant change in the chemical gradients of the ions. How then is this essentially electrical phenomenon, i.e. the action potential, converted into a chemical change, i.e. release of neurotransmitter, at the nerve ending? The crucial event is the opening of **voltage-gated Ca^{2+} channels** in the plasma membrane of the nerve ending. When the action potential reaches the nerve ending, these are triggered by the depolarization to open transiently, and allow Ca^{2+} to enter the cell down its very large concentration gradient (and down the electrical gradient if depolarization is not complete) (Figure 4.20). Although the flux through one of these channels is probably no larger than the small flux of Na^+ through voltage-gated Na^+ channels, the crucial difference is that resting intracellular Ca^{2+} concentration (10^{-7} mol.l^{-1}) is five orders of magnitude lower than that of Na^+ (~10^{-2} mol.l^{-1}). The influx of Ca^{2+} therefore does cause a real increase in the intracellular concentration of Ca^{2+}, probably by at least 10-fold.

There is very good evidence that this increase in Ca^{2+} concentration triggers release of neurotransmitter. The mechanism, although not well understood, is considered in more detail in Chapter 5.

4.2.3 *Excitability in cells other than nerves*

An excitable membrane is one which carries voltage-gated ion channels. Although nerve cells provide the classic examples of excitability, excitable membranes do occur on many other cell types, and probably all cells have at least some voltage-gated ion channels. Important excitable membranes occur on skeletal and cardiac muscle cells. Skeletal muscle is discussed in more detail in Chapter 8. Unlike skeletal muscle, the heart can contract (beat) on its own, without signals from the central nervous system. This is because it contains a "pacemaker" region called the sinu-atrial node which initiates action potentials. The heart muscle cells (myocytes) have both excitable membranes, and gap junctions (Chapter 1) which can pass an electrical signal from one cell to its neighbour. So a wave of depolarization passes through the chamber of the heart, initiating a wave of contraction.

Even the insulin-secreting β-cells of the Islets of Langerhans in the pancreas have excitable regions on the plasma membrane. Stimulation of the cells with an agent causing secretion (in this case, glucose) causes repeated spikes of depolarization in the plasma membrane which can be measured using microelectrodes. This is not directly mediated by glucose itself, but by an unknown metabolite, because inhibitors of glucose metabolism block the effect. The depolarization is initiated by closure of K^+ channels, which in turn causes opening of voltage-gated Ca^{2+} channels. The increase in cytosolic Ca^{2+} then triggers secretion of insulin, by mechanisms discussed in the next chapter.

SUMMARY

All cells strictly regulate their intracellular ionic composition by means of transport systems in the plasma membrane and on internal membranes. Transport can be catalysed by two types of membrane proteins: porters or channels. Normally transport only occurs down electrochemical gradients, but there are a few porters where direct coupling to hydrolysis of ATP can be used to drive transport (e.g. of Na^+ or Ca^{2+}) against a gradient. Gradients of these ions can then be used to transport other molecules against electrochemical gradients by means of co-transport systems.

All cells maintain opposing gradients of Na^+ and K^+ across the plasma membrane by means of a ouabain-sensitive Na^+/K^+-transporting ATPase. Together with the action of K^+ leak channels, this results in a membrane potential of about –70 mV (positive outside). Cells maintain a very large gradient of Ca^{2+} across the plasma membrane by means of a Ca^{2+}-transporting ATPase and several other transport processes. Internal pH is regulated by an amiloride-sensitive Na^+/H^+ co-transport system, and by HCO_3^- transporters.

An excitable membrane such as that of a nerve axon contains voltage-gated Na^+ channels. A small depolarization of the membrane triggers the transient *opening of these channels, resulting in a self-sustaining wave of depolarization, the action potential, which passes down the axon. At the nerve ending the action potential is converted back to a chemical signal via voltage-gated Ca^{2+} channels. Cells other than nerves can also have excitable membranes: these are important, for example, in muscle contraction and insulin secretion.*

FURTHER READING

1. "Ion motive ATPases. Ubiquity, properties, and significance to cell function" – review. Pedersen, P.L. and Carafoli, E. (1987) Trends Biochem. Sci. **12**, 146–150.
2. "Sodium pump: birthday present for digitalis" – minireview. Allen, D.G., Eisner, D.A. and Wray, S.C. (1985) Nature **316**, 674–675.
3. "Amino acid sequence of the catalytic subunit of the (Na^+ + K^+)ATPase deduced from a complementary DNA". Shull, G.E., Schwartz, A. and Lingrel, J.B. (1985) **316**, 691–695.
4. "Intracellular Ca^{2+} homeostasis" – review. Carafoli, E. (1987) Ann. Rev. Biochem. **56**, 395–433.
5. Fluorescence ratio imaging: a new window into intracellular ionic signalling" – review on use of fura-2 for measurement of intracellular Ca^{2+}. Tsien, R.Y. and Poenie, M. (1986) Trends Biochem. Sci. **11**, 450–455.
6. "Amino acid sequence of a Ca^{2+} + Mg^{2+}-dependent ATPase from rabbit muscle sarcoplasmic reticulum, deduced from its complementary DNA sequence". MacLennan, D.H., Brandl, C.J., Korczak, B. and Green, N.M. (1985) Nature **316**, 696–700.
7. "Regulation of cytoplasmic pH by Na^+/H^+ exchange" – review. Moolenaar, W.H. (1986) Trends Biochem. Sci. **11**, 141–143.
8. "The regulation of the intracellular pH in cells from vertebrates" – review. Frelin, C., Vigne, P., Ladoux, A. and Lazdunskji, M. (1988) Eur. J. Biochem. **174**, 3–14.
9. "Primary structure of *Electrophorus electricus* sodium channel deduced from cDNA sequence". Noda, M., Shimizu, S., Tanabe, T., Takai, T., Kayano, T., Ikeda, T., Takahashi, H., Nakayama, H., Kanaoka, Y., Minamino, N., Kangawa, K., Matsuo, H., Raftery, M.A., Hirose, T., Inayama, S., Hayashida, H., Miyata, T. and Numa, S. (1984) Nature **312**, 121–127.

FIVE

CELL BIOLOGY OF FIRST MESSENGER SYNTHESIS AND SECRETION

How are first messengers secreted from the cell in which they are synthesized? Evidence has been accumulating in recent years that this usually, if not always, occurs by the process of **exocytosis**. The essential features of this process are that the messenger is stored inside membrane-bound vesicles in the cytosol. In response to a stimulus for secretion, which might be a nervous impulse, another first messenger, or in the case of the pancreatic β cell, a metabolite (glucose), some of these vesicles fuse with the plasma membrane (Figures 5.1) to release their contents into the extracellular fluid (Figure 5.2). Exocytosis can also be visualized dramatically by electron microscopy of mast cells, which secrete the mediator histamine (see Figure 6.1).

We will consider later in the chapter the mechanism by which vesicle fusion with the plasma membrane takes place. First the question of how the messenger is incorporated into the secretory vesicle must be considered. The mechanism differs for different types of messenger, and we will start by considering peptide messengers.

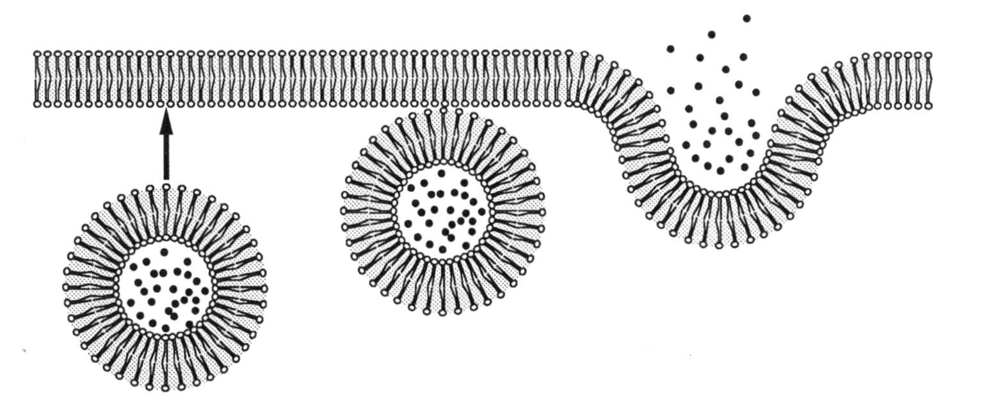

Fig. 5.1 The process of exocytosis involves fusion of bilayers.

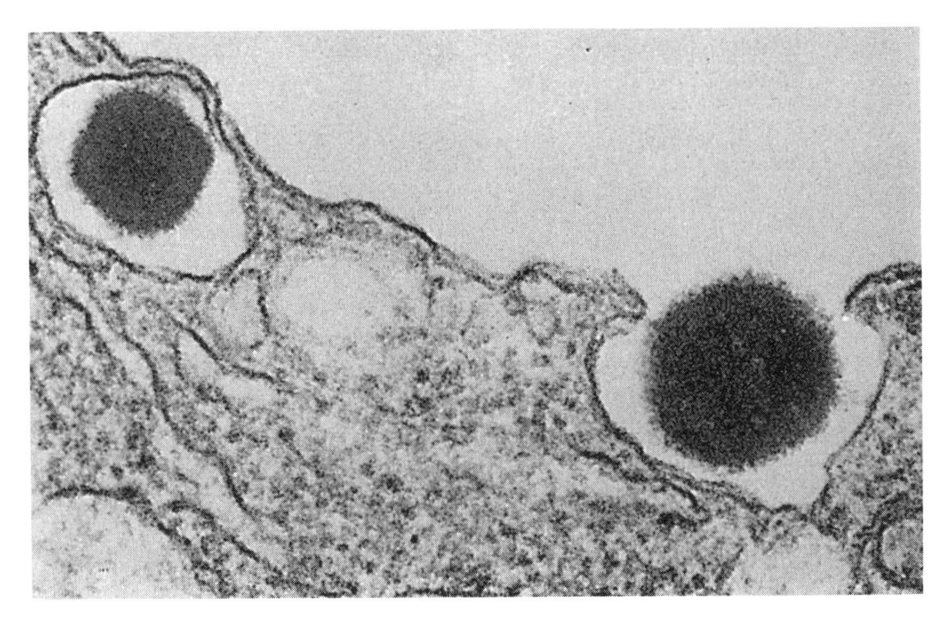

Fig. 5.2 Electron micrograph showing exocytosis of insulin-containing secretory granules from pancreatic β-cells. Courtesy of Lelio Orci.

5.1 *Cell biology of peptide messenger synthesis*

5.1.1 *Targetting of proteins for secretion*

Electron microscopy of the β cells of the Islets of Langerhans in the pancreas, which secrete insulin, shows that they contain features typical of other cells which secrete proteins. They contain numerous electron-dense secretory granules lying close to the plasma membrane, which contain stored insulin (Figure 5.3). In addition, the cells contain around the nucleus parallel arrays of endoplasmic reticulum studded with ribosomes: the so-called rough ER. Since insulin represents 70% of the protein synthesized by these cells, this suggested that the rough ER was the site of insulin synthesis. Indeed it is now clear that all proteins destined for secretion, including insulin, are synthesized on membrane-bound ribosomes.

The problem of how protein synthesis is targetted to free or membrane-bound ribosomes has now been solved in some detail (Figure 5.4). Messenger RNAs coding for either secreted or cytosolic proteins are initially translated in the same way on free ribosomes, but the growing polypeptide for a secreted protein contains at the N-terminus a stretch of hydrophobic amino acids known as the **signal sequence**. When this sequence emerges from the ribosome, it binds to a ribonucleoprotein known as the **signal recognition particle** (SRP). This

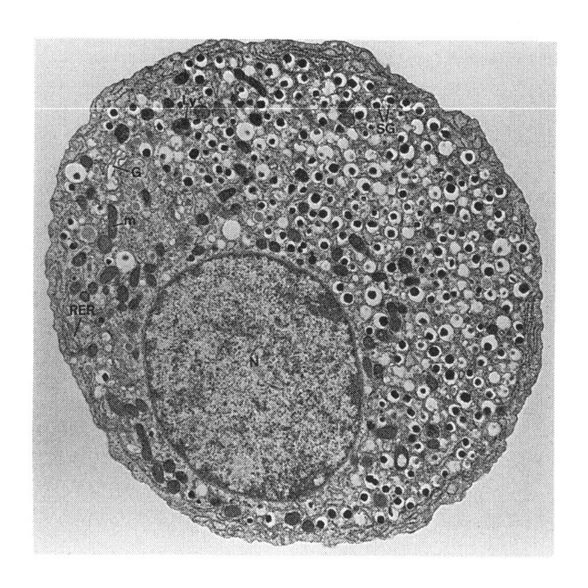

Fig. 5.3 Electron micrograph of insulin-secreting β-cell of the pancreas. The dark granules in the cytoplasm are secretory granules containing stored insulin. Key: SC, secretory granules; RER, rough endoplasmic reticulum; G, Golgi apparatus; Ly, Lysosome; m, mitochondria, N, nucleus. Courtesy of Lelio Orci.

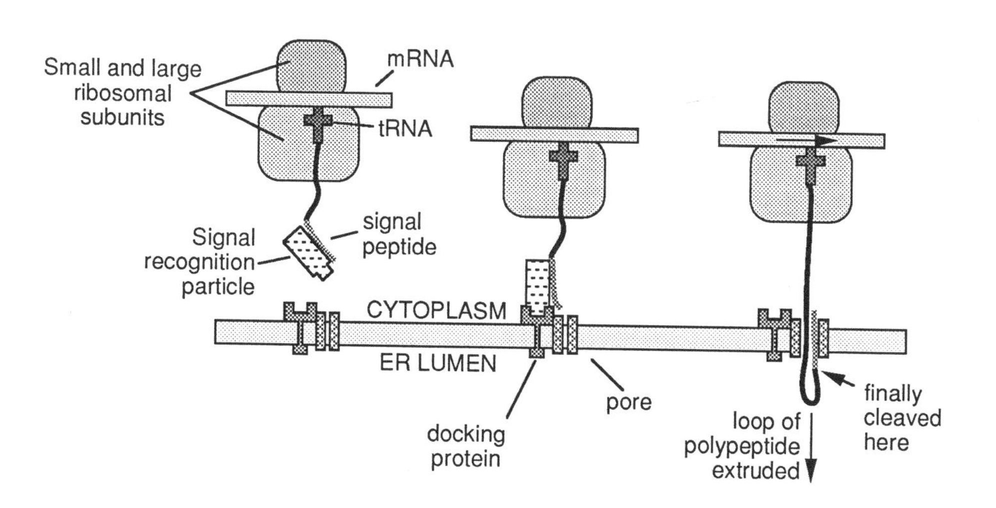

Fig. 5.4 The signal peptide targets a secreted protein to the endoplasmic reticulum.

Preproinsulin: M A R **F L** G **L** C T **W L L** A **L** G P G **L L** A T **V**

Preproenkephalin: M A R **F L** G **L** C T **W L L** A **L** G P G **L L** A T **V**

Preprodynorphin: **M** A WQ G **L L L** AA C **L L V L** P S T **M** A D

Vasopressin precursor: **M** P D A T **L** P A C **F L** S **L L** A **F** T S A

Fig. 5.5 Examples of signal sequences on some prepromessengers. Amino acids are represented using the one letter code (see Appendix). Hydrophobic residues (M = methionine; W = tryptophan; L = leucine; V = valine) are shown in bold type.

particle contains one molecule of RNA and several polypeptides, and appears to have several functions: 1) it recognizes and binds to the signal sequence; 2) when this happens, protein synthesis temporarily halts; 3) it mediates binding to a receptor (termed the **SRP receptor** or **docking protein**) on the surface of the ER. Protein synthesis then continues, but the growing polypeptide is now extruded into the lumen of the ER as a loop of polypeptide, the signal sequence remaining bound at the membrane. Finally, the signal sequence is cleaved off by **signal peptidase**, leaving the mature protein (or, in the case of peptide messengers, the promessenger) in the lumen. Proteins which are destined to span the plasma membrane follow a similar pathway, except that they contain a hydrophobic sequence which is not cleaved by signal peptidase and which remains in the membrane.

This pathway has now been confirmed for many examples of peptide messengers as well as other secreted proteins. The signal sequence for proinsulin was already alluded to briefly in Chapter 3: the initial translation product is called **preproinsulin**, but is cleaved by the signal peptidase into proinsulin. A few examples of signal sequences involved in promessenger synthesis are shown in Figure 5.5.

5.1.2 *The pathway from the rough ER to the secretory vesicle*

How does the protein get from the rough ER to the secretory vesicle? It is now clear that the Golgi apparatus is an obligatory intermediate in this pathway. Initial clues to this came from studies of secreted glycoproteins (recall that some peptide messengers, e.g. lutropin, are glycoproteins). The commonest type of carbohydrate attachment to proteins is via asparagine side chains (**N-linked glycosylation**), which is found exclusively on secreted proteins and the ectoplamic domains of cell surface proteins. The enzymes responsible for the initial attachment of carbohydrate residues to asparagine are found in the ER, but further modification of N-linked carbohydrate, and also glycosylation of serine or

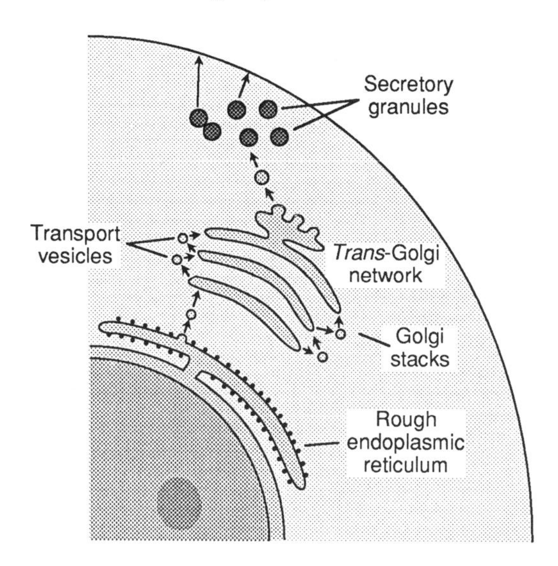

Fig. 5.6 The pathway of a secreted protein from the rough endoplasmic reticulum to the plasma membrane.

threonine residues (**O-linked glycosylation**) appears to occur exclusively inside the Golgi apparatus. Our present model of the overall pathway is summarized in Figure 5.6. The sequence of events is believed to be as follows:

1) Proteins to be secreted are extruded into the lumen of the ER by the mechanisms outlined in Section 5.1.3. In contrast, proteins destined for the plasma membrane remain anchored in the ER via their transmembrane regions, with their ectoplasmic portions on the luminal side.
2) Vesicles containing the protein bud off the ER and are transported to the first (*cis*) cisternum of the Golgi apparatus, which they enter by vesicle fusion.
3) By more budding and fusion of vesicles, the protein gradually works its way through the Golgi cisternae to that on the *trans*-Golgi network, facing the plasma membrane. While this happens, carbohydrate residues are progressively added and removed to and from glycoproteins.
4) At the *trans*-Golgi network proteins are sorted according to their ultimate destination. Vesicles containing first messengers for secretion leave the network and progressively mature to become secretory vesicles.

This pathway has been confirmed for many secreted proteins and peptide messengers, including some which are not glycosylated. The synthesis of insulin in β-cells has been particularly well studied by autoradiography. Cells are incubated for various periods of time with radioactive (^3H) amino acid, which will be incorporated into newly synthesized proinsulin. Thin stained sections of the

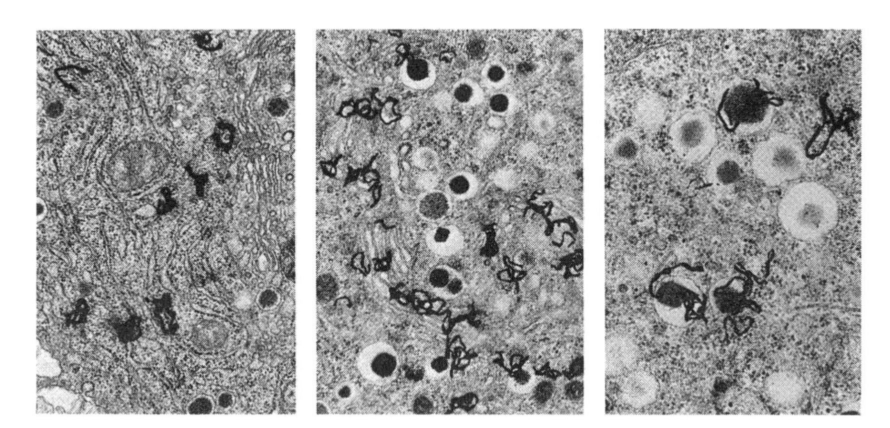

Fig. 5.7 Electron microscope autoradiography of pancreatic β cells that had been incubated with radioactive amino acid to label newly synthesized protein (predominantly insulin) and then "chased" by washing with non-radioactive amino acid. The black twisting tracks are caused by radioactive emission of β particles, and therefore indicate the approximate location of radioactive protein. Five minutes after completion of labelling (left), radioactivity is mainly associated with rough ER; after 10 minutes (centre) it has migrated to the Golgi apparatus, while after 60 minutes (right) it is exclusively associated with secretory granules. Courtesy of Lelio Orci.

cells are prepared, and coated with a photographic emulsion. β-Particles derived from the radioactive protein will reduce silver atoms in the emulsion and leave a characteristic twisting track which can be visualized in the electron microscope, superimposed on top of the cell structures. These experiments show that 5 minutes after adding the radioactive amino acid, most of the radioactivity is associated with rough ER (Figure 5.7). After 10 minutes, it has migrated to the Golgi, while after 1 hour most of the radioactivity is found over the secretory granules.

It now appears that there are at least two pathways from the Golgi to the plasma membrane. One is the so-called **constitutive** pathway followed by proteins whose secretion is not regulated. These proteins migrate in vesicles directly from the Golgi to the plasma membrane (Figure 5.8). The secretion of peptide messengers, on the other hand, must be strictly regulated, and these messengers leave the Golgi via different vesicles which are lined on their cytoplasmic face by a cage-like structures made from the structural protein **clathrin** (Figure 5.9). Shortly after they leave the Golgi, these vesicles lose their clathrin coat (uncoating), and their contents condense to form the mature secretory granule. It is not yet clear how the secretory proteins which follow the "regulated pathway" are directed to the clathrin-coated vesicles: presumably they contain a sorting signal which causes them to be concentrated in the coated vesicle: although aggregation of the secretory

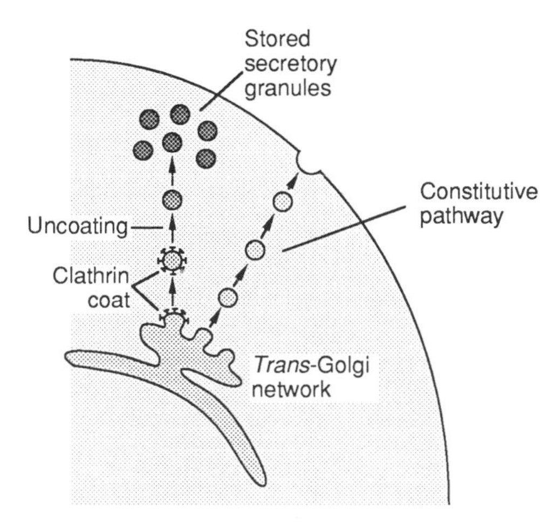

Fig. 5.8 Constitutive and regulated pathways for protein secretion.

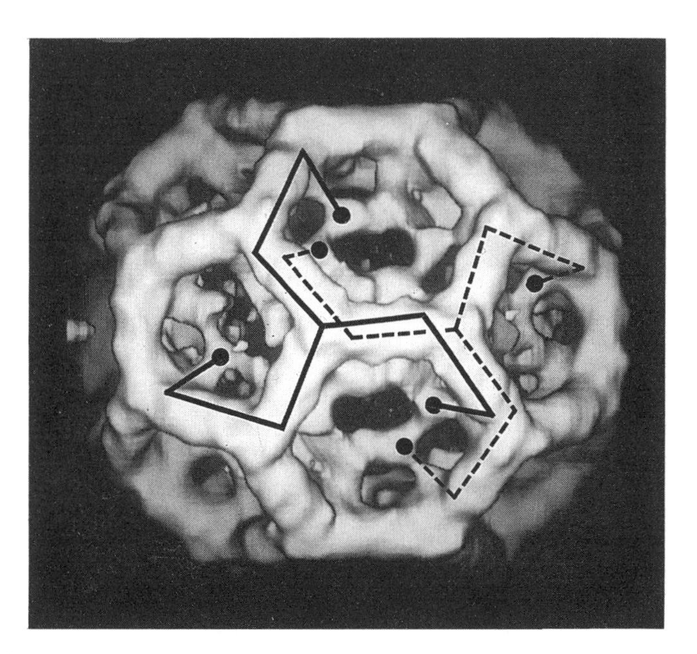

Fig. 5.9 Three-dimensional reconstruction of a clathrin coat which has been reconstituted from purified clathrin into a "cage" formed from 12 pentagons and 8 hexagons. Individual clathrin molecules are known as **triskelions,** each formed from three large and three small polypeptides. Two of these are represented as continuous or dashed lines, showing how they may assemble into the cage structure. Reproduced from Vigers, Crowther and Pearse (1986) EMBO J. **5** 529–534, courtesy of Barbara Pearse.

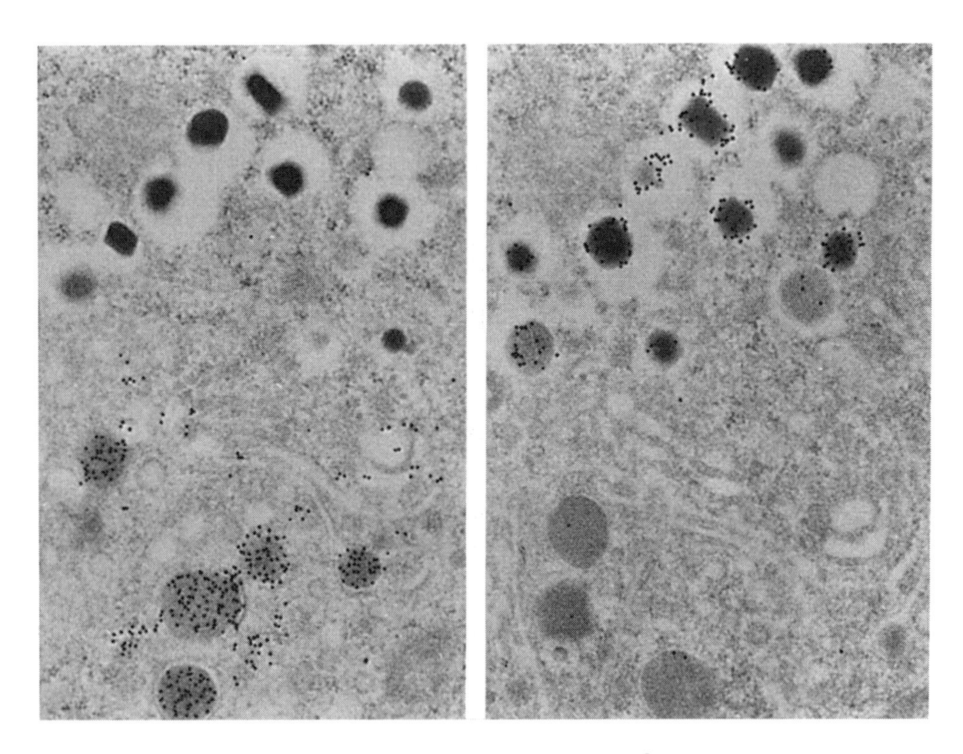

Fig. 5.10 Consecutive sections from a pancreatic β cell stained with antibodies specific for proinsulin (left) or insulin (right). The antibodies were labelled with colloidal gold, and are visible as small black dots. Proinsulin is found only in the Golgi and coated granules (left), whereas insulin is most abundant in the electron-dense, mature uncoated granules (right). Courtesy of Lelio Orci.

protein may also play a role. It is thought that proteins which lack such a signal passively enter the constitutive pathway, which is sometimes therefore called the **default** pathway.

At what stage does proinsulin get converted to insulin? This question was answered by staining thin sections of cells using antibodies specific for proinsulin or insulin. The antibodies were coupled to colloidal gold particles, which being very dense can be visualized in the electron microscope. These experiments show that proinsulin is found in the Golgi, and also in the clathrin-coated vesicles. Insulin, on the other hand, is found in small amounts in the clathrin-coated vesicles, but much more prominently in the mature secretory vesicles, which do not have a clathrin coat (Figure 5.10). Thus proinsulin seems to be converted into insulin as the clathrin-coated vesicle is converted to the uncoated vesicle. While this is happening, there is evidence that the vesicle is also acidified relative to the cytoplasm, probably via mechanisms similar to those described in Section 5.2. It is

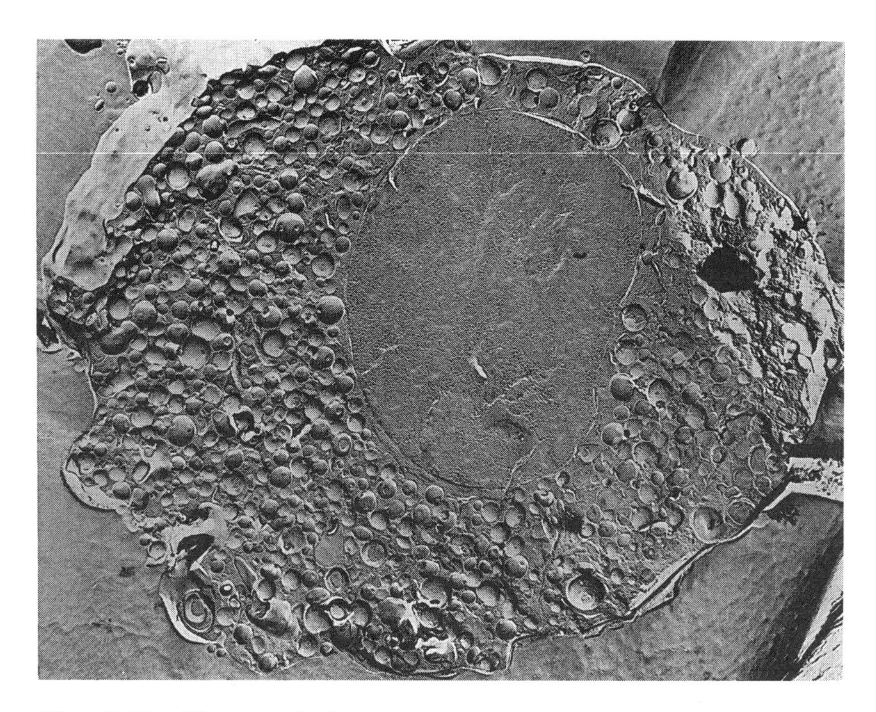

Fig. 5.11 Platinum shadowed electron micrograph of adrenal chromaffin cell showing prominent **chromaffin granules**. Courtesy of Hans Winkler.

likely that this acidification activates the proteolytic enzymes which convert proinsulin to insulin, since incubation of cells with weak bases which prevent this acidification inhibits the conversion. Like most messengers, insulin is less soluble than its precursor (proinsulin), particularly in the low pH of the secretory vesicle, and it crystallizes out to yield the characteristic electron-dense secretory vesicles seen in Figure 5.3. This mechanism allows the endocrine cell to store a large amount of messenger in a small volume, and ensures that when exocytosis does occur, a reasonable extracellular concentration of messenger is achieved even for a hormone which is diluted in the bloodstream.

5.2 *Import of non-peptide messengers into secretory granules*

Unlike peptide messengers, most non-peptide messengers are synthesized in the cytoplasmic compartment of the cell. How then do they become incorporated into the secretory vesicles? This has been studied in most detail in chromaffin cells, the cells which secrete adrenaline (epinephrine) and noradrenaline (norepinephrine) from the adrenal medulla (Figure 5.11). Unlike most neurones, homogeneous

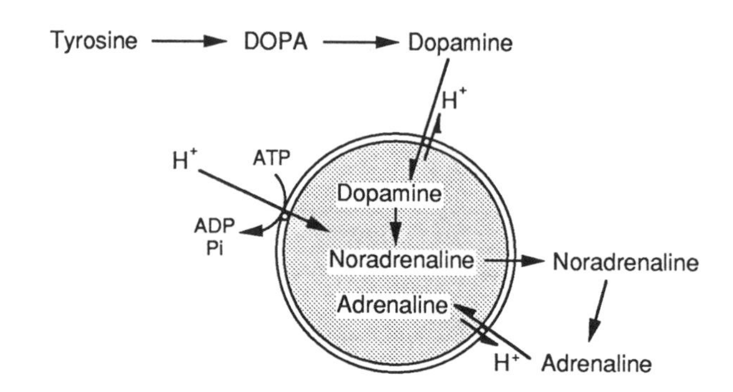

Fig. 5.12 Localization of catecholamine-synthesizing enzymes, and the proton-pumping ATPase, in chromaffin granules.

populations of chromaffin cells can be readily obtained in reasonable amounts, and the secretory granules (known as **chromaffin** granules from the fact that they stain readily with chromium salts) can be isolated by centrifugation.

Adrenaline is synthesized from tyrosine in a four step metabolic pathway (Section 3.1.2). The first two steps, leading to the neurotransmitter dopamine, are carried out in the cytosol, but the third enzyme, which converts dopamine to noradrenaline, is found in the chromaffin granules themselves (Figure 5.12). In about 10% of cells in the adrenal medulla, synthesis ceases at this point, and these cells release noradrenaline only. In the remainder, noradrenaline is converted to adrenaline by the fourth enzyme which, somewhat paradoxically, is found in the cytosol. So noradrenaline has to leak back out of the granules, and then adrenaline is taken up again for final release by exocytosis. The concentration of adrenaline is at least 25,000 times higher in the granule than in the cytosol: how can it be taken up against such a large concentration gradient? It was soon found that isolated vesicles would take up adrenaline, but only in the presence of ATP. This requirement was traced to the presence of an ATP-driven proton pump which acidifies the vesicle interior and thus creates both an electrical and a chemical gradient of protons (Figure 5.12). This electrochemical gradient then drives the uptake of adrenaline and dopamine, presumably via an antiport with protons, although this has not yet been proved.

As well as the enzyme which synthesizes noradrenaline, the chromaffin granule also contains other proteins and peptides, including met- and leu-enkephalins which are derived from the precursor proenkephalin (Section 3.2.2), and are co-released with the catecholamines when the cell is stimulated. These secreted proteins and peptides are of course synthesized on rough ER and transported via the Golgi to the chromaffin granules. The latter therefore have the same type of origin as the secretory granules in other endocrine or neural cells, such as the β–cells which synthesize insulin. You may recall (Section 5.1) that insulin secretory granules are also acidified with respect to the cytosol, and this is presumably because they contain a proton pumping ATPase similar to that of chromaffin granules. Although other systems are less well characterized, it appears that uptake of many other non-peptide neurotransmitters (e.g. serotonin, acetylcholine and glutamate) into secretory granules is also driven by electrochemical gradients created by proton-pumping ATPases.

5.3 *Triggering of exocytosis*

5.3.1 *Role of Ca^{2+} ions*

The mechanisms which trigger exocytosis, and hence secretion, from the cell remain poorly understood, and this could be said to be one of the great unsolved problems of cell biology. What is known is that Ca^{2+} ions play an important role in the process. The evidence for this may be summarized as follows:

1) In many secretory cells including chromaffin cells, the external signal for secretion depolarizes the plasma membrane, causing the opening of voltage-gated Ca^{2+} channels (Section 4.2.2). Secretion does not occur if Ca^{2+} is removed from the extracellular medium. Other cells use a different mechanism involving release of intracellular stores of Ca^{2+} (Section 7.3.1). In these cases removal of external Ca^{2+} does not immediately prevent secretion. However, if the cells are repeatedly stimulated in Ca^{2+}-free medium, their intracellular stores become depleted and they no longer respond.
2) Treatment of secretory cells with Ca^{2+} ionophores (e.g. A23187 or ionomycin) triggers secretion but only when Ca^{2+} is present in the external medium. These ionophores are compounds which bind Ca^{2+} ions and surround them with a lipophilic exterior, rendering them permeable to the plasma membrane.
3) Cells can be made permeable to small molecules by treatments which cause small holes to open in the membrane. The commonest methods are electroporation (treatment with a large electrical discharge), bacterial toxins (e.g. Streptolysin O) and detergents such as saponin or digitonin which bind to cholesterol in the plasma membrane. In most cases treatment of permeabilized cells with Ca^{2+} triggers exocytosis of secretory granule contents.

The permeabilized cell approach is particularly powerful because it enables the Ca^{2+} concentration to be carefully controlled, and also allows the requirement for other small diffusible metabolites to be tested. This has shown that Mg^{2+} and ATP are also necessary for exocytosis, presumably because one step in the pathway requires hydrolysis of ATP.

Exocytosis involves fusion of two lipid bilayers (Figure 5.1). Before this can come about, the extensive shell of water molecules which bind to the phospholipids

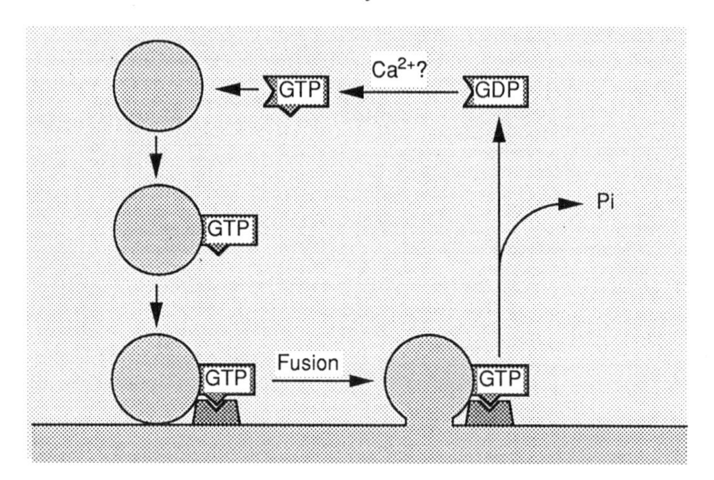

Fig. 5.13 Speculative model for the role of a GTP-binding protein in exocytosis. After Wilschut (1989) Current Opininion in Cell Biology **1**, 639–647.

on the surface of the membrane must be removed. It has been calculated that this would involve a very large activation energy barrier, suggesting that the process must be catalysed to occur at normal temperature. Unfortunately, no reconstituted cell-free system for Ca^{2+}-triggered exocytosis has been developed, so as yet we know very little about the proteins that may be involved in this process. However a family of cytosolic proteins called **annexins**, which bind to secretory granules in the presence of high Ca^{2+} concentrations, have been purified from various secretory cells. One of these, called **annexin II** or **calpactin**, restores Ca^{2+}-dependent secretion in adrenal chromaffin cells which have been made defective by permeabilization and consequent leakage of cell contents. This effect is blocked by anti-calpactin antibody. In unstimulated cells, calpactin is found to be associated with the inner surface of the plasma membrane, and it is possible that it mediates the binding, and perhaps the fusion, of the secretory vesicles with the membrane.

5.3.2 *Role of GTP*

Other experiments suggest that, while Ca^{2+} may regulate the triggering of exocytosis, it may not be involved in the critical final step. Several studies with permeabilized cells, particularly mast cells (Figure 6.1), have indicated that exocytotic secretion can be triggered by GTP analogues such as GTP-γ-S (Figure 5.13) in the absence of any rise in Ca^{2+}. This is consistent with studies of temperature-sensitive mutations in yeast in which, at the restrictive temperature, there is an accumulation of secretory vesicles unable to complete exocytosis. One such mutation has been traced to a gene (SEC4) which encodes a GTP-binding

("G") protein. SEC4 has homologues in higher eukaryotes, and is one of a family of 20–25 kDa, monomeric G proteins, which include an elongation factor (EF-Tu) of protein synthesis, and the *ras* oncogene (Chapter 9). Figure 5.13 shows a model for the role of such a G protein in exocytosis. Presumably the secretory vesicle must contain a targetting molecule on its surface which recognizes a receptor on the plasma membrane. It is proposed that productive binding leading to vesicle fusion occurs only in the presence of the G protein–GTP complex. After fusion the G protein then dissociates and GTP is concomitantly hydrolysed to GDP (this would not occur in the case of GTP-γ-S, which is non-hydrolysable). Possibly, the regulatory event in which Ca^{2+} is involved promotes exchange of GDP for GTP to convert the G protein back to its active form. Hydrolysis of GTP to GDP would ensure that the cycle occurred **in one direction** only. This model is speculative, but it is based on analogies with the well-established role of EF-Tu in protein synthesis, and those of a family of larger G proteins in signal transduction (see Chapter 7).

5.3.3 *Exocytosis of fast-acting transmitters*

The secretory vesicles discussed previously in this chapter are derived from the Golgi apparatus, and generally have an electron-dense appearance. In neurones, the Golgi is found in the cell body, so vesicles derived from them have to be transported along the length of the axon to the nerve ending. A different class of secretory vesicle, whose members are smaller and less electron-dense, appears to be involved in the release of fast-acting transmitters such as acetylcholine, γ-aminobutyrate, glycine and glutamate. Unlike the large, dense type, these small vesicles appear to recycle, and refill with transmitter, locally within the nerve ending without having to return to the cell body. The process has been visualized by electron microscopy of neuromuscular junctions, where acetylcholine is released. In resting cells, the vesicles are seen to be lined up in specialized regions called **active zones** (Figure 5.14), exactly opposite the acetylcholine receptors (Chapter 6) on the target muscle cell. When the presynaptic neurone is stimulated, vesicles fuse with the plasma membrane within a few milliseconds, releasing their contents into the synaptic cleft. Within 10–20 seconds the plasma membrane is seen to contain invaginations lined with clathrin (**coated pits**). These pinch off to form coated vesicles, which rapidly lose their clathrin coat, and are refilled with acetylcholine. This reformation of vesicles from the plasma membrane, the reverse of exocytosis, is called **endocytosis**, and is related to the formation of secretory vesicles at the *trans*-Golgi network (Section 5.1.2), which also involves clathrin.

The small secretory vesicles have now been isolated biochemically and their protein content analysed. Apart from a similar proton-pumping ATPase, their proteins appear to be quite distinct from those of the large, dense vesicles typified by the chromaffin granule. A characteristic feature is the presence of a family of

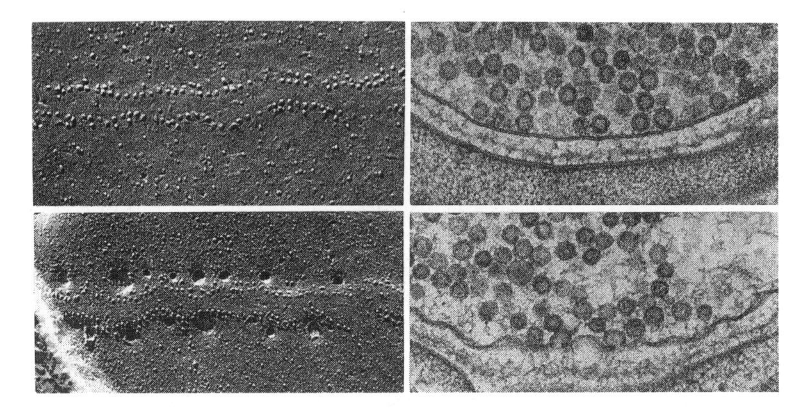

Fig. 5.14 Freeze fracture micrographs (left) and thin sections (right) of neuromuscular junction releasing acetylcholine. The top panels show linear arrays of vesicles lined up in the active zones in the resting state. Five milliseconds after stimulation of the nerve, vesicles are seen to fuse with the membrane (bottom panels). Courtesy of John Heuser, Washington University School of Medicine, St. Louis, Missouri.

peripheral proteins (**synapsins**) on the cytoplasmic surface. These are thought to attach the vesicles to cytoskeletal elements, and a Ca^{2+}-dependent phosphorylation of these proteins may be involved in regulation of exocytosis by causing release of vesicles from the cytoskeleton. This is discussed further in Section 8.2.2.

SUMMARY

Most, if not all, first messengers are released from cells via exocytosis, i.e. the fusion of secretory vesicles with the plasma membrane. Proteins destined for secretion, including all peptide messengers, normally have a hydrophobic signal sequence at the N-terminus which directs the ribosome synthesizing the polypeptide to bind to the endoplasmic reticulum. The polypeptide is extruded into the lumen of the ER, with proteolytic removal of the signal sequence. It passes via vesicles to the Golgi apparatus, and then via clathrin-coated vesicles to the secretory granules. The secretory granules are acidified by a proton-pumping ATPase, and in the case of pancreatic β-cells this appears to activate proteolytic processing of proinsulin to insulin. Non-peptide messengers are synthesized in the cytoplasm and then concentrated in secretory vesicles by transport that is often driven by the pH gradient created by the proton-pumping ATPase.

A variety of evidence indicates that Ca^{2+} triggers exocytosis, but the critical final step may involve a GTP-binding protein without a direct involvement of Ca^{2+}. There are several candidates for proteins involved in exocytosis, but the molecular mechanism remains unclear. The secretory granules which release fast *acting neurotransmitters, such as acetylcholine, differ from the type containing peptide messengers in that they can recycle at the plasma membrane without involvement of the endoplasmic reticulum or Golgi apparatus. In the case of fast-acting neurotransmitters, release may be regulated by Ca^{2+}-dependent phosphorylation of synapsins.*

FURTHER READING

1. "The insulin factory" – review on insulin secretion by the β cell. Orci, L., Vassali, J.D. and Perrelet, A. (1988) Scientific American, September, pp. 50–61.
2. "The adrenal chromaffin cell" – review. Carmichael, S.W. and Winkler, H. (1985) Scientific American, August, pp. 30–39.
3. "A role for calpactin in calcium-dependent exocytosis in adrenal chromaffin cells". Ali, S.M., Geisow, M.J. and Burgoyne, R.D. (1989) Nature **340**, 313–315.
4. "Two roles for guanine nucleotides in the stimulus-secretion sequence of neutrophils". Barrowman, M.M., Cockcroft, S. and Gomperts, B. (1986) Nature **319**, 504–507.
5. "Yeast mutants illuminate the secretory pathway" – minireview on role of GTP-binding proteins in yeast secretion. Burgoyne, R.D. (1988) Trends Biochem. Sci. **13**, 241–243.
6. "Calcium ions, active zones and synaptic transmitter release" – review. Smith, S.J. and Augustine, G.J. (1988) Trends in Neuroscience **11**, 458–464.

SIX

CELL SURFACE RECEPTORS – ANALYSIS AND IDENTIFICATION

What happens when the first messenger has crossed the synaptic cleft, the bloodstream or the extracellular fluid, and arrives at its target cell? The interaction of the messenger with the target cell is normally highly specific, with even closely related compounds (including stereoisomers) usually being ineffective. In addition, the messenger can often act at extremely low concentrations, in the nanomolar or even picomolar range. It has always seemed likely that only proteins could bind messengers with sufficiently high affinity and specificity to account for these observations. A protein which binds a first messenger, and is involved in mediating its biological effects, is known as a **receptor**.

For most first messengers, the available evidence favours the idea that the receptor is located on the outside surface of the cell, and that this is where it exerts its effect. Experiments using radioactive peptide messengers show that they initially bind to the outside surface of the cell, but at normal physiological temperature become rapidly internalized. The internalization is now known to be due to endocytosis through coated pits, and in many cases appears to be a mechanism for degradation of the messenger (Section 6.5). Although hormone–receptor complexes in internalized vesicles may well be active, it appears almost certain that internalization is **not necessary** for the messenger to exert its action. Thus some effects of messenger binding can still be observed at low temperature, when endocytosis does not occur. In addition, several biological effects have been demonstrated for messengers that are covalently attached to large carriers which are too large to enter coated pits. An example is shown in Figure 6.1, in which secretion of histamine has occurred in response to a messenger coupled to the surface of a Sepharose bead. Only the region of the cell next to the Sepharose bead has undergone exocytosis, making it very unlikely that the effect is due to soluble messenger which has become detached from the bead.

From these and similar experiments, the concept has arisen that first messengers act by binding to cell surface receptor proteins. Exceptions to this rule are the steroid and thyroid hormones, and retinoic acid. Unlike most other messengers, these are lipophilic compounds, and autoradiographic experiments show that radioactive steroid and thyroid hormones rapidly enter the cell and accumulate in the nucleus. Consistent with this, it has also been shown that the receptors for these messengers are soluble proteins found in the cell nucleus. The action of these receptors will be discussed in Chapter 10, and the remainder of this

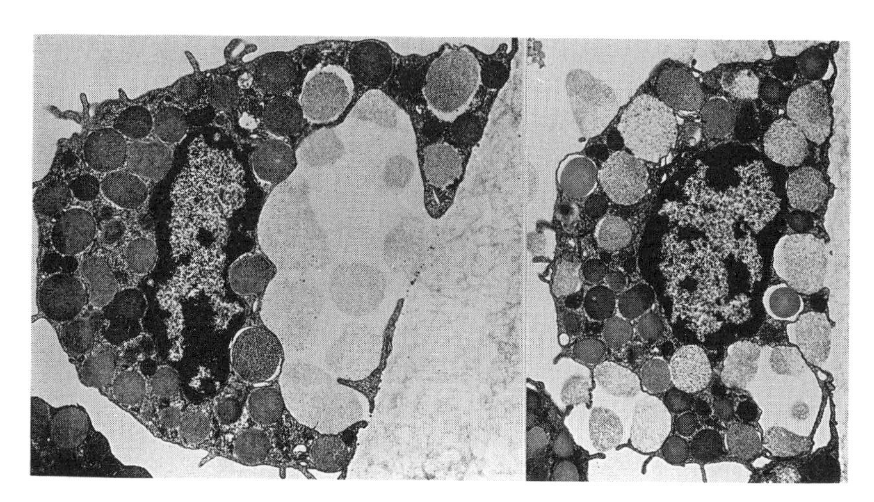

Fig. 6.1 Exocytosis of histamine in mast cells, triggered by agonist attached to the surface of Sepharose, the latter appearing in this electron micrograph as a white bead which is much larger than the cell itself. In the left hand panel, exocytosis only occurs in the region of the cell apposed to the Sepharose bead. In the right hand panel, soluble agonist was also added and exocytosis occurs from other regions. In these experiments the agonist was the lectin concanavalin A, which is not a natural messenger but works by crosslinking messenger receptors. Reproduced from Lawson, Fewtrell and Raff (1978) J. Cell Biol. **79**, 394–400, courtesy of Durward Lawson.

chapter, and the whole of Chapters 7–9, will concentrate on the action of messengers which act at cell surface receptors.

6.1 *Receptor binding studies*

6.1.1 *Quantitative analysis of binding*

When one commences study of a receptor, the only known activity that it exhibits is its ability to bind the messenger. Since binding must be measured at very low messenger concentrations, it is normally essential to radioactively label the messenger. A typical protocol for a binding experiment is shown in Figure 6.2. The labelled messenger is incubated either with whole cells (normally at 0°C to prevent endocytosis), or with isolated membranes, until equilibrium binding is achieved. The free receptor and receptor–messenger complex are then separated. This must be carried out very rapidly to prevent re-equilibration of the messenger,

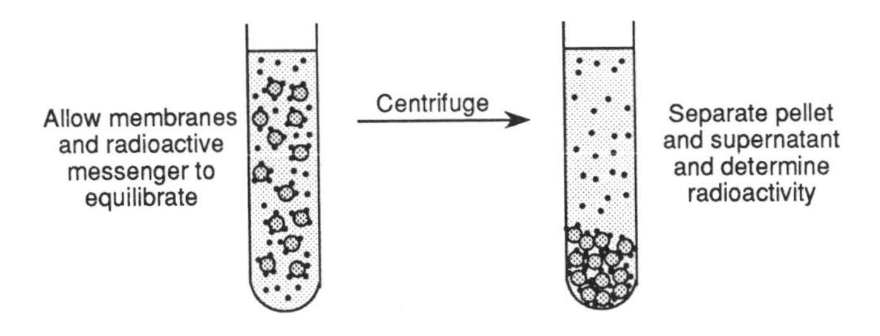

Fig. 6.2 Protocol for a typical messenger-binding experiment on a cell surface receptor.

but with cells or large membrane vesicles this is easily achieved by centrifugation. The concentrations of the free and bound messenger in the original incubation can then be determined from the radioactivity in the supernatant and pellet.

How can one be sure that binding of the messenger to the membranes reflects binding to the receptor which mediates the biological effects of the messenger? Several criteria may be applied:

1) binding should be target-cell specific;
2) since binding is to a protein which is limited in amount, it should be saturable;
3) binding should be specific for the messenger, i.e. analogues (e.g. stereoisomers) which are not biologically active should not displace the bound messenger;
4) binding should have high affinity, consistent with the biologically effective concentrations of the messenger.

The first criterion is easy to apply, but may not be applicable to some messengers (e.g. insulin) where receptors seem to be present on almost all cells. The second criterion implies that binding will reach a limit value when all of the receptor is occupied, directly analogous to the V_{max} for an enzyme reaction. To test this one can study binding of radioactive messenger in the presence and absence of a large (say 10,000-fold) excess of unlabelled messenger. In practice with crude membranes, one finds that the radioactive messenger is not completely displaced (e.g. Figure 6.3). The small non-saturable component is generally assumed to be "non-specific binding", possibly to phospholipids, and results of binding experiments are corrected by subtracting the value obtained in this control experiment, to give "specific" binding.

The easiest way to test criterion (3) is to incubate membranes with a fixed, low concentration of labelled messenger, and examine whether it is displaced by increasing concentrations of the related compound under test. Figure 6.4 shows an example in which membranes from turkey erythrocytes (red blood cells) were incubated with labelled insulin and various other compounds. Labelled insulin can

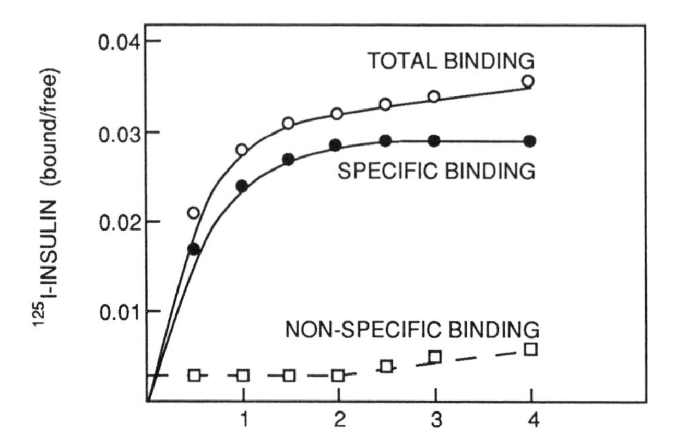

Fig. 6.3 Study showing that most, but not all, of the ^{125}I-insulin bound to turkey erythrocyte membranes (total binding) can be displaced by addition of a large excess of non-radioactive insulin. The residual, non-saturable ^{125}I-insulin binding (non-specific binding) is assumed not to be receptor-mediated and is subtracted to give the "specific binding". Based on Ginsberg, Kahn, and Roth (1976) Biochim. Biophys. Acta **443**, 227–242.

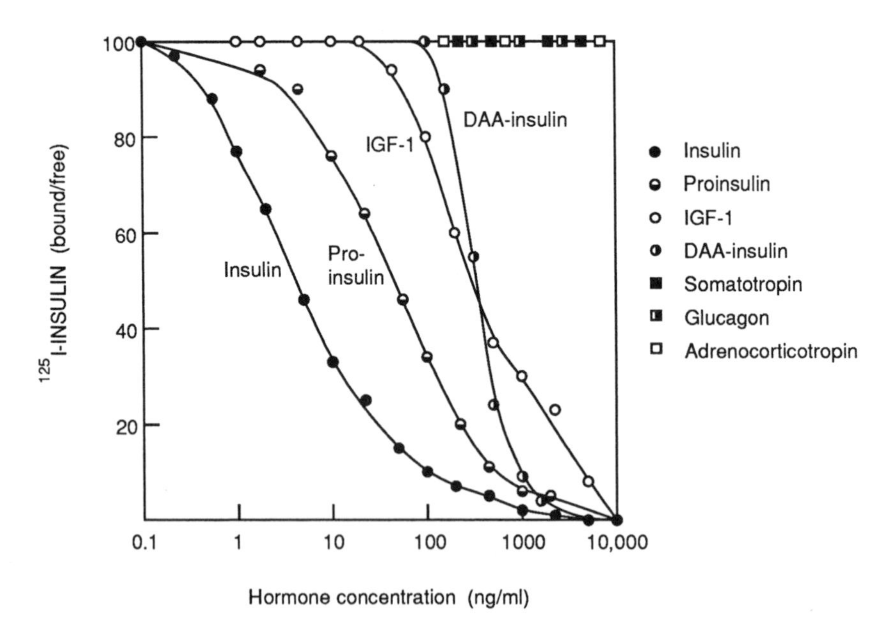

Fig. 6.4 Displacement of ^{125}I-insulin from turkey erythrocyte membranes by non-radioactive insulin and related messengers (IGF-1, insulin-like growth factor-1; DAA-insulin, a proteolytically modified form of insulin). The data have been corrected for non-specific binding (Figure 6.3). Based on Ginsberg, Kahn and Roth (1976) Biochim. Biophys. Acta **443**, 227–242.

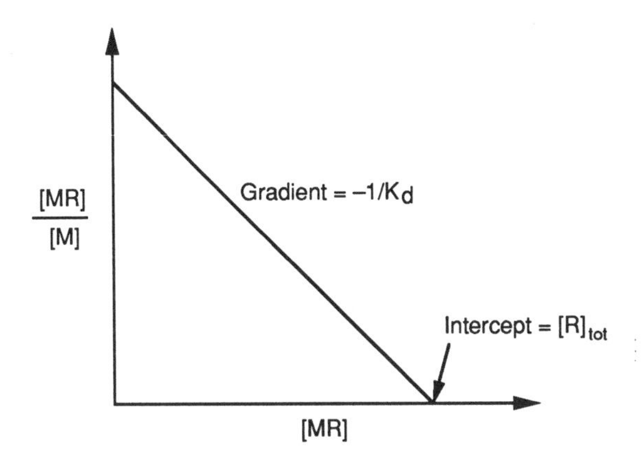

Fig. 6.5 Idealized Scatchard plot for a single class of binding sites.

be displaced by proinsulin, by the related peptide messenger insulin-like growth factor-1, and by a chemically modified insulin, but it is displaced by insulin itself at much lower concentrations. The relative affinity of binding of insulin and these analogues is similar to their relative biological potency. A number of other peptide messengers (e.g. adrenocorticotropin, glucagon, somatotropin) do not displace radioactive insulin at all.

To apply criterion (4), it is necessary to measure the affinity of binding. This is usually carried out as in Figure 6.4, using a fixed, low concentration of labelled messenger and increasing concentrations of unlabelled messenger. Although the specific radioactivity decreases and must be recalculated for each incubation, this method is economical on labelled messenger, which is often expensive to produce. A qualitative assessment of the affinity of binding can be made from a plot similar to that in Figure 6.4, but for quantitative results it is usual to convert the data into a form which gives a linear plot: the **Scatchard plot** (Figure 6.5). Although the analysis is almost invariably now performed by computer, the results are often still displayed as a Scatchard plot, so it is important to understand what it means.

For a messenger M, and single class of receptor (R), binding is described by a simple equilibrium:

$$MR \leftrightarrow M + R$$

At equilibrium, the dissociation constant, $K_d = \dfrac{[M][R]}{[MR]}$ .... (1)

One can readily measure the concentration of free messenger, [M], and the concentration of bound messenger, [MR], but measuring the free receptor, [R], is not so easy. However it can be calculated by subtracting the concentration of bound messenger from the total receptor concentration ($[R]_{tot}$):

$$[R] = [R]_{tot} - [MR]$$

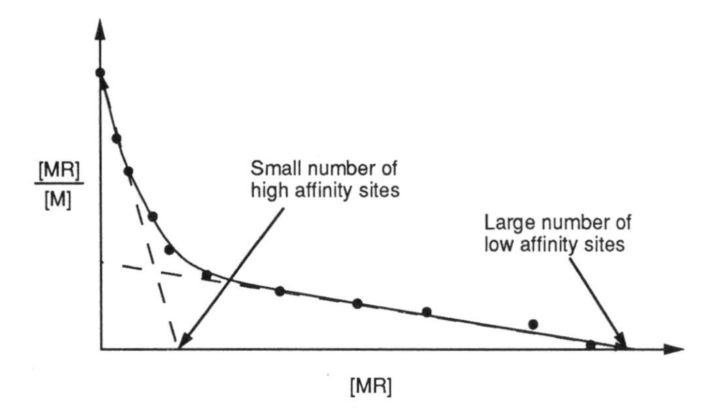

Fig. 6.6 Scatchard plot for binding of insulin to turkey erythrocyte membranes. The plot is derived from the data shown in Figure 6.4.

Substituting this expression for [R] in equation (1):

$$K_d = \frac{[M]\,([R]_{tot} - [MR])}{[MR]}$$

Rearranging:

$$\frac{[MR]}{[M]} = \frac{-1}{K_d}[MR] + \frac{[R]_{tot}}{K_d} \quad \ldots\ldots\ldots\ldots\ldots\ldots (2)$$

This equation has the form y = –mx + c, so a plot of bound messenger over free messenger ([MR]/[M]) against bound messenger [MR] (a **Scatchard** plot) will give a straight line with a negative slope, gradient $-1/K_d$, from which the dissociation constant K_d can be determined (Figure 6.5). Also, as [MR]/[M] approaches zero, the equation reduces to $[MR] = [R]_{tot}$. Hence the intercept on the [MR] axis, when [MR]/[M] = 0, is equal to the total receptor concentration. The reason for this can be grasped more intuitively by considering that if the messenger concentration is increased indefinitely, [M] can also increase indefinitely, but [MR] can only increase to the point where the receptor is saturated. At this point, [MR]/[M] will be effectively zero, and all the receptors will have bound messenger, so $[MR] = [R]_{tot}$.

In practice, one often finds that a single cell type may contain two or more types of receptor of different affinity for a particular messenger. In this case, the plot will appear curved (Figure 6.6). It is essential to use a computer to fit the data to models involving two or more sites, because fitting by eye becomes notoriously unreliable.

It may be worth mentioning that the binding of a single substrate to an enzyme is directly analogous to the binding of messenger to a receptor. In fact an analogue

Nicotine
From the tobacco plant. Mimics the effect of acetylcholine at some receptors, such as those found on skeletal muscle:

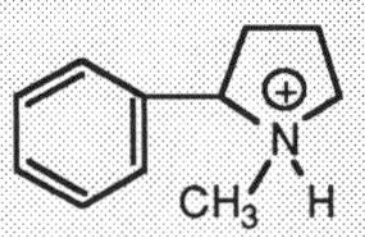

Muscarine
Extracted from the poisonous toadstool *Amanita muscarina.* Mimics the effect of acetylcholine at some receptors, such as those found at the termini of parasympathetic nerves:

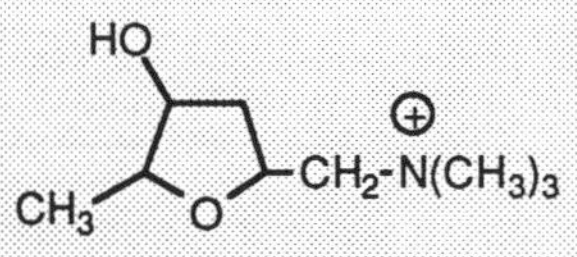

Curare
A plant extract, originally used by South American Indians to poison blowgun darts for hunting. Antagonizes the action of acetylcholine at nicotinic receptors, and so paralyses the prey. Used clinically to paralyse muscles for surgery. The main active ingredient is tubocurarine:

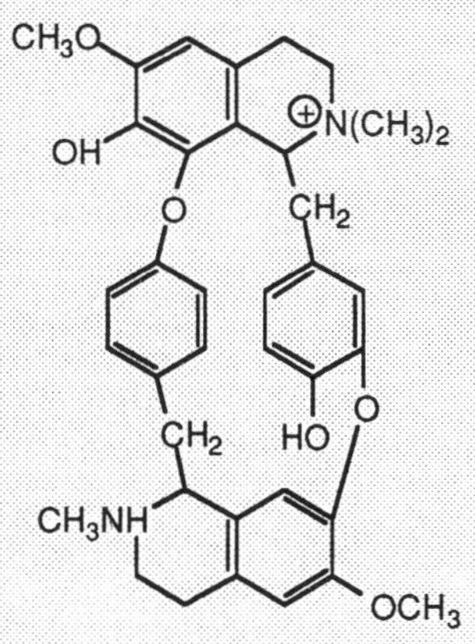

Atropine
Extracted from the poisonous plant *Atropa belladonna,* the Deadly Nightshade. Blocks the action of acetylcholine at muscarinic receptors:

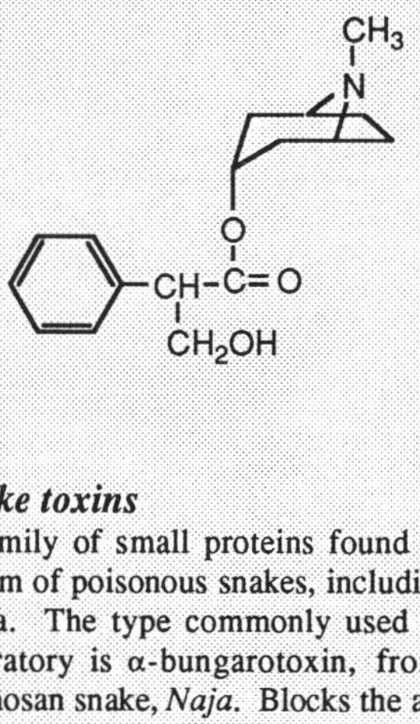

Snake toxins
A family of small proteins found in the venom of poisonous snakes, including the cobra. The type commonly used in the laboratory is α-bungarotoxin, from the Formosan snake, *Naja.* Blocks the actions of acetylcholine at nicotinic receptors, thus causing paralysis of the victim of a snake bite.

Box 6.1 Agonists and antagonists acting at acetylcholine receptors

of the Scatchard plot (the Eadie and Hofstee plot) can be used to determine K_m (analogous to K_d) and V_{max} (which is related to the total number of enzyme molecules) for a simple enzyme reaction.

Having obtained the total receptor concentration in mol.l^{-1} from the Scatchard plot, and knowing Avogadro's number and the number of cells per unit volume, it is trivial to calculate the number of receptors per cell. This generally lies in the

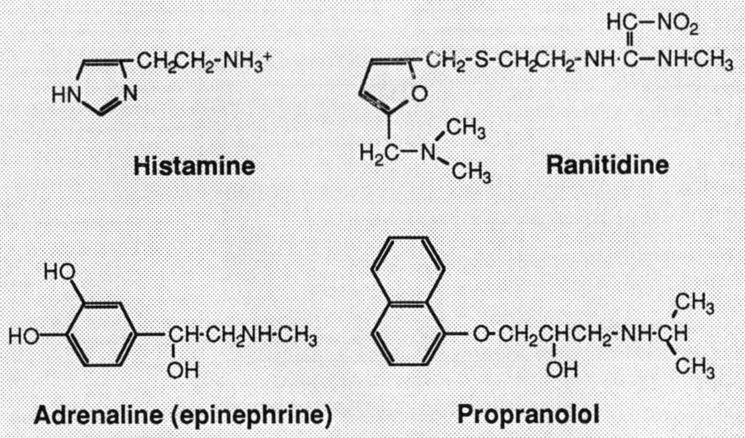

Histamine

Ranitidine

Adrenaline (epinephrine)

Propranolol

Box 6.2 Structures of ranitidine and propranolol, two of the world's largest selling drugs, which were developed as analogues of histamine and adrenaline.

range 10^4 to 10^5. Calculation of the dissociation constant (K_d) from the slope of the plot allows one to test criterion (4) above, i.e. that binding is of sufficiently high affinity to account for the biological effects of the messenger.

6.1.2 *Labelling of messengers and pharmacological studies of receptor binding*

In practice, the messenger itself is often not used for receptor binding studies. This is because of two problems:

1) it may be difficult to achieve a high enough specific radioactivity to measure the messenger at very low concentrations;
2) the target cell often contains very active mechanisms for breakdown of the messenger.

For larger polypeptide messengers, the first problem can often be overcome by labelling tyrosine residues with ^{125}I, using a chemical procedure, or using a bacterial peroxidase which catalyses a reaction similar to that in which tyrosine residues are iodinated during synthesis of thyroid hormones (Section 3.1.4). One does of course end up with a chemically modified form of the messenger, and it is important to check that this does not interfere with biological activity.

The second problem, degradation of the messenger, is particularly serious for messengers like acetylcholine, for which there are very active breakdown mechanisms in the tissues. Fortunately, pharmacologists have over the years found many compounds other than the natural messengers which bind to receptors, some of which are resistant to breakdown. Some of these compounds activate the receptor (**agonists**), while others bind in an unproductive manner and block binding of the messenger (**antagonists**). They have been found by at least three different approaches:

1) As components of naturally occurring toxins, many from exotic sources and often originally used in primitive herbal medicine. Examples (Box 6.1) are atropine, curare, and protein toxins found in snake venom, e.g. cobra toxin. All are antagonists of different types of acetylcholine receptor (see below). One of the snake venom toxins, α-bungarotoxin, has been particularly useful for binding measurements of the acetylcholine receptor, because it can be iodinated to high specific radioactivity using ^{125}I. Another example would be heroin, the active ingredient of the opium poppy, which is now known to be an agonist at opioid peptide receptors (Section 3.2.2).
2) As synthetic analogues of the messenger, which may be agonists or antagonists (Box 6.2). Some of these have become the world's biggest selling drugs, e.g. the "β-blocker" propranolol which antagonizes catecholamine receptors on the heart, and ranitidine, which antagonizes one class of histamine receptor and is used for treatment of stomach ulcers.
3) As synthetic compounds found to relieve symptoms, and only later discovered to interact with messenger receptors. Examples of these are the anxiety-suppressing benzodiazepine drugs (e.g. Librium, Valium, Mogadon) which are now believed to activate the γ-aminobutyrate receptor, by binding at a site distinct from the messenger-binding site.

As well as often providing more practical alternatives to the messengers themselves for receptor binding assays, these pharmacological studies have revealed heterogeneity in messenger receptors. This was first shown as long ago as 1914, when Dale showed that responses to acetylcholine could be subdivided into two types, nicotinic and muscarinic, depending on whether they were mimicked by the agonists nicotine or muscarine (Box 6.1). Nicotinic receptors are blocked by curare and by α-bungarotoxin, and are found, among other places, at the synapse between a motor nerve and a skeletal muscle (a **neuromuscular junction**). This explains how both South American Indians and venomous snakes are able to paralyse their prey! Muscarinic receptors are found at terminal synapses in the parasympathetic nervous system, and are blocked by atropine. More recent biochemical studies have revealed that muscarinic receptors are themselves heterogeneous, and can be divided into further sub-types (Section 6.3.2). This is a recurring theme in modern studies of receptors.

Another important case of receptor heterogeneity has been found in catecholamine receptors, which have now been divided into four functional classes (termed the α_1-, α_2-, β_1- and β_2-**adrenergic** receptors) by means of agonists

Table 6.1 Classification of adrenergic receptors according to potency of action of catecholamines [adrenaline (Adr), and noradrenaline (Nor)], agonists [e.g. isoproterenol (Iso)], and antagonists. After Lefkowitz *et al* (1983) Ann. Rev. Biochem. **52**, 159–186.

Type	Agonists (Order of potency)	Agonists (specific)	Antagonists (specific)	Selected effects
α_1	Adr≥Nor>Iso	Phenylephrine	Prazosin	Vascular constriction Glycogen breakdown (liver)
α_2	Adr≥Nor>Iso	Clonidine	Yohimbine	Vascular constriction
β_1	Iso>Nor≈Adr		Atenolol Practolol	Increased contractility (heart) Lipolysis (adipose tissue)
β_2	Iso>Adr>>Nor		Butoxamine	Vascular relaxation Bronchial relaxation

and antagonists which are more specific for these receptor subclasses than the catecholamines themselves (Table 6.1). The development of these specific compounds has been of great value to the researcher in receptor biochemistry, as well as making the compounds more useful clinically. Thus propranolol (a β-antagonist) has been very successful as a therapeutic drug because it blocks the effects of adrenaline on the heart (which contains mainly β_1-receptors) without affecting many other actions of the catecholamines. More specific β_1-antagonists have been developed subsequently.

These classifications of receptor sub-types, based originally on the specificities of pharmacological agents, have now been provided with a biochemical basis by isolation and structural analysis of the receptor proteins. We will now discuss how this important next step has been taken.

6.2 *Identification and characterization of receptor proteins*

Ideally the analysis of receptor proteins involves purification to homogeneity as a prerequisite step. However, this is often a formidable task, for at least two reasons:

1) All cell surface receptor proteins are integral membrane proteins. Before purification can commence, they must be solubilized using detergent. Unfortunately, detergents tend to interfere with many conventional protein purification procedures.
2) An individual receptor protein often constitutes a very small proportion of total membrane protein (typically $<0.01\%$). One is looking for the proverbial needle in a haystack.

Fortunately, some information about the receptor polypeptide can often be obtained without purification, by exploiting the very high affinity binding of the messenger to its receptor.

6.2.1 *Identification of receptor polypeptides – crosslinking and affinity labelling*

These two techniques are variations on a theme: the objective is to identify the receptor polypeptide while it is still an integral component of the membrane. They can also be used with purified receptor proteins to identify the messenger binding site with more precision.

1) *Crosslinking*

This method is particularly applicable to receptors for peptide messengers, although it can also be used with other receptors if a protein agonist or antagonist exists (e.g.α-bungarotoxin for the nicotinic acetylcholine receptor). Membrane preparations containing the receptor are incubated with labelled ligand until equilibrium binding is achieved (a ligand is any molecule which binds reversibly to a protein, and in the case of a receptor could be the messenger, or an agonist or antagonist). One then adds a crosslinking agent, which is a "double-headed" reagent which reacts with amino acid side chains, typically lysine (Figure 6.7). Such reagents will crosslink at random any reactive amino acid side chains which are a suitable distance apart, and could therefore crosslink the labelled ligand to any protein in the preparation. However, the rationale behind the method is that only the receptor will bind the ligand with sufficient affinity to form a detectable number of crosslinks. The crosslinked preparation is then analysed by polyacrylamide gel electrophoresis in the presence of the detergent sodium dodecyl sulphate, in which non-covalent interactions are disrupted and polypeptides separate according to their molecular weight. Although many proteins in the preparation will be crosslinked, only ligand–receptor crosslinks will be radioactive, and so are detectable by autoradiography.

2) *Affinity labelling and photoaffinity labelling*

In the affinity labelling technique, rather than using a double-headed crosslinking reagent, the crosslinker is an integral part of the ligand itself. An example is the

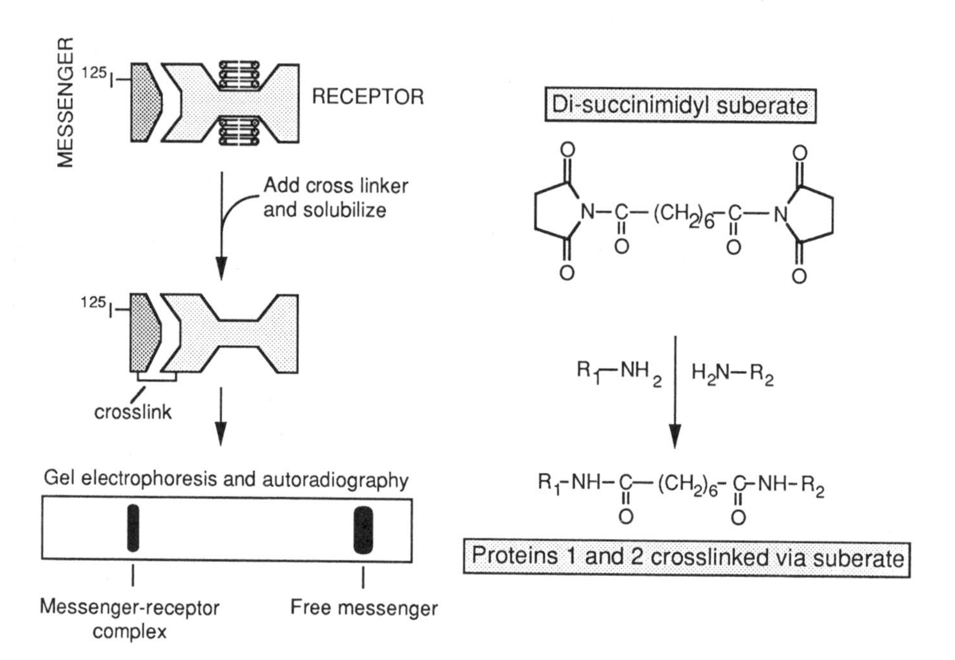

Fig. 6.7 Principle of a ligand–receptor crosslinking experiment (left) and a typical crosslinking agent (right), which can crosslink lysine side chains ($R—NH_2$) on the messenger and the receptor.

use of bromoacetylcholine, an analogue of acetylcholine which binds to the receptor, but has a reactive bromoacetyl group which can alkylate cysteine side chains (Figure 6.8). If the bromoacetylcholine is radioactive, the receptor subunit which binds acetylcholine will now have a radioactive tag and can be identified by polyacrylamide gel electrophoresis and autoradiography. A more controlled reaction is obtained with the technique of **photoaffinity labelling**, in which the ligand is reactive only after exposure to light. An example is the use of 4-azido-2-nitrophenyl-^{125}I-insulin (Figure 6.8). This insulin analogue is allowed to bind to membranes containing the receptor, which are then exposed to a strong source of ultraviolet light. In a photolytic reaction, the azido group is converted to a nitrene free radical, a highly reactive species which will react with almost any convenient amino acid side chain.

For multisubunit receptors, these methods only identify the ligand-binding subunit (and with purified receptors may be used for exactly that purpose). However it may be possible to obtain information about other subunits if they are attached to the ligand-binding subunit by disulphide bridges (see example of the insulin receptor in Section 6.3.3).

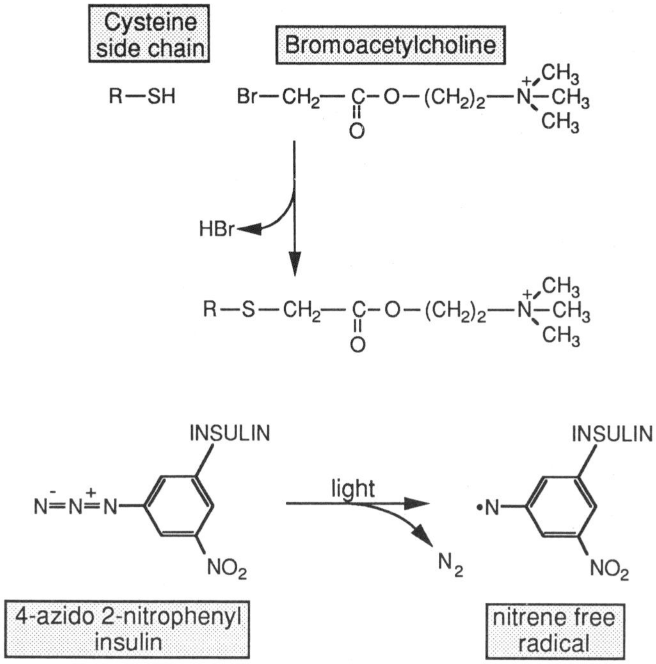

Fig. 6.8 Structure and action of two receptor affinity labelling reagents.

6.2.2 *Purification of receptor proteins*

1) *Choice of tissue*

Although useful information can be obtained by crosslinking, there is ultimately no substitute for purification of the receptor. However, given the low abundance of receptors in a typical membrane, a very important decision is the choice of starting material for purification. This can be illustrated in the case of the nicotinic acetylcholine receptor. Although the receptor is present in high concentration at the neuromuscular junction, these structures are anatomically very small, and impossible to dissect out in reasonable quantity. Very little progress was made in biochemical characterization of the nicotinic receptor until it was found that they are present in great abundance in the electroplax organ of electric rays (*Torpedo* spp.) and electric eels (*Electrophorus* spp.). These species stun their prey using electrical discharges generated in the electroplax organ, which appears to have evolved from muscle cells. The cells have asymmetrical surfaces, one of which is innervated and is excitable, containing very abundant nicotinic acetylcholine receptors, as well as voltage-gated Na^+ channels. The other surface is not excitable. In the resting state, both membranes will have the normal membrane potential of –90 mV, but because these are in opposite directions there is no net potential difference across the cell.

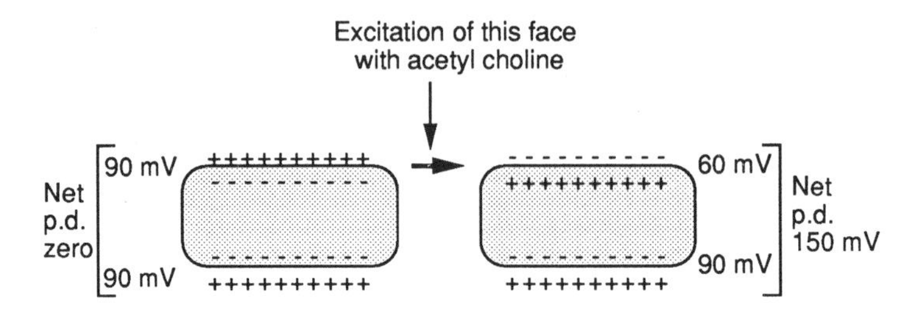

Fig. 6.9 Mechanism of voltage generation, involving nicotinic acetylcholine receptors, in the electroplax organs of electric fish.

However, when the excitable face is stimulated with acetylcholine, that membrane is depolarized through mechanisms discussed in Chapter 7, and the new membrane potential of +60 mV combines with the –90 mV across the non-excitable membrane to produce a net potential difference across the whole cell of 150 mV (Figure 6.9). Since several thousand of these cells are arranged in series in the electroplax organ, it can pack a total "punch" of around 750 volts!

The electroplax organ is a uniquely rich source of nicotinic acetylcholine receptor, but cells or tissues which show useful, if less spectacular, enrichment of receptors can be found for other messengers. For reasons which are not clear, insulin receptors are rather abundant in placenta, and human placenta has proved to be a valuable source for receptor purification. In the case of the epidermal growth factor (EGF) receptor, a cell line derived from a human epidermal cancer (A431 cells) was found by Scatchard analysis to contain unusually large numbers of receptors ($>10^6$ per cell, cf. $\sim10^4$ per cell in normal epidermal cell). It is now known that this is due to DNA rearrangements which led to amplification of the receptor gene, and which may have been instrumental in the development of the cancer (Chapter 9). Through this chance discovery, the EGF receptor remains one of the best characterized of all peptide messenger receptors.

2) *Purification and solubilization of plasma membrane fractions*

Once the cells are homogenized, membrane fractions enriched in plasma membrane can be obtained by various forms of centrifugation, which separate according to sedimentation rate and/or density. However, procedures which result in a large enrichment are usually associated with a low yield, and most receptor purification schemes begin with a relatively crude membrane fraction. The next step is solubilization, normally carried out by resuspension of the membrane fraction in a solution of detergent. Detergents are amphipathic molecules which differ from phospholipids in that in aqueous solution they tend to form micelles rather than bilayers. They readily form mixed micelles with phospholipids and integral membrane proteins (Figure 4.15), in which the hydrophobic transmembrane segments of the latter are shielded from the solvent by detergent molecules.

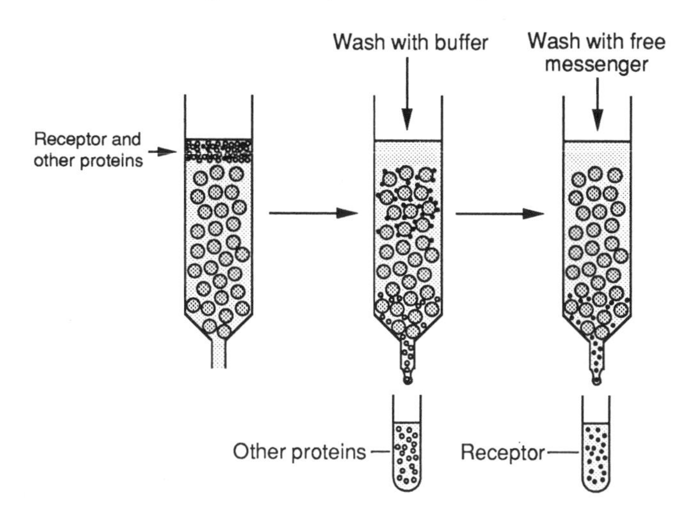

Fig. 6.10 Purification of a messenger receptor on immobilized messenger column.

3) *Purification methods*

Although most general protein purification methods (e.g. gel filtration, ion exchange, etc.) can be used with solubilized receptors, it is generally necessary to use highly selective techniques, due to the low abundance of the receptor in the preparation. Some of those commonly used are:

a) *Chromatography on immobilised lectins*
Lectins are plant proteins which bind the carbohydrate residues on glycoproteins. Different lectins differ in their specificity, e.g. concanavalin A (from jack beans) binds mannose, whereas wheat germ lectin binds N-acetylglucosamine. Most cell surface receptors are glycosylated on their ectoplasmic domains, and useful purification can be achieved using lectins immobilized on Sepharose or other matrices.

b) *Chromatography on immobilized messenger*
Generally the most selective method and very successful for larger polypeptide messengers such as insulin (Figure 6.10). The receptor is usually eluted using free messenger, but one problem can be that the messenger–receptor complex formed is so tight that it may be difficult to regenerate the free receptor.

c) *Chromatography on immobilized toxin, agonist or antagonist*
With some small messengers, it may not be possible to couple the messenger itself to a matrix without losing receptor binding activity. It may be possible instead to use an immobilized toxin or other pharmacological agent. An additional advantage is that these are often more specific for receptor subtypes than the receptor itself. For example, the acetylcholine receptor can be purified

on immobilized α-bungarotoxin, while the β_1- and β_2-adrenergic receptors can be purified on columns of immobilized alprenolol. The latter is a β-antagonist that is not specific for β_1- or β_2-receptors. However, binding can be made selective by carrying it out in the presence of more specific β_1- or β_2-antagonists.

6.3 *Receptor superfamilies*

Using the methods outlined above, a large number of cell surface receptors for first messengers have now been identified and purified. Amino acid sequence information from the purified proteins has then been used to design synthetic oligonucleotides with which to screen cDNA libraries. Sequencing of cDNA has usually been the quickest route to obtaining complete sequences for the receptor polypeptides. As well as allowing informative models of the receptors to be developed, this effort has also shown that the receptors for different messengers are similar in sequence and may be derived from common ancestors. Thus the bewildering number of known receptors may be reduced to a much smaller number of receptor classes or superfamilies. We will finish this chapter by considering the major classes of receptor that have been defined by these studies.

6.3.1 *Multi-subunit receptors – the nicotinic acetylcholine receptor family*

Due largely to its great abundance (up to 70% of total membrane protein) in the electroplax organs of electric fish (Section 6.2.2), the nicotinic acetylcholine receptor is one of the best understood of all receptors. After solubilization in detergent, the electroplax receptor can be readily purified by affinity chromatography on columns of immobilized α-bungarotoxin. The purified protein contains four polypeptides from 40 to 60 kDa designated α, β, γ and δ. The relative proportions of these suggested a stoichiometry of $\alpha_2\beta\gamma\delta$ which, making allowance for glycosylation of all four subunits, is consistent with the observed minimal native molecular weight of 290,000. The 290 kDa form exists in equilibrium with a dimeric form in which the two pentamers are linked by disulphides (see Figure 6.13). Crosslinking studies using α-bungarotoxin, and affinity labelling using radioactive acetylcholine analogues such as bromoacetylcholine (Figure 6.8), suggest that both the messenger and the toxin bind to the α subunits, consistent with the observed stoichiometries of binding of 2 molecules per $\alpha_2\beta\gamma\delta$.

As is the case for most integral membrane proteins, it has not proved possible to obtain crystals of the receptor suitable for X-ray crystallography. However, if membrane vesicles from electroplax organs are stored for some time, they aggregate into so-called "tubular crystals" in which the receptors are present in

Fig. 6.11 Density map of the electroplax nicotinic acetylcholine receptor, reconstructed from negatively-stained electron micrographs of tubular crystals. The view is at right angles to the plane of the membrane, looking straight down the central "pore". The five subunits, in pseudosymmetrical orientation, are clearly seen. Reproduced from Brisson and Unwin (1985) Nature **315**, 474–477, courtesy of Nigel Unwin.

regular arrays. Electron microscope images of these arrays can be processed using the methods discussed for gap junctions in Section 1.3.1, to generate a three-dimensional image of an "average" receptor. The reconstructed image, viewed down an axis at right angles to the plane of the membrane, shows that the receptor consists of five regions of heavy staining arranged in a pseudosymmetrical fashion around a central lightly staining region or "pore" (Figure 6.11). A side view shows that all of the five structures span the membrane (Figure 6.12). These five structures correspond to the five subunits ($\alpha_2\beta\gamma\delta$), because reconstructed electron microscope images of receptors coupled to subunit-specific antibodies, or crosslinked to α-bungarotoxin, allow identification of the subunits (Figure 6.13).

Since symmetrical protein oligomers are usually formed from identical subunits, it was surprising that four different subunits could form a pseudo-symmetrical structure. However, the explanation for this became apparent when it was found that all four subunits are related in sequence. This was suggested initially by amino acid sequencing of the N-termini of the four subunits, and sequencing of full length cDNA coding for the four subunits confirmed that they

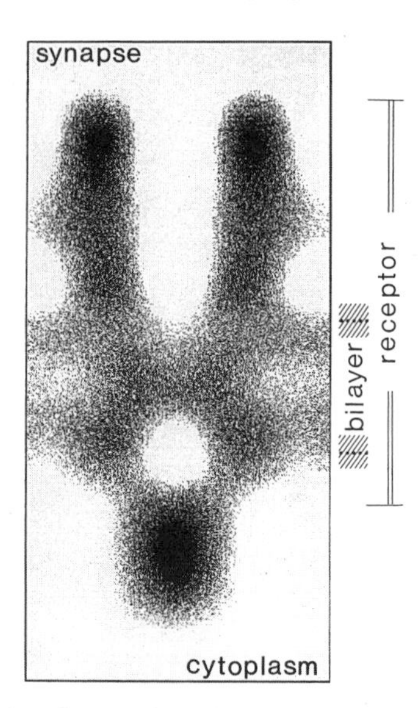

Fig. 6.12 Map of the electroplax nicotinic acetylcholine receptor, reconstructed from electron micrographs of frozen tubular crystals viewed at different angles. The dark mass apparently blocking the cytoplasmic exit of the channel is a distinct peripheral protein whose function is currently unknown. Reproduced from Toyoshima and Unwin (1988) Nature **336**, 247–250, courtesy of Nigel Unwin.

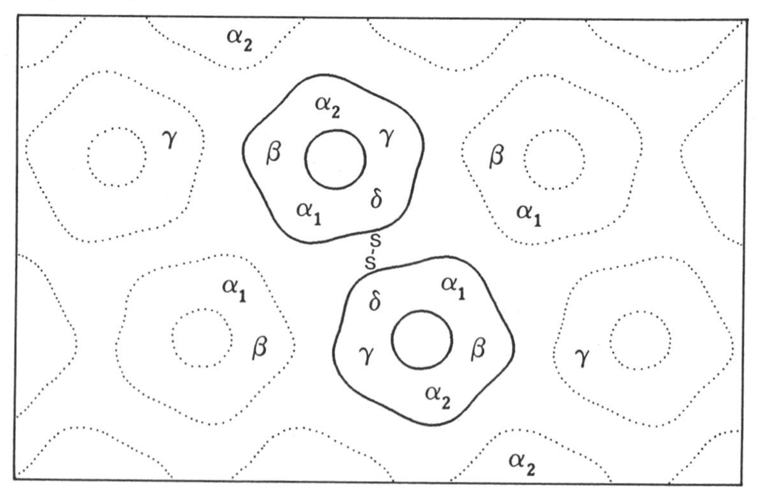

Fig. 6.13 Identification of subunits of the nicotinic acetylcholine receptor, based on reconstructed electron micrographs of receptors coupled to snake toxins and subunit-specific antibodies. Reproduced from Kubalek, Ralston, Lindstrom, and Unwin (1987) J. Cell Biol. **105**, 9–18, courtesy of Nigel Unwin.

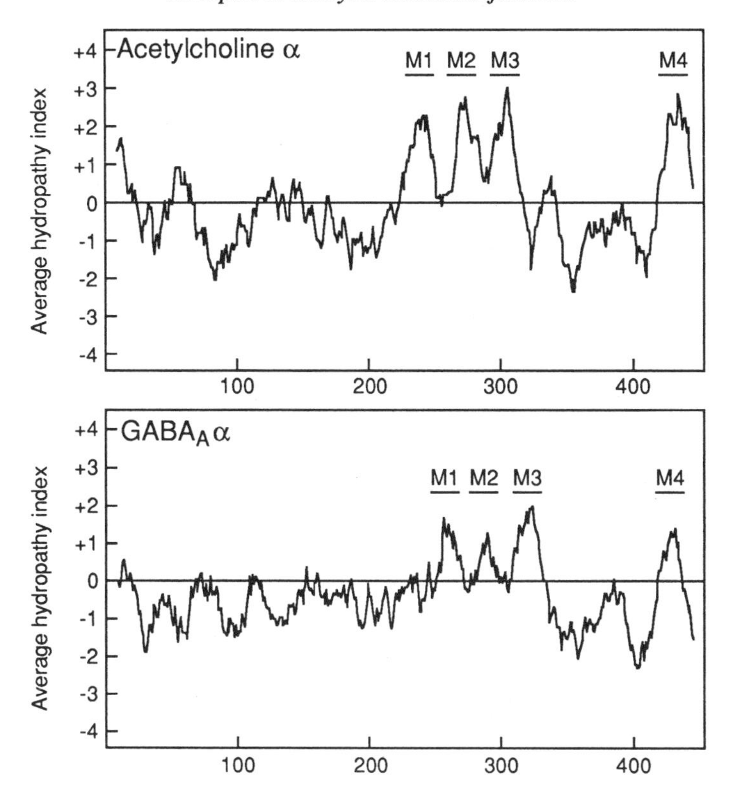

Fig. 6.14 Hydropathy plots of the α subunits of the nicotinic acetylcholine and GABA$_A$ receptors, indicating the identical locations of hydrophobic segments (M1 to M4). See Figure 1.12 for an explanation of these plots.

were related. The cDNAs from the electric fish electroplax were also used to screen cDNA libraries from mammals, where four very similar receptor subunits were found in a more conventional physiological setting at the neuromuscular junction. Nicotinic receptors in the central nervous system are less well characterized, but appear to be constructed from only two subunits, which are isoforms of the α and β subunits found at the neuromuscular junction.

A hydropathy plot (see Section 1.3.1) for the α subunit of the acetylcholine receptor is shown in Figure 6.14, and suggests four hydrophobic segments, M1 to M4. Similar results are obtained from hydropathy profiles on the other three subunits, leading to the models shown in Figure 6.15. This remains of course a working model, but it is consistent with other observations. Thus, for example, glycosylation of all four subunits occurs on the N-terminal region, which the model predicts to be on the extracellular (**ectoplasmic**) side of the membrane, whereas phosphorylation of serine residues, which would be expected to occur only on the cytoplasmic side, occurs on the loop between helices M3 and M4.

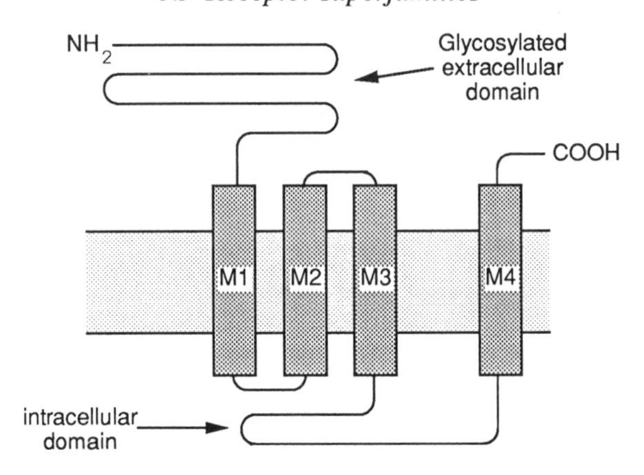

Fig. 6.15 Model for the membrane topology of one subunit of the nicotinic acetylcholine receptor.

Acetylcholine is an **excitatory** neurotransmitter because it causes depolarization of the target cell membrane (Chapter 7), promoting the formation of an action potential. As briefly discussed in Chapter 2, there are also **inhibitory** neurotransmitters, which antagonize the action of excitatory transmitters acting on the same target cell, by resisting depolarization. Two examples are γ-aminobutyrate (GABA), the most important inhibitory transmitter in the brain itself, and glycine, found in the brain stem and spinal cord and important in inhibition of motor nerves. The importance of these "negative" transmitters can be seen by the effects of the plant alkaloids picrotoxin and strychnine, which are antagonists at one class of GABA receptor ($GABA_A$) and glycine receptors respectively. Both cause uncontrolled firing of nerves, leading to convulsions and death. Conversely, $GABA_A$ receptors are activated by the barbiturate and benzodiazepine drugs, which inhibit firing of nerves, and at increasing dose levels induce loss of anxiety, sleep, anaesthesia and ultimately death. The benzodiazepines (e.g. Librium, Valium, Mogadon) appear to act more specifically on $GABA_A$ receptors than the barbiturates, and are much less likely to cause death due to overdose.

Both the $GABA_A$ and glycine receptors have now been purified by affinity chromatography on columns of immobilized benzodiazepine and immobilized strychnine respectively. The $GABA_A$ receptor appears to consist of two subunits with $\alpha_2\beta_2$ stoichiometry, while the glycine receptor appears to have two subunits of 48 and 58 kDa which are related in sequence, plus a 93 kDa subunit that is a peripheral protein on the cytoplasmic side. cDNAs coding for both subunits of the $GABA_A$ receptor, and the 48 kDa subunit of the the glycine receptor, have now been cloned and sequenced using methods similar to those used for the nicotinic acetylcholine receptor. Remarkably, the four subunits of the nicotinic acetylcholine receptor, the two subunits of the $GABA_A$ receptor and the 48 kDa (and, presumably, the 53 kDa) subunits of the glycine receptor all show significant

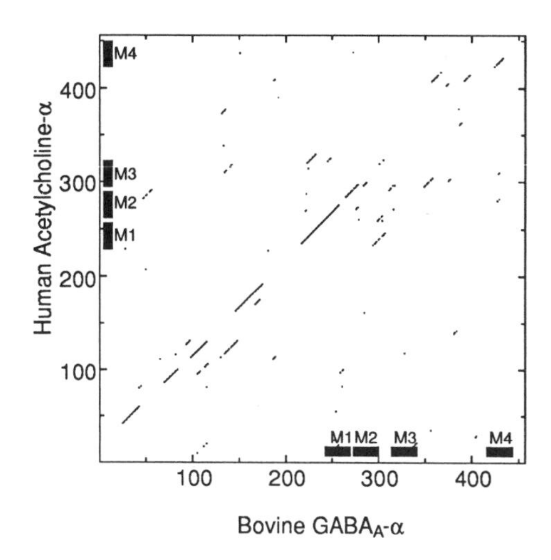

Fig. 6.16 Dot matrix plots showing sequence similarities between the α subunits of the nicotinic acetylcholine and $GABA_A$ receptors. The diagonal lines (see Figure 3.15) indicate that the two sequences are particularly related in the N-terminal regions (residues 0–200) and in the regions of transmembrane helices M1/M2.

sequence similarity. Hydropathy plots suggest that four transmembrane helices would form in analogous positions of all of these subunits (e.g. Figure 6.14), so that the model shown in Figure 6.15 for the acetylcholine receptor subunits can also be applied to the other two receptors. The sequence similarity is also shown by the dot matrix comparison of the acetylcholine and $GABA_A$ receptors in Figure 6.16. This shows that the similarity is most noticeable in the N-terminal ectoplasmic domain, thought to be the messenger-binding region, and in the regions of the proposed transmembrane helices M1/M2.

Thus the receptors for these three rapid-acting neurotransmitters, both excitatory and inhibitory, appear to fall into a **superfamily** which has probably arisen by duplication and divergence of an ancestral neurotransmitter receptor. The great significance of this finding is that knowledge gained about the function of any one of these receptors (discussed in the next chapter) may, with caution, be applied to the other members of the family.

6.3.2 *"Seven-pass" receptors – the rhodopsin/β-adrenergic receptor family*

Paradoxically, the first member of this family to be characterized was not a messenger receptor at all, but a visual pigment. A slight digression is necessary at this point to consider the mechanism of light absorption in the eye. The lens of the

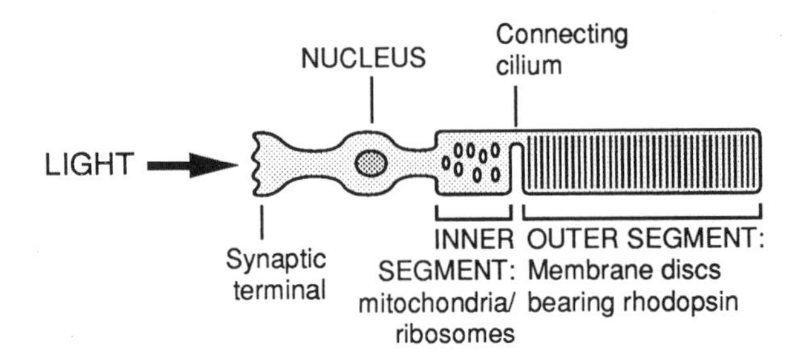

Fig. 6.17 Schematic structure of retinal rod cell.

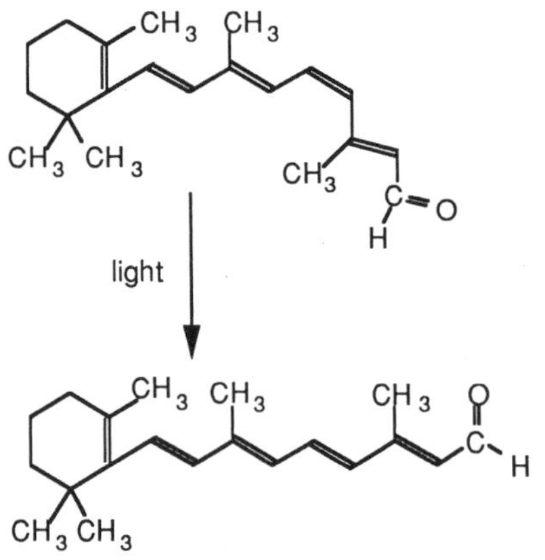

11-*cis*-retinal

light

all-*trans*-retinal

Fig. 6.18 *Cis* ↔ *trans* isomerization of retinal, which occurs when rhodopsin absorbs a photon.

mammalian eye focuses light onto the retina, which in turn converts the light signal into action potentials in the optic nerve. The retina contains two types of photoreceptor cells, i.e. cones, which are responsible for colour vision in normal light, and rods, which give black and white vision at low light levels. The structure of a rod cell is shown diagrammatically in Figure 6.17. Photons of light are absorbed by the pigment rhodopsin in the membrane discs of the outer segment. Rhodopsin consists of a protein, **opsin**, covalently attached to a chromophore, **retinal**, and absorption of a single photon of light causes isomerization of retinal from the 11-*cis* to the all-*trans* form (Figure 6.18). Retinal is derived by oxidation of retinol (vitamin A – Section 2.3.6), so there is some truth in the rumour that eating carrots helps you to see at night! Apart from differences in size and shape, cone cells are functionally similar to rods. The major difference is that where rods have a single rhodopsin with a broad absorption spectrum covering most of the visible region, there are three types of cone cells,

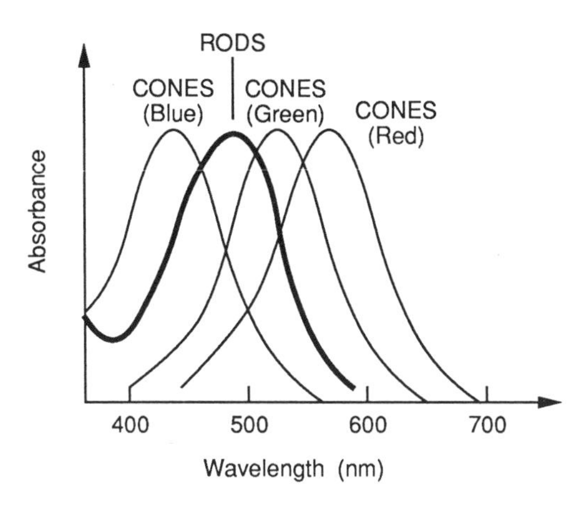

Fig. 6.19 Absorption spectra of rhodopsins in rods and the three types of cones.

each containing a different opsin. The three opsins modify the absorption spectrum of their retinal chromophores to produce three pigments with overlapping spectra (Figure 6.19).

Rhodopsin is the major component of the membrane discs of the rod cell outer segment, and so could easily be purified for sequencing at the amino acid level. Hydropathy plots (Figure 6.20) suggested that it contained seven transmembrane helices (hence the term "seven-pass"), suggesting the model shown in Figure 6.21, which has been confirmed by physical measurements. The retinal is covalently attached via a Schiff's base to a lysine residue in the centre of transmembrane helix 7. The other transmembrane helices appear to form a pocket which surrounds the retinal in the membrane. Isomerization of retinal from the *cis* to the *trans* form, which takes picoseconds, involves a movement of 5 Å of the Schiff base linkage with respect to the ring portion, and this triggers conformational changes of the opsin component on the millisecond time scale. The mechanism by which this is converted into an action potential is discussed in Chapter 7.

As discussed in Section 6.1.2, catecholamines like adrenaline (epinephrine) bind to at least four classes of receptor, β_1, β_2, α_1 and α_2. Initial analysis of β-adrenergic receptors involved photoaffinity labelling using [^{125}I] *p*-azidobenzoylpindolol and [^{125}I]cyanopindolol-azide (Figure 6.22), these being derivatives of the β-antagonist pindolol. These reagents labelled membrane polypeptides of around 60 kDa. The β_2-receptors were purified by repeated affinity chromatography on the β-antagonist alprenolol coupled to Sepharose. Very small amounts were obtained, but they were sufficient to obtain partial amino acid sequences. Using synthetic oligonucleotides based on these sequences, cDNA coding for the receptors was cloned and sequenced. At this point a remarkable analogy was noticed with the structure of rhodopsin. Hydropathy plots revealed

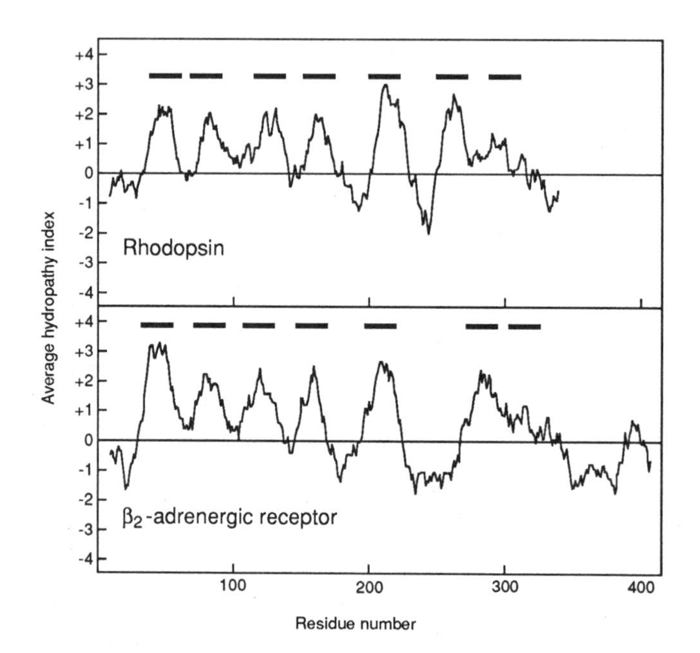

Fig. 6.20 Hydropathy plots of rhodopsin/β_2-adrenergic receptor, showing similar locations of hydrophobic segments (black bars). See Figure 1.12 for an explanation of these plots.

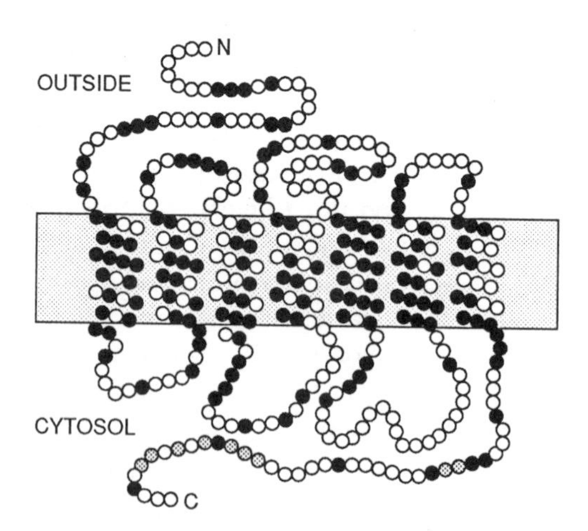

Fig. 6.21 Model for membrane topology of rhodopsin. Hydrophobic residues (val, leu, ile, met, phe, trp, tyr) are shown in black. Serine and threonine residues thought to be phosphorylated by rhodopsin kinase in the C-terminal tail are shaded.

Adrenaline (epinephrine)

p-azidobenzyl pindolol

Cyanopindolol azide

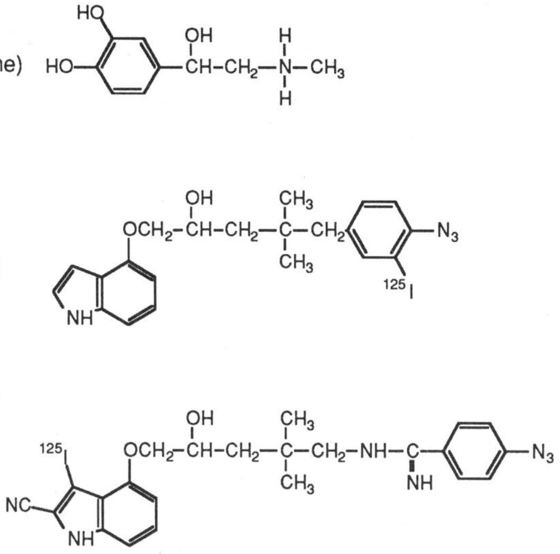

Fig. 6.22 Photoaffinity analogues of adrenaline (epinephrine) used for identification of β-adrenergic receptor proteins.

seven likely transmembrane helices in positions very similar to those found in rhodopsin (Figure 6.20). There was also some similarity in amino acid sequence, chiefly within the transmembrane helices. The rhodopsin-like model for the topography of the β_2-receptor shown in Figure 6.23 has been largely confirmed by proteolysis studies, and by analysis of sites of covalent modification. Thus, for example, treatment of intact cells with trypsin causes cleavage between helices 4 and 5, this loop therefore being ectoplasmic. Glycosylation occurs in the N-terminal region, which must therefore be ectoplasmic, whereas phosphorylation (Section 6.4) occurs on the C-terminal tail and the loop between helices 5 and 6, which must therefore be cytoplasmic.

Since the original sequencing of rhodopsin and the β_2-adrenergic receptor, a large variety of other receptors have been shown to be members of this family and to conform to this "seven-pass" model, including all known catecholamine receptors, and the receptors for several peptide messengers (Table 6.2). In some cases the sequences have been obtained after purification by methods similar to those used for the β_2-adrenergic receptor. A powerful alternative approach has been the use of low-stringency hybridization with existing cDNA probes to detect related DNA sequences. In order to identify novel receptors cloned by this approach, the cloned cDNA can be converted to RNA using a viral RNA polymerase, and then expressed by injection of the RNA into *Xenopus* oocytes (Figure 4.16). The oocytes translate the RNA into protein, which is then inserted into the membrane. The ligand binding properties of the expressed protein can then be examined.

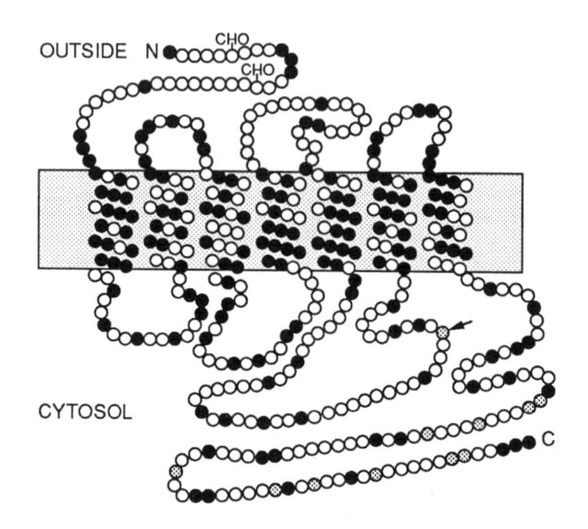

Fig. 6.23 Model for the membrane topology of the β_2-adrenergic receptor. Hydrophobic residues (val, leu,ile, met, phe, trp, tyr) are shown in black. "CHO" signifies asparagine residues in the N-terminal region that may be glycosylated. Serine and threonine residues thought to be phosphorylated by β-adrenergic receptor kinase in the C-terminal tail are shaded. A probable site for phosphorylation by cyclic AMP-dependent protein kinase in the 3rd cytoplasmic loop is arrowed.

Table 6.2 Some messenger receptors shown to be of the "seven-pass" type. These receptors have been characterized in mammals unless stated otherwise.

Messenger	Messenger type	Receptor types
(Nor)adrenaline	Catecholamine	α_1, α_2, β_1, β_2
Dopamine	Catecholamine	D_2
Acetylcholine	Amine	Muscarinic (M1-M5)
Serotonin	Amine	1A, 1C, 2
Cyclic AMP	Nucleotide	(*Dictyostelium discoideum*)
Lutropin	Peptide	
Thyrotropin	Peptide	
Mating factor	Peptide	a, α (*Saccharomyces cerevisiae*)
Substance K	Peptide	
Angiotensin	Peptide	

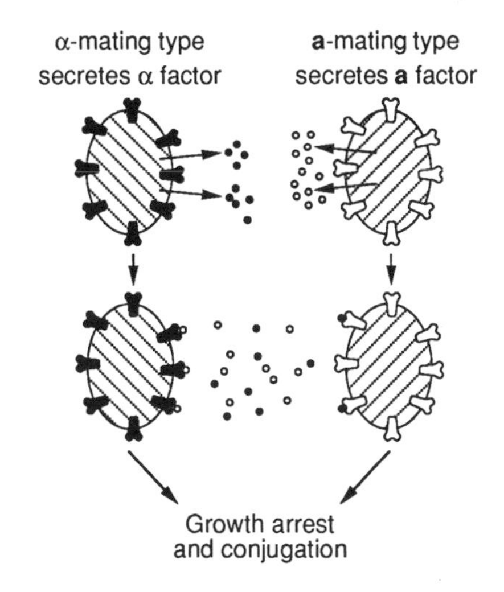

Fig. 6.24 Function of yeast mating factors.

These approaches have revealed a hitherto unsuspected degree of heterogeneity in many of these receptors. Previous pharmacological studies had suggested that muscarinic acetylcholine receptors could be divided into two subtypes, but at least five have now been found by DNA cloning. All are related to the β_2-adrenergic receptor and conform to the "seven-pass" model. No less than seven serotonin receptors have been postulated from pharmacological studies, and to date three of these have been shown to be members of the "seven-pass" family. Intriguingly, members of the family are also found in unicellular eukaryotes, such as yeasts and slime moulds. The yeast *Saccharomyces cerevisiae* normally exists as a haploid form which divides mitotically by budding. However, if nutrients become limiting, cells of opposite mating type conjugate to form diploids which then sporulate. To trigger this process, the two mating types release the a and α mating factors, which are recognized by receptors on the surface of the opposite mating type (Figure 6.24). Although the intracellular mechanisms are not understood, the a and α mating factor receptors have now been cloned and shown to be members of the "seven-pass" family. An analogous intercellular communication process occurs in the slime mould *Dictyostelium discoideum*. In rich medium this organism exists as a unicellular amoeba-like form, but on starvation the cells aggregate and differentiate to form a spore body. This is triggered by the secretion of cyclic AMP from the cells, and the cyclic AMP receptor is also a member of the "seven-pass" family.

By analogy with the location of the retinal in rhodopsin, one might expect the ligand-binding site on the receptor proteins to be within the transmembrane helices. Consistent with this idea, various affinity labelling reagents for β, α_2 and M_1-muscarinic receptors have been shown to react covalently with one or other of

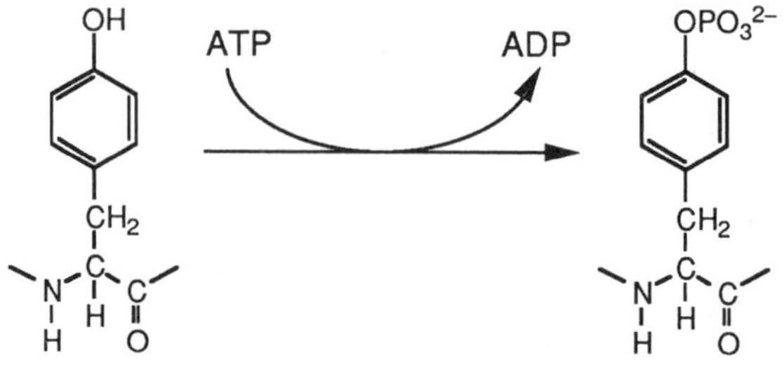

Fig. 6.25 Phosphorylation of a tyrosine side chain.

the transmembrane regions, including helices 2, 3, 4 and 7. Site-directed mutagenesis studies, in which point mutations or small deletions are made in the DNA which is then expressed by transfection, also suggest that the transmembrane regions are involved in ligand binding. An exception to this may be receptors for the larger polypeptide hormones such as that for lutropin and choriogonadotropin. These messengers are members of the glycoprotein hormone family (Section 3.3.1) which have molecular masses of 28 to 38 kDa, and would appear to be much too large to fit within the helices of a seven-pass receptor. Sequence analysis shows that the common receptor for lutropin and choriogonadotropin does belong to the seven-pass family, but with a much larger N-terminal, ectoplasmic domain which appears to be involved with hormone binding.

Although the "seven-pass" topology is highly conserved between the members of the family, actual sequence similarities are generally limited and confined to the transmembrane regions. One might ask the question whether the similarities between rhodopsin and the receptors have arisen by divergence from a common ancestor, or by convergent evolution towards a common function. Although this type of question can never be answered with certainty, the mechanisms of signal transduction at these receptors via G proteins (Section 7.2.4), and their regulation by phosphorylation (Section 6.4), are also very closely related. It therefore seems likely that the members of this class of receptor for first messenger, and the sensory receptor, rhodopsin, are derived from a common ancestor. It is also worth noting that both taste and smell receptors in mammals, although poorly characterized, are believed to be mediated by the second messenger cyclic AMP (Section 7.2.1). All other receptors coupled to cyclic AMP production are of the "seven-pass" type. Therefore other sensory systems may operate in the same way. One possibility is that the "seven-pass" class of receptors for first messengers have evolved from more ancient sensory receptors.

6.3.3 *Single pass receptors – the tyrosine kinase family*

The first member of this family to be well characterized at the protein level was the

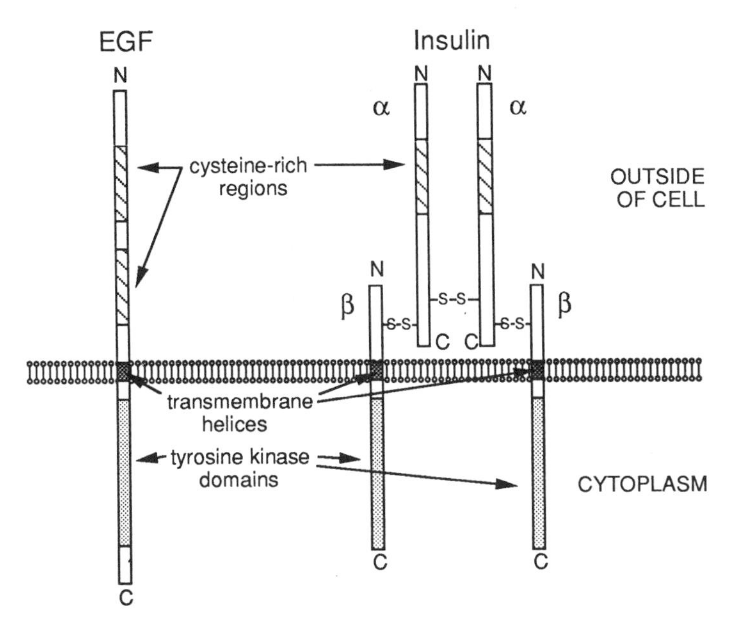

Fig. 6.26 Model for the membrane topology of the epidermal growth factor and insulin receptors.

epidermal growth factor (EGF) receptor, due largely to its abundance in the A431 tumour cell line (Section 6.2.2). The receptor was purified on immobilized EGF, and shown to be a glycosylated single polypeptide of apparent molecular mass 170 kDa. Associated with the purified receptor was an EGF-dependent protein kinase activity which phosphorylated the receptor itself as well as exogenous substrates. This protein kinase was subsequently shown to be specific for phosphorylation of **tyrosine** residues (Figure 6.25). The function of this tyrosine kinase activity is considered in Chapter 7.

Purification of the EGF receptor, in amounts large enough for partial amino acid sequencing, opened the way to cDNA cloning and complete sequence determination. The human receptor contains 1186 residues, and a single stretch of amino acids suitable for a transmembrane helix (Figure 6.26). The N-terminal, ectoplasmic domain contains all of the glycosylation sites and must contain the EGF-binding site. It exhibits two regions of internal sequence similarity, which appear to have resulted from duplication of a cysteine-rich sequence of ~300 residues. Consistent with the identification of the receptor as an EGF-activated tyrosine kinase, the C-terminal cytoplasmic domain has a region of ~250 residues that is similar to the sequence of $pp60^{v\text{-}src}$, a tyrosine kinase encoded by the Rous sarcoma virus (Section 9.2.1). There is also a C-terminal tail which contains the tyrosine residues phosphorylated on the receptor itself in response to EGF (the **autophosphorylation sites**).

Because of its major importance in the disease diabetes, the insulin receptor has received much attention. Purification of the insulin receptor on insulin-Sepharose in the early 1970s was among the first attempted applications of affinity chromatography. Progress was initially slow due to the low abundance of the protein, but eventually a preparation was obtained that on reduction of disulphides exhibited a major polypeptide of 125 kDa, with variable amounts of a second polypeptide at ~45 kDa. At about the same time, crosslinking and photoaffinity labelling studies were initiated, either using ^{125}I-insulin and the crosslinking agent disuccinimidyl suberate, or using photoactivatable azide analogues of ^{125}I-insulin (Figure 6.8). These revealed that the receptor consists of a 125 kDa α subunit and a 90 kDa β subunit, joined in a β–α–α–β layout by disulphide bridges (Figure 6.26). This tetrameric structure could be reduced under mild conditions to α–β heterodimers, which then separated in higher concentrations of reducing agent to the individual subunits. Only the α subunit was labelled using crosslinking or photoaffinity analogues, suggesting that it contained the insulin-binding site. The 45 kDa polypeptide observed in the original purified preparations is now known to be a proteolytic fragment of the β subunit.

At first sight there would seem to be little similarity between the structures of the EGF and insulin receptors. However, a twist in the story came when insulin receptors were labelled biosynthetically using radioactive amino acids, and then immunoprecipitated. This showed an initial synthesis of a 180 kDa precursor that was converted to a 200 kDa form by addition of carbohydrate. The 200 kDa form was then further converted to the mature 125 and 90 kDa subunits. Cloning and sequencing of insulin receptor cDNA has now completely confirmed that the α and β subunits are synthesized as a single precursor with the α sequence at the N-terminus, which is then cleaved at a sequence of four basic residues to the mature subunits. Only the β-subunit contains a sequence suitable for a transmembrane helix, and the α subunit is believed to be entirely extracellular (Figure 6.26). It is a remarkable coincidence that both insulin and its receptor are synthesized as single chain precursors that are converted to twin polypeptide structures linked by disulphides.

Shortly after the elucidation of the β–α–α–β structure, it had been shown that insulin activated the phosphorylation of the β subunit on tyrosine residues. It was therefore no surprise that the sequencing revealed a putative tyrosine kinase domain in the β subunit in a very similar position to that found in the EGF receptor (Figure 6.26). More surprising was the finding of significant similarity in sequence between cysteine-rich regions in the α subunit and in the ectoplasmic domain of the EGF receptor. Crosslinking studies had already suggested that insulin binds to the α subunit, and it is likely that these regions have diverged from a common ancestral peptide messenger-binding domain.

In the last few years, several more members of this tyrosine kinase receptor superfamily have been defined by cloning and sequencing. In all cases where the ligand is known, it is a peptide growth factor: this is true of insulin itself, which is

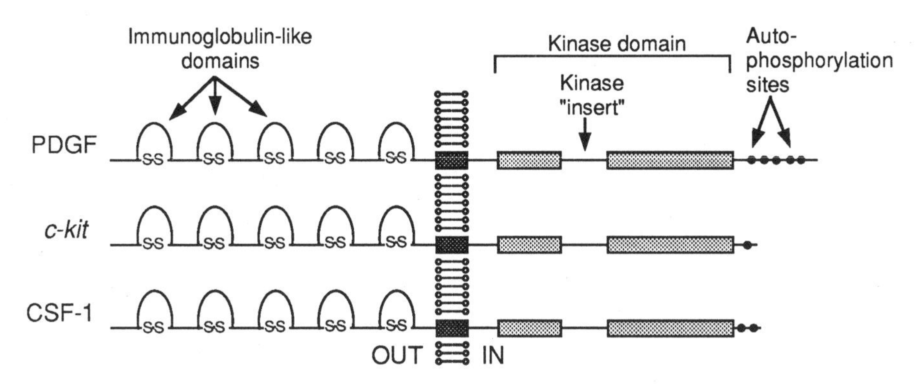

Fig. 6.27 General model of the platelet-derived growth factor/*fms*/*kit* family of growth factor receptors.

a growth factor for many cell types in culture. There is good evidence that tyrosine phosphorylation is involved in control of cell growth by these factors (Chapter 9). The superfamily may be divided into at least three subfamilies:

1) *The EGF receptor family*

The EGF receptor itself binds both EGF and transforming growth factor-α. A close relative of the EGF receptor, the *neu* gene product, was originally isolated from a neural tumour and identified as an **oncogene** (cancer-causing gene) by its ability to transform cells in culture (Section 9.3.2). The oncogene form does this due to a single point mutation in the transmembrane region, and the wild type *neu* protein is assumed to be a growth factor receptor, although its natural ligand is not yet known.

2) *The insulin receptor family*

This type, with a subunit structure of β–α–α–β, is represented by the receptors for insulin and insulin-like growth factor-1 (IGF-1). Like the peptide messengers themselves, these two receptors are very similar, and in fact insulin binds to the IGF-1 receptor, and IGF-1 binds to the insulin receptor, although the affinities are lower than with their own receptors.

3) *The PDGF receptor family*

This family is exemplified by the two known isoforms of the receptor for platelet-derived growth factor (PDGF). Other examples are the *c-fms* and *c-kit* genes (Figure 6.27). The latter two were cloned through hybridization with the *v-fms* and *v-kit* oncogenes, cancer-causing genes of tumour viruses (Chapter 9). The *c-fms* protein is now known to be the receptor for colony stimulating factor-1, a growth factor for macrophages. The ligand for the *c-kit* gene product has not been identified. Receptors in this class are characterized by an insertion of ~100

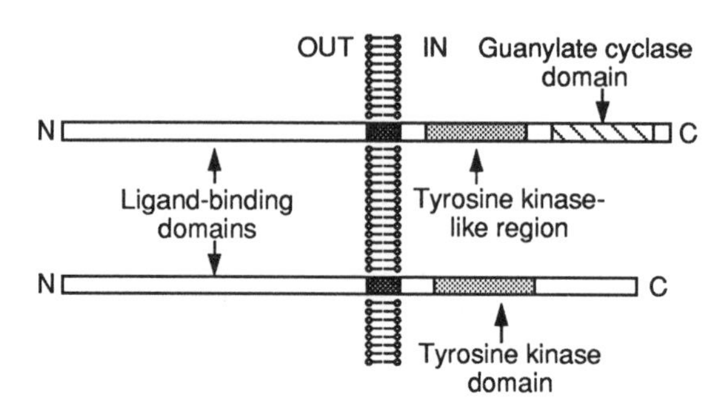

Fig. 6.28 Model for the resact/atrial natriuretic peptide family of receptors (top) and comparison with a protein (tyrosine) kinase receptor.

residues in the centre of the tyrosine kinase domain. In addition, the extracellular domains contain five internally homologous sequence repeats, each with a size and disposition of cysteine residues similar to that in an immunoglobulin domain.

6.3.4 *Single-pass receptors – the guanylate cyclase family*

The receptors for atrial natriuretic peptide (ANP) have now been purified, cloned and sequenced from the human and the rat. A closely related protein has also been cloned and sequenced from spermatozoa of sea urchins, and identified as a receptor for a chemotactic peptide (**resact**) secreted by the egg cells. All three sequences suggest a protein of ~1000 residues, with a single transmembrane helix in the centre (Figure 6.28). The cytoplasmic domain is well conserved and has two distinct regions. One shows some sequence similarity with the tyrosine kinase domains of the growth factor receptors described above. However, several residues which are thought to be essential for kinase activity, are not found in the ANP/resact receptor sequence. The function of this region is therefore unclear. The other region is homologous with part of the sequence of mammalian adenylate cyclase (which catalyses cyclic AMP synthesis) and is likely to be the guanylate cyclase domain that is activated by binding of messenger to the ectoplasmic domain (Section 7.2.7).

6.3.5 *Single-pass receptors for other peptide messengers*

The single-pass receptors for peptide messengers described in the previous two sections have ligand-activated enzyme activities (tyrosine kinase/guanylate cyclase) as their cytoplasmic domains. A final group of sequenced peptide messenger receptors also contain single transmembrane regions, but their cytoplasmic regions seem too small to contain any catalytic activity. This group includes the receptors for somatotropin and prolactin, two related polypeptide hormones (Section 3.3.1) which also have closely related receptors, and the receptors for interleukins 1, 2 and 6, all lymphocyte growth factors. In view of the very small cytoplasmic regions for some of these receptors, it seems likely that their effects must be transduced via interaction with other membrane proteins, but in most cases the mechanism of signal transduction is not understood. The ectoplasmic portions of the receptors for interleukins 1 and 6 contain immunoglobulin-like domains similar to those in the PDGF receptor class.

6.4 *Desensitization of receptors*

If cells are continuously exposed to a first messenger, a common finding is that they become increasingly resistant to the action of that messenger. This is clearly a type of feedback regulation, and takes two forms: a rapid, acute effect which involves changes in the function of existing receptor molecules (**desensitization**), and a longer-term, chronic effect which involves a decrease in the number of receptor molecules. The latter phenomenon is known as **down-regulation**, and is dicussed in Section 6.5.

Desensitization can take two forms:

1) **homologous desensitization** occurs when incubation of cells with one messenger produces a loss of sensitivity specifically to that messenger;
2) **heterologous desensitization** occurs when incubation of cells with one messenger produces a loss of sensitivity to that messenger, and to other messengers acting through different receptors.

It is now clear that the mechanisms of desensitization involve protein phosphorylation (a topic discussed in more detail in Chapter 8), and they are best understood in the case of the β_2-adrenergic receptor. It has been found that the receptor can be phosphorylated on serine and threonine residues by a protein kinase termed **β-adrenergic receptor kinase**. The exact number and location of the site(s) of phosphorylation are not known, but they are all in the C-terminal tail, which contains at least 10 serine and threonine residues (Figure 6.23). An important feature is that this only occurs when agonist is bound to the receptor, suggesting that the phenomenon could be responsible for homologous desensitization. In confirmation of this, when these serine and threonine residues are abolished by site-directed mutagenesis, neither phosphorylation nor desensitization occurs in response to agonist. An exactly analogous mechanism seems to operate for rhodopsin, which is phosphorylated at multiple residues in the C-terminal tail by a specific **rhodopsin kinase**. This prevents the interaction of rhodopsin with the next component in the signal transduction pathway (Chapter 7).

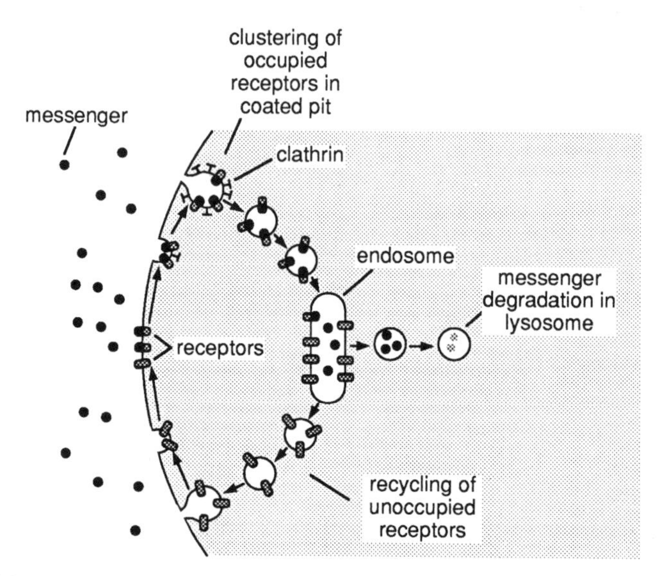

Fig. 6.29 Model of the pathway for receptor-mediated endocytosis.

A consequence of this mechanism of homologous desensitization is that only the individual receptor molecule to which agonist is bound becomes desensitized: others would be unaffected. It would operate efficiently at synapses where catecholamine was released, since high concentrations of messenger would occur, and the receptors would be readily saturated. In agreement with this idea, β-adrenergic receptor kinase is found in highest concentrations in brain and in tissues which receive sympathetic innervation. By contrast, at receptors where catecholamine is acting as a hormone or local mediator, it appears that heterologous desensitization predominates.

As discussed in Section 7.2, the β-adrenergic receptor mediates its effects through cyclic AMP and cyclic AMP-dependent protein kinase. The β_2-adrenergic receptor is also phosphorylated by cyclic AMP-dependent protein kinase, probably at a serine residue which occurs in the loop between transmembrane helices 5 and 6 (Figure 6.23). As expected, replacement of this residue with alanine by site-directed mutagenesis reduces desensitization. Since many other first messengers also increase cyclic AMP in their target cells, this would explain the phenomenon of heterologous desensitization.

Although the mechanisms of desensitization are not so well understood in other cases, several other receptors are known to be phosphorylated on their cytoplasmic domains in intact cells, and it appears likely that this is involved in regulating receptor activity. In some cases this may be truly heterologous, and allow receptors acting through completely different signal transduction systems to interact. A possible example is the phosphorylation of the γ and δ subunits of the nicotinic acetylcholine receptor (which is definitely **not** coupled to cyclic AMP production) by cyclic AMP-dependent protein kinase, which appears to increase the rate of desensitization of the receptor by acetylcholine.

6.5 *Down-regulation of receptors*

If radioactive peptide messengers are incubated with intact cells at 0°C, the messenger becomes bound to receptors on the cell surface, but the messenger is usually still accessible to degradation by added proteinases. If however the cells are incubated at physiological temperature, the messenger becomes resistant to external proteinases. Autoradiography has shown that this is associated with the migration of the messenger into the interior of the cell (**internalization**). It is now clear that this happens by the process of **receptor-mediated endocytosis**, which is related to the recycling of acetylcholine vesicles discussed in Section 5.3.3. This pathway (Figure 6.29) was originally delineated by studies in human fibroblasts of the receptor for **low-density lipoprotein** (LDL), which is involved in cholesterol uptake into the cells in the form of its plasma carrier, LDL. In cells from normal individuals, labelled LDL binds to a plasma membrane receptor which is concentrated in invaginations in the plasma membrane, **coated pits**, which are coated with the protein, clathrin (Section 5.1.2). The coated pits pinch off and form coated vesicles, which rapidly lose their clathrin coats in an ATP-requiring process. The vesicles then fuse with an intracellular compartment called the **endosome**, which is acidified with respect to the cytoplasm due to a proton-transporting ATPase similar to that in secretory vesicles (Section 5.1.2). The low pH causes the LDL to dissociate from the receptor: the latter returns in vesicles to the plasma membrane while the cholesterol esters from the LDL pass into progressively more acidic compartments (pre-lysosomes and lysosomes), where acid lipases hydrolyse them to release free cholesterol. The importance of this pathway was confirmed from studies with fibroblasts from patients with the inherited condition **familial hypercholesterolaemia**. These patients cannot take cholesterol up into cells efficiently: their plasma cholesterol is very high and they have a marked propensity to heart attacks. In some cases the inherited defect is due to a single point mutation (tyr→cys) in the cytoplasmic tail of the LDL receptor, which prevents it from being concentrated in the coated pits.

Although the LDL receptor is involved with nutrient uptake rather than signal transduction, it is now clear that most messenger–receptor complexes are internalized by the same pathway. What is the function of receptor-mediated endocytosis in the case of messenger receptors, which are not concerned with nutrient uptake? It has been argued that it may be necessary for signal transduction, but there is little evidence for this. More important may be its role in degradation both of the messenger and the receptor. As in the case of the LDL receptor, the messenger-receptor complex normally dissociates in the low pH of the endosome, and while the receptor recycles to the surface, the messenger is routed to lysosomes. The latter contain a variety of degradative enzymes including proteinases, and this is a major mechanism by which the target cell degrades the messenger, particularly for peptide messengers.

A difference between the LDL receptor and many messenger receptors such as the epidermal growth factor (EGF) receptor is that whereas the former recycles continuously in the presence or absence of ligand (i.e. **constitutively**), the latter only appears to cluster in coated pits in the presence of its ligand (EGF). Incubation of cells with the messenger will therefore cause a loss of receptors from the cell surface as they move to endosomes. Although some receptors may recycle to the surface, a certain proportion fail to dissociate from the messenger and are routed to lysosomes, where they are degraded. Thus continuous exposure to high concentrations of messenger leads to net loss of receptor protein, an effect known as **down-regulation**. It should be noted that this is a different phenomenon from **desensitization**, a more short-term mechanism in which the affinity of existing receptor molecules is modulated (Section 6.4). However down-regulation and desensitization both lead to a loss in sensitivity of target cells to messenger. These phenomena may explain, at least in part, the problems of habituation and withdrawal in drug addicts, and are probably also involved in conditions such as Type II (insulin-independent) diabetes, where insulin levels may be normal or even elevated, but cells become insulin-resistant.

SUMMARY

First messengers work by binding to proteins called receptors and, with the notable exceptions of steroid and thyroid hormones and retinoic acid (discussed in Chapter 10), these receptors are found on the cell surface. Cell surface receptors may be studied initially by binding studies using radioactive ligand, which can be the messenger itself, or a pharmacologically defined toxin, agonist or antagonist. The receptor proteins can be identified biochemically by crosslinking or affinity labelling using radioactive ligands. Purification, often involving affinity chromatography, eventually allows amino acid sequencing and cDNA cloning.

Sequencing of receptor cDNA clones has reduced the large number of pharmacologically defined receptors to a relatively small number of receptor families, while at the same time allowing recognition of previously unsuspected subtypes within each family. These receptor families include: (i) the ligand-gated ion channels, which form a channel made from five subunits arranged in pseudosymmetrical fashion; (ii) "seven-pass" receptors containing seven transmembrane helices, which include the catecholamine, muscarinic acetylcholine, and many peptide messenger receptors; (iii) three classes of "single pass" receptors for peptide messengers, i.e. those displaying tyrosine kinase activity, those displaying guanylate cyclase activity, and those displaying no known enzyme activity.

Receptors are regulated by acute and chronic mechanisms, i.e. desensitization and down-regulation respectively. In the few cases where it is understood, desensitization involves phosphorylation of cytoplasmic domains of the receptor. Down-regulation is a loss of receptor protein, and probably results from ligand-induced endocytosis through coated pits, followed by the routing of a proportion of the internalized receptor to lysosomes.

FURTHER READING

1. "The insulin receptor of the turkey erythrocyte: characterization of the membrane-bound receptor" – example of receptor-binding study. Ginsberg, B.H., Kahn, C.R. and Roth, J. (1976) Biochim. Biophys. Acta **443**, 227–242.
2. "Electrophoretic resolution of three major insulin receptor structures with unique subunit stoichiometries" – crosslinking study. Massague, J., Pilch, P.F. and Czech, M.P. (1980) Proc. Natl. Acad. Sci. USA **77**, 7137–7141.
3. "Adenylate cyclase-coupled beta-adrenergic receptors: structure and mechanisms of activation and desensitization" – review on pharmacological studies and purification of adrenergic receptors. Lefkowitz, R.J., Stadel, J.M. and Caron, M.G. (1983) Ann. Rev. Biochem. **52**, 159–186.
4. "Structural homology of *Torpedo californica* acetylcholine receptor subunits". Noda, M., Takahashi, H., Tanabe, T., Toyosato, M., Kikyotani, S., Furutani, Y., Hirose, T., Takashima, H., Inayami, S., Miyata, T. and Numa, S. (1983) Nature **302**, 528–532.
5. "Quarternary structure of the acetylcholine receptor" – electron microscope study. Brisson, A. and Unwin, P.N.T. (1985) Nature **315**, 474–477.
6. "Sequence and functional expression of the $GABA_A$ receptor shows a ligand-gated receptor super-family". Schofield, P.R., Darlison, M.G., Fujita, N., Burt, D.R., Stephenson, F.A., Rodriguez, H., Rhee, L.M., Ramachandran, J., Reale, V., Glencorse, T.A., Seeburg, P.H. and Barnard, E.A. (1987) Nature **328**, 221–227.
7. "Adrenergic receptors. Models for the study of receptors coupled to guanine nucleotide regulatory proteins" – minireview of more recent work. Lefkowitz, R.J. and Caron, M.G. (1988) J. Biol. Chem. **263**, 4993–4996.
8. "Rhodopsin and bacteriorhodopsin: structure and function relationships" – review. Ovchinnikov, Y.A. (1982) FEBS Lett. **148**, 179–190.
9. "The genes for colour vision" – review. Nathans, J. (1989) Scientific American, February, pp.28–35.
10. "Lutropin-choriogonadotropin receptor: an unusual member of the G protein-coupled receptor family". McFarland, K.C., Sprengel, R., Phillips, H.S., Köhler, M., Rosemblit, N., Nikolics, K., Segaloff, D.L. and Seeburg, P.H. (1989) Science **245**, 494–499.

11. "Conservation of a receptor/signal transduction system" – minireview on receptors for yeast mating factors. Herskowitz, I. and Marsh, L. (1987) Cell **50**, 995–996.
12. "Affinity labelling of the protein kinase associated with the epidermal growth factor receptor in membrane vesicles from A431 cells" – conclusive evidence that the EGF receptor has an integral protein (tyrosine) kinase. Buhrow, S.A., Cohen, S. and Staros, J.V. (1982) J. Biol. Chem. **257**, 4019–4022.
13. "Human epidermal growth factor receptor cDNA sequence and aberrant expression of the amplified gene in A431 epidermoid carcinoma cells". Ullrich, A., Coussens, L., Hayflick, J.S., Dull, T.J., Gray, A., Tam, A.W., Lee, J., Yarden, Y., Libermann, T.A., Schlessinger, J., Downward, J., Mayres, E.L.V., Whittle, N., Waterfield, M. and Seeburg, P.H. (1984) Nature **309**, 418–425.
14. "Human insulin receptor and its relationship to the tyrosine kinase family of oncogenes" – cDNA cloning and sequencing of the insulin receptor. Ullrich, A., Bell, J.R., Chen, E.Y., Herrera, R., Petruzzelli, L.M., Dull, T.J., Gray, A., Coussens, L., Liao, Y.C., Tsubokawa, M., Mason, A., Seeburg, P.H., Grunfeld, C., Rosen, O.M. and Ramachandran, J. (1985) Nature **313**, 756–761.
15. "Cell surface receptors for polypeptide hormones, growth factors and neuropeptides" – review. Panayatou, G. and Waterfield, M.D. (1989) Current Opinion in Cell Biology **1**, 167–176.
16. "The guanylate cyclase/receptor family of proteins" – review. Schulz, S., Chinkers, M. and Garbers, D.L. (1989) FASEB J. **3**, 2026–2035.

SEVEN

CELL SURFACE RECEPTORS – SIGNAL TRANSDUCTION

In this chapter we will consider how the effect of binding of a messenger to a protein receptor on the outside (**ectoplasmic**) surface of the cell is transmitted into a chemical change on the inside (**cytoplasmic**) surface of the membrane – the problem of **signal transduction**. Very considerable progress has been made in this area during the last decade, to the extent that we can now make a fairly confident statement that most, if not all, signal transduction at cell surface receptors occurs by one of three basic mechanisms (Figure 7.1):

1) **Ion channel systems**, in which binding of the messenger on the ectoplasmic side of the membrane causes opening of an ion channel. The best characterized examples of this are **ligand-gated ion channels**, in which the ion channel is an integral part of the receptor which binds the messenger (the **ligand**). However recently it has been found that some ion channels may be coupled indirectly to receptors via G proteins (see under (2) below).
2) **Second messenger systems**, in which binding of the (**first**) messenger on the ectoplasmic side activates an enzyme on the cytoplasmic side, which increases the concentration of a **second messenger** inside the cell. The receptor and second messenger generating enzyme are coupled indirectly via GTP-binding proteins (G proteins). Some ion channels are also coupled to receptors via G proteins, without the involvement of a second messenger. Since these appear to utilize the same family of G proteins, they will be discussed with the second messenger systems (Section 7.2.6).
3) **Receptors with integral enzyme activity**, in which binding of the messenger on the ectoplasmic domain directly activates an enzyme on the cytoplasmic domain of the receptor polypeptide. In at least one case, this results in production of a second messenger (cyclic GMP).

7.1 *Ligand-gated ion channels*

7.1.1 *The nicotinic acetylcholine receptor*

This is the classic example of a ligand-gated ion channel, and due to its great abundance in the electroplax organ of electric rays and eels is one of the best

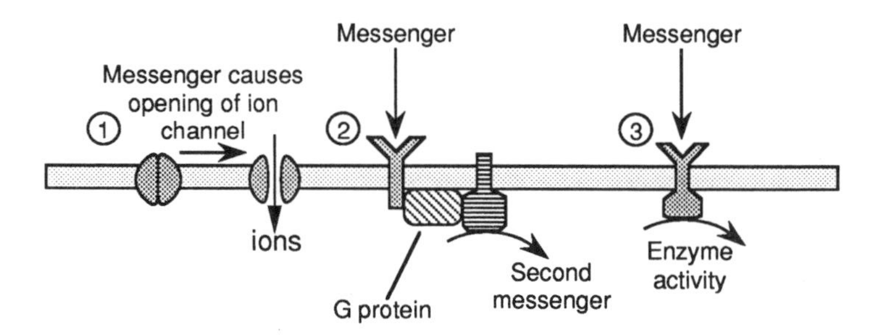

Fig. 7.1 Three basic mechanisms for signal transduction by cell surface receptors: (1) messenger causes opening of ion channel; (2) messenger is coupled to a second messenger generating enzyme via a G protein; (3) receptor has a messenger-activated enzyme as its cytoplasmic domain.

understood of all receptors. Electrophysiological studies show that treatment of electroplax membranes with acetylcholine results in the transient opening of cation channels in the membrane. These channels are relatively unselective for cation, but because of the large electrochemical gradient of Na^+ ions across the plasma membrane of resting cells (Chapter 4), the major effect of channel opening is a large influx of Na^+ and a consequent depolarization of the membrane. In this way the receptor converts a chemical signal (the neurotransmitter acetylcholine) into an electrical signal (depolarization), the reverse function to the voltage-gated Ca^{2+} channels in nerve endings (Section 4.2.2). Thus acetylcholine **excites** the target cell, and if the latter is a neurone will tend to initiate a new action potential. The further consequences of depolarization by ligand-gated ion channels are discussed more fully in Chapter 8.

These electrophysiological experiments did not answer the question as to whether the ion channel was an integral function of the receptor itself, or a separate protein which interacted with it. Resolution of this uncertainty required purification and reconstitution of the receptor. As discussed in Chapter 6, the acetylcholine receptor consists of five subunits ($\alpha_2\beta\gamma\delta$), each of which is a transmembrane glycoprotein arranged in a pentameric ring around a central pore (Figures 6.11–6.13). Could this pore be the ion channel? To answer this question it was necessary to reconstitute the receptor back into a lipid bilayer membrane.

7.1.2 *Reconstitution of receptor activity*

The original approach to reconstitution involved taking mixed micelles of ionic detergent and phospholipid, and of ionic detergent and purified receptor, and dialysing slowly to remove the detergent. This proved to be a laborious task, with the nature of the detergent, the phospholipids, and the conditions of dialysis all being critical for success. However it eventually proved possible to obtain vesicles

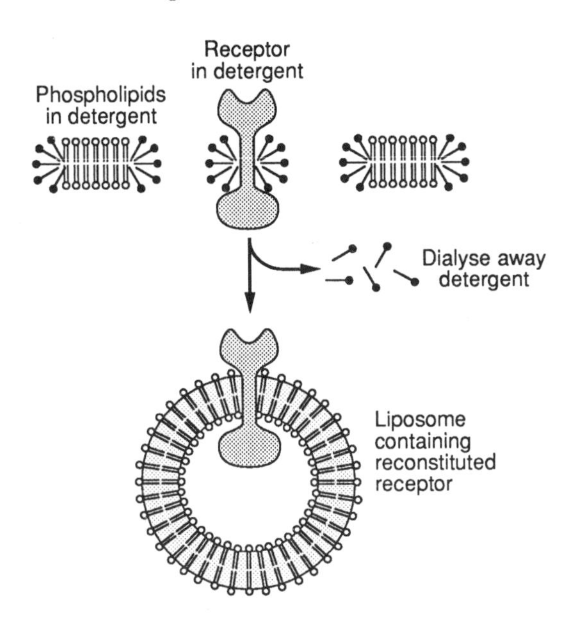

Fig. 7.2 Reconstitution of a cell surface receptor into an artificial liposome membrane. To allow removal by dialysis, an ionic detergent (e.g. deoxycholate) must be used.

in which receptors had incorporated into a phospholipid bilayer with the acetylcholine-binding site facing out (Figure 7.2). The ion channel activity of the reconstituted receptors could be examined by measuring $^{22}Na^{+}$ uptake into the vesicles in response to added agonists. This confirmed that acetylcholine caused opening of Na^{+} channels which were blocked by α-bungarotoxin. The response to acetylcholine was sigmoidal, with a Hill coefficient of close to 2. This is also observed with the native receptor, and probably results from there being two binding sites for the transmitter on the two α subunits (Box 7.1). These experiments provided the first direct evidence that the ion channel was integral to the receptor rather than being a separate protein that interacted with the receptor. Using this approach it is possible to make vesicles of defined lipid composition, and thus study the interaction between the receptor and different membrane lipid components. However, the small size of the vesicles precludes electrophysiological analysis of channel kinetics.

A different and potentially more powerful approach to reconstitution, applicable also to many other receptor types, became possible with the cloning of the four receptor subunits. cDNA coding for each of the four subunits were ligated to promoters from the bacteriophage SP6, so that SP6 RNA polymerase could be used to transcribe them *in vitro* into mRNA. The mRNAs were then injected into the very large egg cells (**oocytes**) of the toad *Xenopus laevis*, which do not

Cooperativity can occur when a protein has more than one binding site for a small molecule (or *ligand*), usually because of identical sites on subunits in an oligomer. Cooperativity occurs when binding of ligand to one site increases the affinity of the remaining sites. Imagine a protein P with n binding sites for a ligand L, *for which the increase in affinity on binding the first molecule of L is very large.* This means that P will only exist in the form of free P and PLn:

$$PL_n \leftrightarrow P + nL$$

$$K_d \text{ (dissociation constant)} = \frac{[P][L]^n}{[PL_n]}$$

$$\% \text{ saturation with ligand} = \frac{[PL_n]}{[P] + [PL_n]}$$

$$= \frac{[P][L]^n / K_d}{[P] + [P][L]^n / K_d}$$

$$= \frac{[L]^n}{K_d + [L]^n}$$

This is known as the *Hill equation,* and n is the *Hill coefficient.* The graph below shows that when n is >1 (more than one binding site), one obtains an S-shaped or sigmoid curve rather than the usual hyperbola for single, non-interacting sites ($n = 1$). If n >1 the system is very sensitive to small changes in ligand concentration over a certain range:

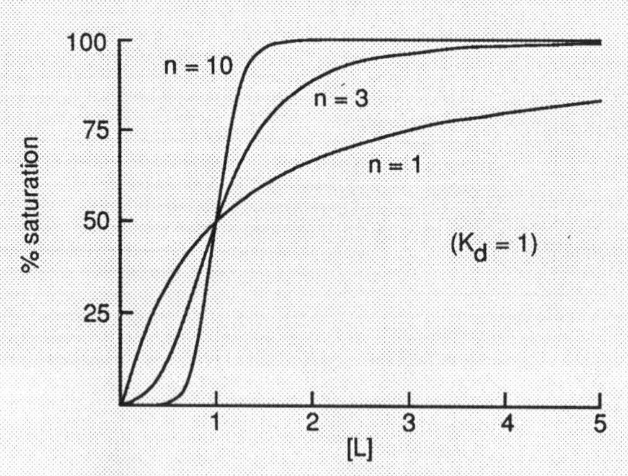

The simple treatment given above is in fact an oversimplification. Although binding data can generally be fitted to the Hill equation, in practice n is usually less than the number of binding sites, and is often not an integer (e.g. for binding of oxygen to the 4 sites on haemoglobin, $n = 2.6$). This is because the assumption that there is an infinitely large increase in affinity is often not valid. Other, more complex treatments do not make this assumption. Nevertheless, the Hill coefficient remains a useful empirical measure of cooperativity which gives a *minimum* estimate of the number of interacting binding sites.

Box 7.1 Cooperativity and the Hill coefficient

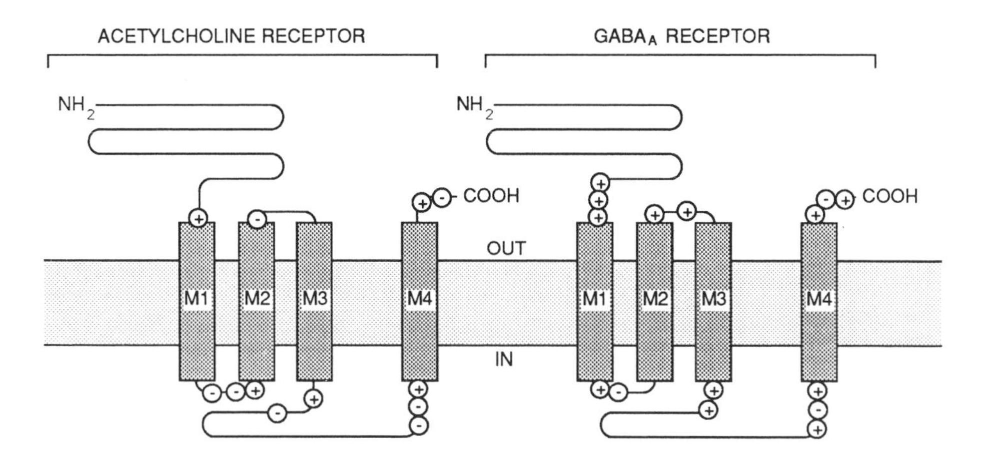

Fig. 7.3 Models for the membrane topology of the nicotinic acetylcholine and $GABA_A$ receptors, showing clusters of positively charged residues around the ends of the transmembrane helices in the latter.

normally express acetylcholine receptors (see Figure 4.16). Shortly after microinjection, electrophysiological recording made using patch clamping (Figure 4.13) showed that the cell membrane contained acetylcholine-gated ion channels identical to those in cells in which the acetylcholine receptor is normally expressed. All four subunits were required for functional channels, although weak α-bungarotoxin binding was obtained using combinations of three subunits, provided that the α subunit was present. This method of reconstitution has several advantages:

1) DNA cloning assures absolute purity, and rules out the possibility, always possible with protein preparations, that the observed functional effect is due to a minor contaminant.
2) Patch clamping (and hence, observation of individual channel molecules) is technically easier with *Xenopus* oocytes than with artificial membrane systems.
3) Most important of all, use of cloned DNA introduces the possibility of investigating structure-function relationships by **site-directed mutagenesis**, the intentional mutation of particular amino acid residues.

7.1.3 *Modelling of the receptor and site-directed mutagenesis*

Before meaningful site-directed mutagenesis studies on protein function could be carried out, it was necessary to have a working model of the protein structure so that intelligent choices could be made about which residue to modify. The model shown in Figure 7.3 was devised from hydropathy plots and biochemical studies of the protein topography, as discussed in the last chapter (Section 6.3.1). Site-directed mutagenesis can now be performed to test hypotheses and hence refine the model. Initial studies were carried out by expressing in *Xenopus* oocytes hybrid receptors in which three subunits came from one species and one from another. The receptors from *Torpedo* electroplax and mammals do in fact show subtle differences in function:

1) the mean opening time of the *Torpedo* receptor is 0.6 msec, as against 7–8 msec for the mammalian type.
2) at low Ca^{2+} the *Torpedo* receptor has a higher K^+ conductance than mammalian receptors.

Hybrid receptors in which the $\alpha\beta\gamma$ subunits came from the human but the δ subunit came from *Torpedo* behaved like the *Torpedo* receptor, emphasizing the importance of the δ subunit in controlling gating behaviour. Further refinement was possible by combining the mammalian $\alpha\beta\gamma$ with a recombinant hybrid δ subunit in which only helix M2 and part of the M2–M3 surface loop came from the *Torpedo* sequence. Once again this behaved like the Torpedo receptor, emphasizing the importance of helix M2.

Further analysis of point mutations and small deletions suggest that the N-terminal, ectoplasmic domain is the region of the α subunit which binds acetylcholine, consistent with a variety of other data. Deletions in any of the proposed transmembrane helices usually lead to loss of activity, while some deletions in the proposed cytoplasmic loop between M3 and M4 appear to be acceptable.

7.1.4 *Other ligand-gated ion channel receptors*

How do $GABA_A$ and glycine receptors inhibit the firing of action potentials in the target cell? Electrophysiological studies show that both are ligand-gated channels permeable to small anions, chiefly Cl^-. Cl^- concentration in a resting cell is normally about 10 times higher outside than in, but the chemical gradient for re-entry is roughly balanced by the resting membrane potential which opposes it. Opening of Cl^- channels will therefore maintain the normal resting potential (or even cause slight hyperpolarization). More importantly, opening of Cl^- channels will resist any tendency to depolarization caused by opening of cation channels (Figure 7.4). Injection of mRNAs coding for the two subunits of the $GABA_A$ receptor into *Xenopus* oocytes reconstitutes a functional GABA-activated Cl^- channel which is activated by benzodiazepines and blocked by picrotoxin. These results suggest that, unlike the four subunit nicotinic acetylcholine receptor, only two subunits may be necessary for the function of the $GABA_A$ receptor, although it is likely that they form tetrameric or pentameric structures in the membrane. More

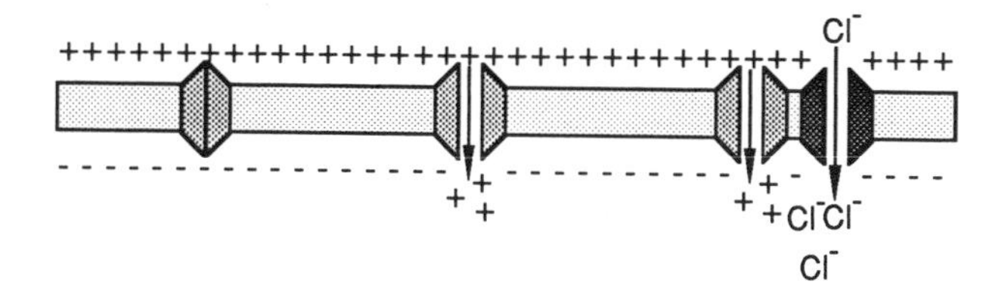

Fig. 7.4 Antagonistic effects of anion and cation channels on membrane depolarization. In a resting cell (left), the cation channels are closed and there is a membrane potential of –70 mV. When stimulated by an excitatory transmitter, the cation channels open, causing a local collapse of the membrane potential (centre). However, if a Cl^- channel also opens in response to an inhibitory transmitter, Cl^- will enter down its concentration gradient and antagonize depolarization (right).

recently, cDNAs encoding additional $GABA_A$ receptor subunits have been cloned, but it is not yet clear whether these are distinct subunits or tissue-specific isoforms.

Interesting speculations have been made based on detailed comparison of the models of the nicotinic acetylcholine, $GABA_A$ and glycine receptors (Figure 7.3), taking into account the sequences of more than 20 nicotinic acetylcholine receptor subunits from different species:

1) All contain a conserved proline residue in the middle of helix M1. This would introduce a bend in the helix, and it has been proposed that this may be involved in the gating mechanism.
2) In every case helices M1 and, particularly, M2 contains several side chains bearing hydroxyl functions (serine and threonine). It is proposed that these may line the hydrophilic pore through which the ions pass. The importance of helix M2 has already been demonstrated by mutagenesis studies on the acetylcholine receptor (see above).
3) Close to where the transmembrane helices would enter the membrane on both the ectoplasmic and cytoplasmic sides are clusters of charged residues, which are predominantly positive in the $GABA_A$ and glycine receptor models, and negative in the acetylcholine receptor model (Figure 7.3). This could help to explain the selectivities of these receptors for anion or cation.

Thus it seems likely that all of the well-characterized ligand-gated ion channel receptors have arisen from a common ancestral channel protein and form a protein "family". This theme will be repeated when we consider the other classes of signal transduction mechanism.

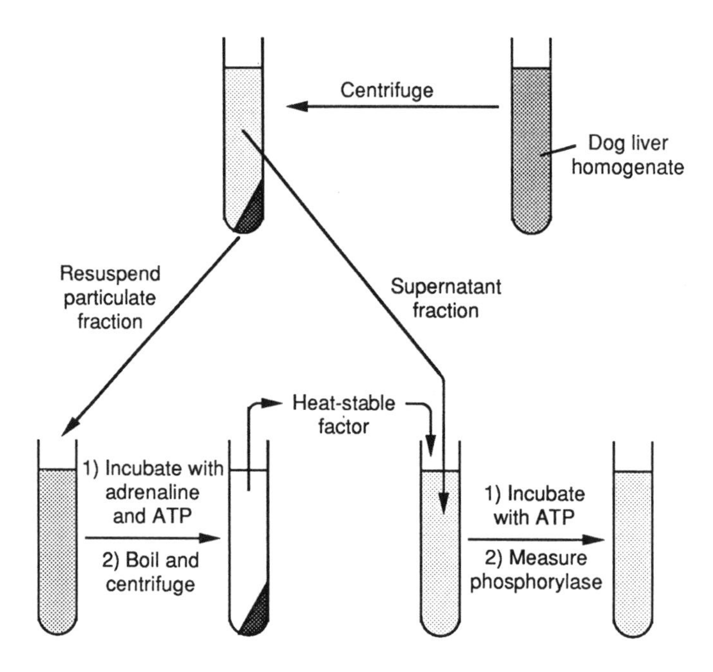

Fig. 7.5 Experiments which led to the discovery of cyclic AMP by Earl Sutherland. The heat-stable factor turned out to be cyclic AMP.

7.2 *Second messenger systems – cyclic nucleotides*

Although others have now been identified, the archetypal second messenger is **cyclic AMP**. The identification of cyclic AMP by Earl Sutherland in 1957 was a great milestone in the whole area of intercellular signalling, so we will describe it in some detail.

7.2.1 *Discovery of cyclic AMP - the second messenger hypothesis*

In the 1950s it was known that adrenaline (epinephrine) caused breakdown of glycogen in liver. In dog liver, which Sutherland studied, this occurs via what we now call β-adrenergic receptors. It was known that phosphorylase catalysed the key step in the breakdown of glycogen, and that adrenaline converted it into an active form due to a covalent modification (later shown to be phosphorylation – Chapter 8). Sutherland was able to demonstrate that adrenaline still activated phosphorylase in broken cells as long as ATP was present. Having made this crucial breakthrough of developing a cell-free system, he began to fractionate it to investigate the components of the mechanism (Figure 7.5). He eventually showed

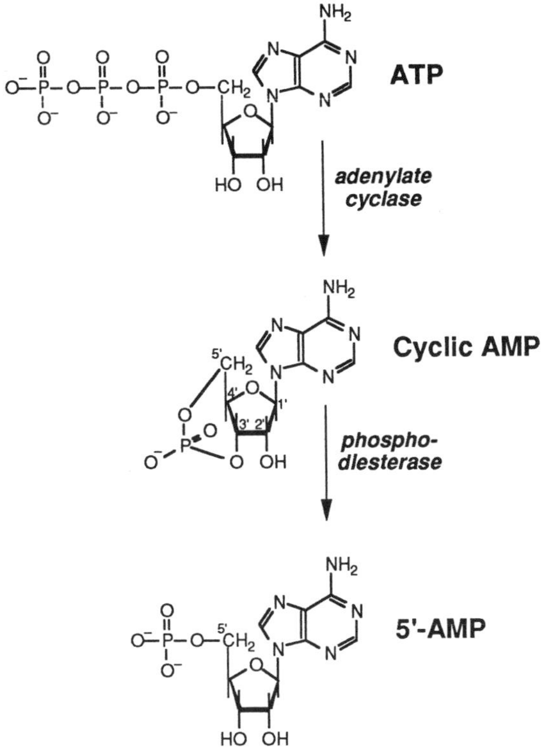

Fig. 7.6 Structure of cyclic AMP, and mechanisms of its formation and breakdown.

that adrenaline acted on the particulate fraction (which, as we now know, contains the membrane-bound receptor), and produced a compound that was stable to boiling and that activated phosphorylase in the soluble fraction. ATP was required both for the production of the compound in the particulate fraction, and for its activation of phosphorylase in the soluble fraction. Purification of this factor showed it to be an unusual nucleotide, soon identified as cyclic AMP (Figure 7.6). Its full name is cyclic adenosine-3',5'-monophosphate, meaning that it is a cyclic ester between phosphoric acid and the hydroxyl functions on the 3' and 5' carbons on the ribose ring. RNA contains 3'→5' phosphate ester linkages, and cyclic AMP is like a little bit of RNA that has formed a cyclic ester with itself. Just as the adenosine residues in RNA are derived from ATP, cyclic AMP is synthesized from ATP by the enzyme **adenylate cyclase**. It is broken down by **phosphodiesterases** which hydrolyse the 3'-phosphate ester to give the common metabolite, 5'-AMP (Figure 7.6).

From this classical work arose the concept of the **second messenger**. This

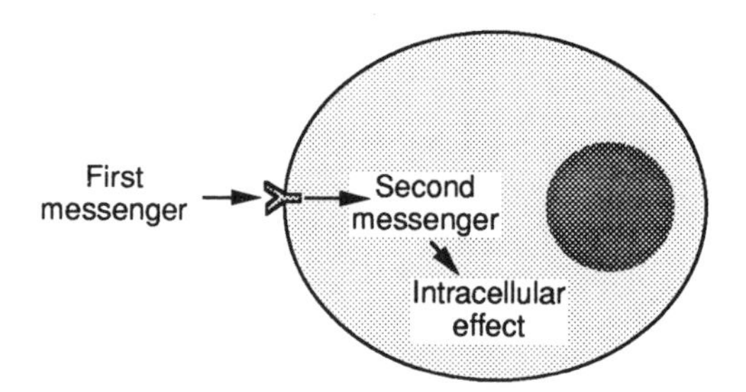

Fig. 7.7 The second messenger concept.

concept (Figure 7.7) can be stated as a sequence of three events, which happen to correspond to the topics discussed in Chapters 6, 7 and 8 respectively:

1) the hormone, neurotransmitter or mediator (the **first messenger**) does not enter the cell, but binds to a receptor on the cell surface;
2) this causes the formation of a different chemical signal, the second messenger, inside the cell;
3) the second messenger activates some system which modulates a variety of cell functions, leading to the physiological end-result.

This mechanism has obvious advantages over the alternative mechanism whereby the first messenger diffused into the cell and directly activated the target enzyme. Like all biosynthetic pathways, formation of extracellular messengers requires energy. Particularly in the case of hormones, which are diluted in the large volume of the bloodstream, it would be very costly to produce them at high concentration, and hormones are typically effective at nanomolar concentrations. Yet they have to activate enzymes which may be present at very high concentration in the cell. For example, phosphorylase in skeletal muscle, which as in liver is activated by occupation of β-adrenergic receptors, is present at concentrations of around 10^{-4} mol.l^{-1}. By the second messenger mechanism, the hormone need only be at very low concentration in the bloodstream, and higher concentrations of the second messenger are produced **selectively** inside those cells which express the cell surface receptor (i.e. target cells). The second messenger is produced at high concentration because its production is catalytic rather than stoichiometric. Further amplification of the signal is obtained at points downstream of the second messenger, a topic which is discussed in the next chapter.

7.2.2 *Cyclic AMP is the second messenger for many different extracellular (first) messengers*

How could one determine whether a particular second messenger like cyclic AMP was involved in the response to an extracellular (first) messenger? Two initial

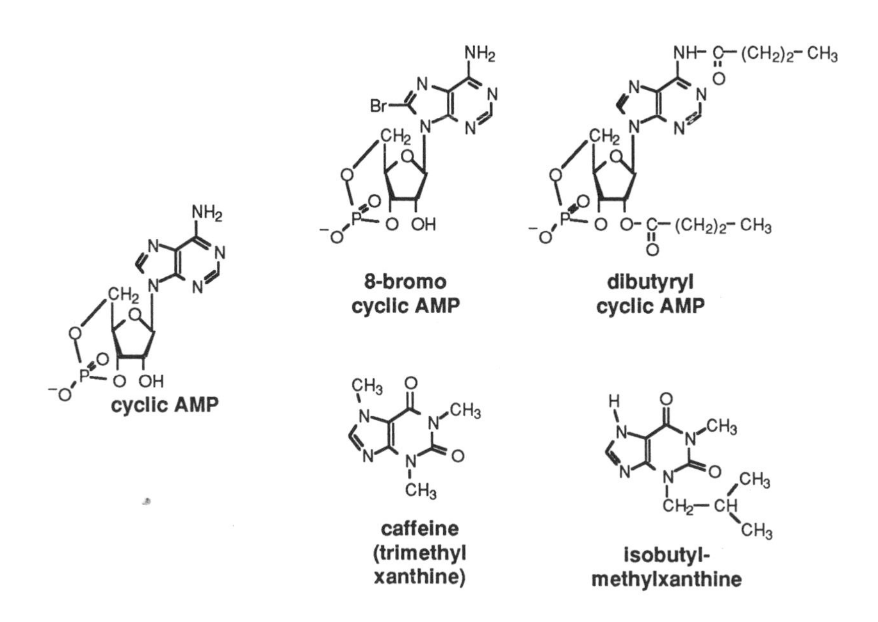

Fig. 7.8 Comparison of the structure of cyclic AMP (left) with that of cell-permeable analogues of cyclic AMP (top right), and inhibitors of cyclic AMP phosphodiesterase (bottom right).

criteria could be put forward:

1) the first messenger should increase the concentration of the second messenger inside the target cell;
2) increasing the concentration of the second messenger inside the cell should mimic the effect of the first messenger.

In the two decades after Sutherland's discovery, numerous laboratories tested with great success the idea that cyclic AMP could mediate the effects of first messengers other than adrenaline acting at β-receptors. Since cyclic AMP is found at rather low concentration even in unstimulated cells (around 10^{-6} mol.l^{-1}), testing of the first criterion required development of sensitive assays. A technical problem with examination of the second criterion is that the plasma membrane is not readily permeable to cyclic AMP. However, a number of less polar analogues of cyclic AMP have been developed, which are not only more cell permeable but are also resistant to breakdown by phosphodiesterases (Figure 7.8). There are also several pharmacological agents which elevate cyclic AMP itself inside the cell. An example is **caffeine** (Figure 7.8), the mild stimulant drug found in tea, coffee and cocoa, which, along with other actions, inhibits the breakdown of cyclic AMP by

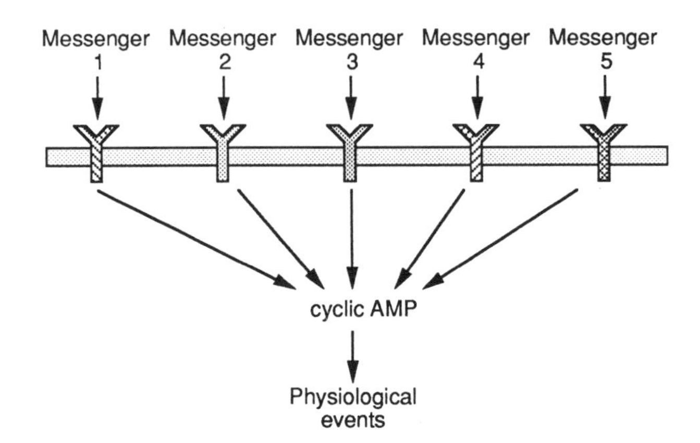

Fig. 7.9 Several first messengers can converge on a single second messenger.

phosphodiesterases. **Isobutylmethylxanthine** (Figure 7.8) is another, more potent phosphodiesterase inhibitor. Another drug which increases intracellular cyclic AMP is the plant terpenoid **forskolin**, which directly activates adenylate cyclase.

Using these approaches, evidence was obtained that a very large number of first messengers exert their effects via the second messenger cyclic AMP. Table 7.1 shows a list which is far from exhaustive. This illustrates another advantage of the second messenger mechanism: many different first messengers can converge onto a single second messenger (Figure 7.9). The target enzyme or protein does not have to be recognized by every different first messenger, but only by the system which mediates the actions of the common second messenger.

While many first messengers elevate cyclic AMP in their target cell, it was subsequently recognized that there are also messengers which decrease it. These are analogous to the inhibitory neurotransmitters (e.g. GABA), which dampen down the effect of excitatory neurotransmitters (Section 7.1.4). The net effect on cyclic AMP levels depends on the balance between those messengers which stimulate adenylate cyclase (and hence increase cyclic AMP) and those which inhibit it. A list of some inhibitory messengers is shown in Table 7.2.

7.2.3 *Generation of cyclic AMP involves three separate components*

The simplest model to explain how binding of the messengers in Table 7.1 regulates adenylate cyclase would be that the latter is an integral component of the receptor polypeptide. The first clear indication that this is not the case came from some elegant cell fusion experiments. Turkey erythrocytes have β-adrenergic receptors which activate adenylate cyclase, but the endogenous adenylate cyclase

Table 7.1 Some first messengers whose effects are mediated by cyclic AMP

Messenger	Target cell	Physiological effects
Glucagon	Liver	{Glycogen breakdown {Gluconeogenesis
Adrenaline	Adipose (β_1 receptor)	{Lipolysis {Inhibits fatty acid synthesis
Adrenaline	Heart (β_1 receptor)	{Increase in heart rate {Increase in contractile force
Vasopressin	Kidney (V_2 receptor)	Na^+/water reabsorption
Lutropin	Ovary	Progesterone synthesis
ACTH	Adrenal cortex	Glucocorticoid synthesis
Thyrotropin	Thyroid gland	Thyroid hormone synthesis
Parathormone	Bone	Bone resorption

Table 7.2 Some first messengers which **inhibit** adenylate cyclase

Messenger	Target cell	Physiological effects
Adrenaline	{Vascular smooth muscle {(α_2 receptor)	Contraction
Adenosine	{Adipose tissue {A_1 purinergic receptor)	Inhibits lipolysis
Acetylcholine	{Heart muscle {(M2 receptor)	Relaxation
Enkephalin	Neurones (δ receptor)	Behavioural effects
Somatostatin	Anterior pituitary	Inhibits ACTH release

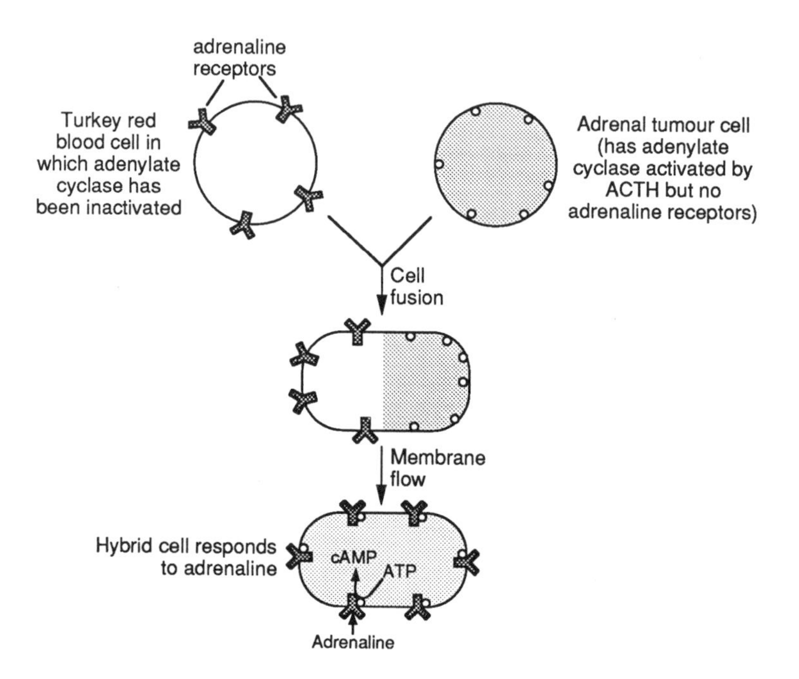

Fig. 7.10 Cell fusion experiment showing that the β-adrenergic receptor and adenylate cyclase are discrete proteins which can migrate within the plane of the membrane.

can be inactivated by treatment with a thiol-modifying reagent. On the other hand, an adrenal tumour cell line had adrenocorticotropin receptors which activated adenylate cyclase, but no β-adrenergic receptors. These two cell types were fused using a virus, and the hybrid cells now responded to adrenaline (epinephrine) by increased cyclic AMP levels, while unfused controls did not (Figure 7.10). The most reasonable explanation was that the receptor and adenylate cyclase are separate components which can diffuse independently within the plane of the membrane. These experiments also suggested that different receptors could interact with the same adenylate cyclase catalytic unit.

The third component of the system is a GTP-binding, or "G" protein. Early experiments using solubilized membranes had produced the puzzling observation that adenylate cyclase, as well as requiring ATP as substrate, was greatly stimulated by GTP, and even more dramatically by non-hydrolysable analogues such as GPP(NH)P (Figure 7.11). In addition, hormone-stimulated adenylate cyclase activity was depleted by passage through a GTP-agarose column, but could be restored by re-addition of fractions eluted from the column by GTP, which in themselves had no adenylate cyclase activity. The existence of a distinct GTP-binding component became much clearer with the isolation of a mutant (cyc^-) of the

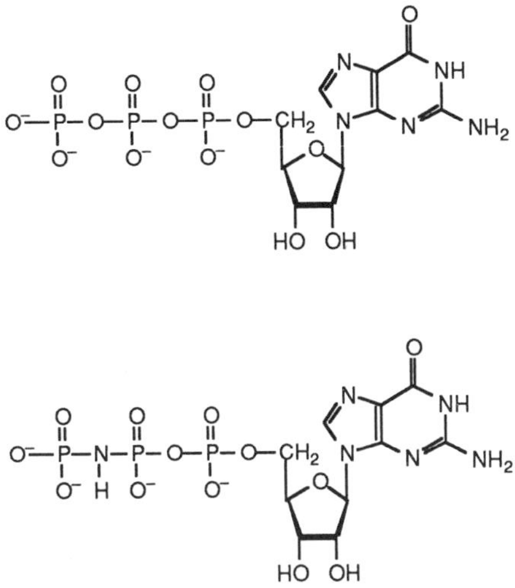

Fig. 7.11 Structure of GTP and the non-hydrolysable analogue GPP(NH)P.

S49 lymphoma cell line, which failed to respond to adrenaline (epinephrine) by increasing cyclic AMP, despite exhibiting normal binding of β-agonists, and having measurable adenylate cyclase activity. Normal hormone-stimulated adenylate cyclase could be reconstituted by adding a detergent extract of wild-type cells to cyc– membranes.

The reconstitution of hormone-sensitive adenylate cyclase in cyc^- membranes provided for the first time a practical assay for this third component. The protein, which is now called Gs (designating a GTP-binding, stimulator of adenylate cyclase) has been purified and characterized in detail. It consists of three subunits, α (~50 kDa), β (35 kDa) and γ (8 kDa), and it is the α subunit that is missing in the cyc^- mutation. We now have a detailed model of the manner in which the receptor and Gs are coupled to adenylate cyclase. Development of this model was greatly aided by the discovery of a remarkable analogy to the mechanism by which the absorption of light in the retina is converted into an action potential.

7.2.4 *Signal transduction and photoreception involve analogous GTP-binding (G) proteins*

It was known from electrophysiological studies that absorption of light in retinal rod cells causes hyperpolarization of the outer segment membrane. The membrane contains Na^+ channels which are open in the dark, and close upon illumination. Since the membrane discs containing rhodopsin are not continuous with the plasma membrane (Figure 6.18), it seemed that there must be an intracellular messenger responsible for carrying the signal between the two membrane systems. It would

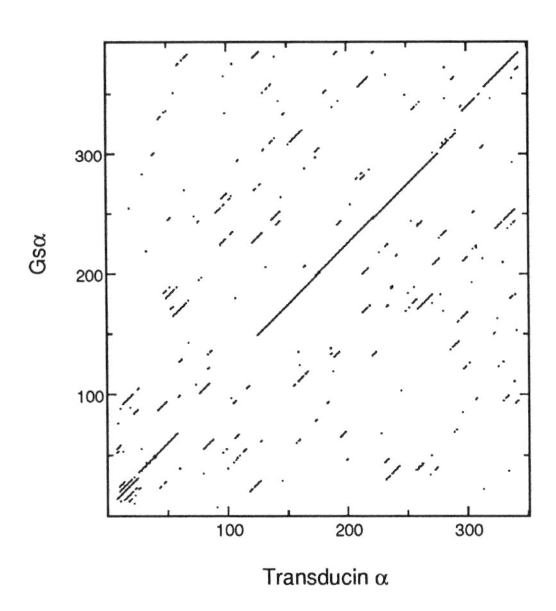

Fig. 7.12 Dot matrix comparisons of the α subunits of Gs and transducin. The distinct diagonal lines indicate that the two proteins are closely related (see Figure 3.15).

also seem that this messenger would have to be produced catalytically, since absorption of a single photon of light can result in blocking of flux of $>10^6$ Na^+ ions. For many years the rival claims of cyclic GMP and Ca^{2+} to fulfil this role were debated, but it now appears that both are involved. The plasma membrane contains cyclic GMP-gated cation channels, and absorption of light activates a phosphodiesterase which hydrolyses cyclic GMP, so the channels close. One effect of this is to reduce influx of Ca^{2+} into the cell. This in turn stimulates resynthesis of cyclic GMP, because Ca^{2+} either inhibits guanylate cyclase, or activates phosphodiesterase, or both. Thus a decrease in cyclic GMP mediates the initial response to light, while a decrease in Ca^{2+} exerts a feedback effect to switch off the response.

How is absorption of light by rhodopsin coupled to activation of cyclic GMP phosphodiesterase? Because the disc membranes and their protein components can be readily purified, it has been possible to study this in detail, and it has become clear, unexpectedly, that they are coupled via a G protein (**transducin**) very similar to that (Gs) which couples receptors to adenylate cyclase. The remarkable similarities between rhodopsin and messenger receptors such as the β_1- and β_2-adrenergic receptors have already been discussed in Chapter 6. Because of the ease with which rhodopsin and transducin can be purified from the disc membranes of the retinal rod, the mechanism which couples the receptor to the effector (i.e. cyclic GMP phosphodiesterase) was originally elucidated in the transducin system.

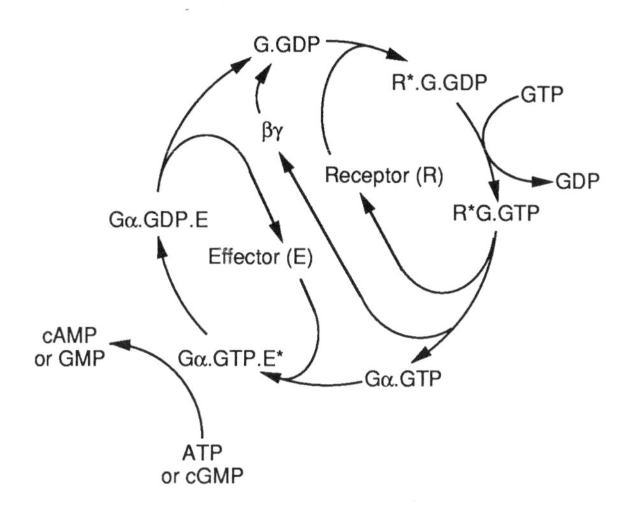

Fig. 7.13 Model for the function of Gs and transducin. R and E are the receptor and effector respectively: asterisks indicate activated forms.

However, the system involving Gs and adenylate cyclase appears to be entirely analogous, so they will be described together. Both transducin (which can be called Gt) and Gs consist of very similar α, β and γ subunits. The βγ subunits may in fact be identical, while the α subunits are different, although related in sequence (Figure 7.12). There are two isoforms of Gsα produced by alternative splicing of a single gene, as well as distinct isoforms of Gtα in rod and cone cells. Reconstitution experiments show that it is the α subunits which bind GTP and provide the specificity for both the receptor and effector proteins.

A widely accepted model for the action of transducin and Gs is shown in Figure 7.13. In the resting state, the G proteins are present as αβγ complexes containing bound GDP (G.GDP). When rhodopsin absorbs a photon, or the messenger binds to its receptor, the activated receptor (R*) interacts with the respective G protein, causing dissociation of GDP. GTP then binds to the Gα subunit, causing Gα.GTP to dissociate from βγ and bind to and activate the effector protein (E*: phosphodiesterase or adenylate cyclase): at least in the former case, this is because α.GTP displaces an inhibitory subunit.

It is obviously important that activation of the effector protein (phosphodiesterase or adenylate cyclase) is reversed when the stimulus is no longer present. This is achieved by means of a GTPase activity which is intrinsic to the α subunit. The GDP that is formed remains bound to the α subunit, which then reassociates with βγ so that the cycle can begin again. The crucial importance of the GTPase activity is shown by the action of cholera toxin (Box 7.2). The B subunit of the toxin is an enzyme which irreversibly transfers ADP-ribose from cellular NAD^+ to a cysteine residue on both Gtα and Gsα. In the case of Gs, this blocks the cycle shown in Figure 7.13 with the α subunit in the active GTP-bound

Cholera is caused by the bacterium *Vibrio cholerae* which infects water supplies. The organism secretes a protein toxin which binds to the epithelial cells lining the gut. The toxin is taken up by endocytosis, and after proteolytic activation a fragment of the A subunit enters the cytoplasm where it specifically ADP-ribosylates an arginine residue on the α subunit of Gs:

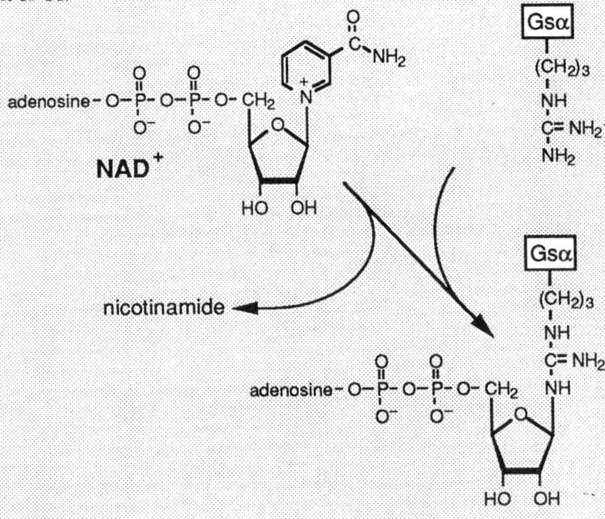

This irreversible covalent modification locks the enzyme in the active, GTP-bound state. There are massive and persistent rises in cyclic AMP in the cells lining the gut. This causes excessive secretion of sodium ions and water, and hence debilitating diarrhea. Cholera has been eradicated in Western countries by the provision of efficient sewage systems and clean water supplies. Although cholera can be treated relatively easily, it is still a major killer in the Third World, particularly of children.

Box 7.2 Cholera and cholera toxin

form, leading to massive and persistent increases in cyclic AMP. Infection with the cholera bacterium thus causes uncontrolled secretion of Na^+ from the cells lining the gut, leading to debilitating and potentially lethal diarrhoea.

An additional mechanism for terminating the response is desensitization of the receptor by phosphorylation, a mechanism which has already been discussed for the β_2-adrenergic receptor (Section 6.4). Rhodopsin (which has to be in the light activated form) is phosphorylated by a rhodopsin kinase at multiple residues in the C-terminal, cytoplasmic tail, and in this form no longer activates transducin (Figure 6.21). This is directly analogous to the phosphorylation of the hormone-bound β_2-adrenergic receptor by β-adrenergic receptor kinase. A further analogy is that the

Whooping cough *(pertussis)* is caused by the bacterium *Bordetella pertussis*. The organism produces a toxin containing two components. One component is an enzyme which, in direct analogy to cholera toxin, ADP-ribosylates a cysteine residue in the inhibitory GTP-binding protein Gi (Fig. 7.14). This prevents Gi from inhibiting the host cell adenylate cyclase. The other component is a bacterial adenylate cyclase enzyme. Their combined effect is a large increase in cyclic AMP in affected cells.

Despite their similarities, the clinical effect of pertussis is very different to that of cholera because it is an air-borne infection which attacks the epithelial cells lining the airways. The consequence is a massive secretion of fluid and mucus from these cells, causing paroxysmal coughing followed by a "whooping" noise when air is drawn in.

Box 7.3 Whooping cough and pertussis toxin

desensitization of rhodopsin requires the presence of a soluble 48 kDa inhibitor protein termed **arrestin**. It has recently been found that the inhibitory effect of phosphorylation of the β_2-adrenergic receptor is lost during purification, but can be restored by addition of arrestin purified from the retina.

Coupling of the receptor and its **effector** (i.e. phosphodiesterase or adenylate cyclase) via an intermediary G protein relies on interaction between proteins which are mobile in the plane of the membrane, rather than formation of a static complex. It is therefore referred to as a **collision-coupling mechanism**. What is the advantage of such a complicated and indirect mechanism, which involves apparently wasteful hydrolysis of GTP? One answer probably lies in the large degree of amplification achieved. Each molecule of activated rhodopsin or receptor can catalyse exchange of GDP for GTP on at least ten molecules of G protein before becoming inactivated by mechanisms discussed above. In order to permit this first stage of amplification, Gs is normally present in large molar excess over the receptor (this is not true of rhodopsin, where a high concentration of the 'receptor' is necessary to trap light over a wide range of intensity). A second stage of amplification comes from the fact that the rate constant for the GTPase activity (~13 min^{-1} for Gs at 37°C) is much lower than the turnover number for the effector enzyme (~1100 min^{-1} for adenylate cyclase). The overall amplification is thus ~1000-fold in the case of Gs/adenylate cyclase.

7.2.5 *G proteins can also inhibit adenylate cyclase*

An additional advantage of the indirect coupling mechanism is that many receptors can interact with a single effector. As mentioned earlier, messengers can either activate or inhibit adenylate cyclase. A number of receptors which inhibit adenylate cyclase have been shown to have receptors of the seven transmembrane helix type related to rhodopsin and the β_2-adrenergic receptor. How are these coupled to adenylate cyclase? A key observation was the finding that **pertussis toxin**, produced by the whooping cough bacterium (Box 7.3), abolishes the effect

```
Gs:  arg-met-his-cys-[arg]-gln-tyr-glu-leu-leu
Go:  ala-asn-asn-leu-arg-gly-[cys]-gly-leu-tyr
Gi1: lys-asn-asn-leu-lys-asp-[cys]-gly-leu-phe
Gi2: lys-asn-asn-leu-lys-asp-[cys]-gly-leu-phe
Gi3: lys-asn-asn-leu-lys-glu-[cys]-gly-leu-tyr
```

Fig. 7.14 C-terminal sequences of G protein α subunits, showing the arginine residue which is altered in the *unc* mutation of Gs, and the cysteine residues which are ADP-ribosylated by pertussis toxin in Go/Gi.

of these inhibitory messengers. Pertussis toxin was found to cause ADP-ribosylation of a novel αβγ-type G protein termed Gi. Use of pertussis toxin and radioactive NAD to specifically label the α subunit of this protein was a great help in its identification and purification. cDNA cloning and sequencing has now shown that there are in fact at least three isoforms of Gi, plus a closely related protein, Go. The functional significance of this heterogeneity is not yet clear. Gi_1, Gi_2, Gi_3 and Go all contain βγ subunits which are similar or identical to those in transducin and Gs, but differ in their α subunits, which are ADP-ribosylated by pertussis toxin at a cysteine residue close to the C-terminus (Figure 7.14). This prevents interaction of the G protein with its receptor, and it is likely that this region, which is the most variable in sequence between different G proteins, is the part of the protein that provides the necessary specificity of interaction with the receptor. This is also suggested by analysis of the *unc* (uncoupled) mutation of S49 lymphoma cells, in which mutation of an arginine to a proline near the C-terminus of Gsα (Figure 7.14) prevents interaction with the receptor.

How do the various isoforms of Gi inhibit adenylate cyclase? Early work on the separation of the α and βγ subunits suggested that only the latter produced significant inhibition. Since the βγ subunits of Gs and Gi are identical or very similar, the simplest hypothesis was that βγ released from Gi could combine with, and inactivate, free Gsα (Figure 7.15). While this is plausible, it seems unlikely that it is the whole story. For example, the model cannot explain how hormones can still inhibit adenylate cyclase in the cyc^- mutant of S49 cells, which lack Gsα altogether.

7.2.6 *G proteins also directly regulate ion channels*

Acetylcholine released by the parasympathetic nervous system reduces both the strength and rate of contraction of the heart. It does this via muscarinic receptors, and at least part of their action is explained by lowering of cyclic AMP via Gi proteins. This antagonizes the stimulatory effects of cyclic AMP-elevating agents

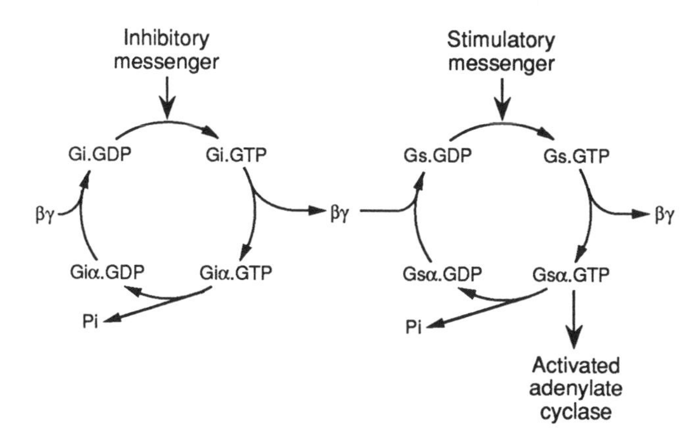

Fig. 7.15 Proposed mechanism of adenylate cyclase inhibition by Gi. Activation of Gi releases a large pool of βγ subunits which displaces the equilibrium for stimulatory systems in favour of the inactive Gs.GDP complex.

(e.g. adrenaline acting at β1-receptors), the mechanisms for which are discussed in Section 8.2.4. However, muscarinic agonists also relax the heart directly by causing opening of K^+ channels, and thus hyperpolarization, in cells in the pacemaker region of the atrium. It has been shown that addition of a preparation of G proteins from red blood cells, in the presence of GTP-γ-S, will cause opening of single K^+ channels in patch clamp experiments. Further experiments showed that it was the α, rather than the βγ, subunits which were responsible. Since ATP is not added in these experiments, it is very unlikely that the effect is mediated indirectly via changes in the level of cyclic AMP or any other second messenger, and it is therefore thought that there is a direct protein–protein interaction between the G protein and a component of the ion channel system. The effect is sensitive to pertussis toxin, and it was originally proposed that there may be a novel G protein involved in this effect, but analysis of the effects of $Gi_1\alpha$, $Gi_2\alpha$ or $Gi_3\alpha$ expressed from cloned DNA in *Escherichia coli* showed, surprisingly, that all of them were effective in opening of K^+ channels. The isoforms of Gi may therefore work via multiple effectors, causing both lowering of cyclic AMP and opening of ion channels.

7.2.7 *Another cyclic nucleotide second messenger – cyclic GMP*

Shortly after the discovery of cyclic AMP, researchers began to look for other cyclic nucleotides, and soon found cyclic-3',5'-guanosine monophosphate (cyclic GMP). At first it seemed that this would be very quickly established as another

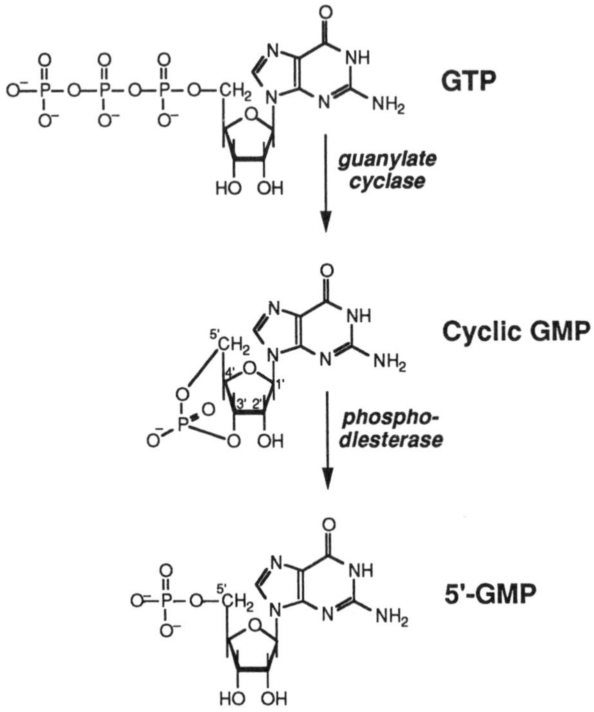

GTP

guanylate cyclase

Cyclic GMP

phosphodiesterase

5'-GMP

Fig. 7.16 Structure of cyclic GMP, and mechanisms of formation and breakdown. Compare with Figure 7.6.

second messenger, since all the enzymes analogous to those involved with cyclic AMP metabolism and action were also found, i.e. guanylate cyclase and cyclic GMP phosphodiesterase (Figure 7.16) and a protein kinase analogous to cyclic AMP-dependent protein kinase which mediates the actions of cyclic GMP (see Chapter 8). However a frustrating period of several years then ensued in which no system could be found where a first messenger modulated the concentration of cyclic GMP. This situation has now been rectified. The first clues came from observations that certain drugs used to lower blood pressure (e.g. nitroprusside) and related compounds (e.g. nitric oxide) activated the **soluble** guanylate cyclase in arterial smooth muscle and that this caused relaxation of the muscle. The physiological significance of this is unclear, although there is evidence that nitric oxide may be a natural mediator released by endothelial cells. More recently it has been found that atrial natriuretic peptides, released by the atria of the heart in response to high blood pressure (Section 2.2.1), activate **particulate** guanylate cyclase in arterial smooth muscle, causing relaxation of the blood vessel wall. It is now known that the particulate guanylate cyclase which is activated is an integral component of the atrial natriuretic peptide receptor (Section 6.3.4). A closely related system is involved in the response of sea urchin spermatozoa to the chemotactic peptide, **resact**, released by eggs. An additional system in which cyclic GMP is involved as a second messenger, but in a distinctly different manner, is in visual transduction in the retina (Section 7.2.4).

7.3 *Calcium-mobilizing messengers utilize a novel second messenger system*

If Sutherland had used rat liver rather than dog liver for his experiments, he might never have discovered cyclic AMP, because rat liver (especially that of adult males) lacks β-adrenergic receptors, and the effects of adrenaline (epinephrine) are mediated entirely by α_1-receptors, which do not increase cyclic AMP. Yet in these cells, the hormone still causes phosphorylation of phosphorylase and consequent activation of glycogen breakdown. A variety of evidence indicated that this process involved an increase in cytosol Ca^{2+}:

1) the effect was mimicked by addition of Ca^{2+} ionophores, which are drugs which make the plasma membrane permeable to the cation;
2) prior depletion of Ca^{2+} from cells by incubation with EGTA rendered them insensitive to adrenaline;
3) experiments with fluorescent Ca^{2+} indicators such as fura-2 (Section 4.1.2) showed that an increase in cytosol Ca^{2+} from the resting level of ~0.1 $\mu mol.l^{-1}$ to ~1 $\mu mol.l^{-1}$ occurred.

Although the effect of adrenaline could be mimicked via entry of Ca^{2+} catalysed by ionophores, it was also clear that the initial response to the hormone was not mediated by external Ca^{2+} and was due to release from intracellular stores, probably the endoplasmic reticulum. How then could binding of adrenaline to the α_1-receptor at the plasma membrane trigger release at internal sites? Although it was not realized at the time, many clues to identification of the missing second messenger had already been found.

7.3.1 *A novel second messenger - inositol trisphosphate*

In 1953, Hokin and Hokin reported the curious observation that in pancreatic cells acetylcholine (acting at muscarinic receptors) stimulated the incorporation of radioactivity from [^{32}P]phosphate into the minor phospholipid, phosphatidylinositol, but not into the other major phospholipids. A number of other first messengers were subsequently shown to produce this "phosphatidylinositol response" (Table 7.3) but for a long time its significance was not understood. Then in 1975 Michell pointed out that messengers which activate

Table 7.3 Some first messengers which activate phosphoinositide breakdown

Messenger	Target cell	Physiological effects
Adrenaline	Liver (α_1 receptor)	Glycogen breakdown
Vasopressin	Liver (V_1 receptor)	Glycogen breakdown
ATP	Liver (P_2 receptor)	Glycogen breakdown
Acetylcholine	{Exocrine pancreas {(M1/M3 receptor)	Secretion
Acetylcholine	{Smooth muscle {(M3 receptor)	Contraction
Thrombin	Platelets	Aggregation
PDGF	Fibroblasts	Cell proliferation

the phosphatidylinositol response (including adrenaline acting at α_1-receptors) invariably also cause increases in cytosolic Ca^{2+} concentration, and suggested a causal link between the two events. The nature of this link became clearer with the discovery that the primary event stimulated by the messengers was not the turnover of phosphatidylinositol (PI) itself, but of a minor derivative, phosphatidylinositol-4,5-bisphosphate (PIP_2; Figure 7.17). Phosphatidylinositol is mainly found in the inner leaflet of the plasma membrane, and a small proportion of the phospholipid is phosphorylated to the 4-phosphate and 4,5-bisphosphate by specific kinases. These inositol lipids are known collectively as **phosphoinositides**. The first messengers are now known to activate a phosphoinositide-specific phospholipase C (**phosphoinositidase C**), causing breakdown of PIP_2 (Figure 7.17) to two products: a water soluble product, inositol-1,4,5-trisphosphate, and a lipophilic component, diacylglycerol, which remains associated with the membrane (Section 7.3.2). Diacylglycerol can be phosphorylated to phosphatidate, and then phosphatidylinositol can be reformed, via the intermediate CMP-phosphatidate, to complete the cycle (Figure 7.17).

Since the initial phase of increase in cytosol Ca^{2+} produced by the messengers shown in Table 7.3 was derived from endoplasmic reticulum, there was a need to postulate a water soluble messenger which could cross the gap between the plasma membrane and the endoplasmic reticulum. Inositol-1,4,5-trisphosphate was an obvious candidate for this role. The first direct evidence came from studies with pancreatic acinar cells, which secrete digestive enzymes in response to stimulation

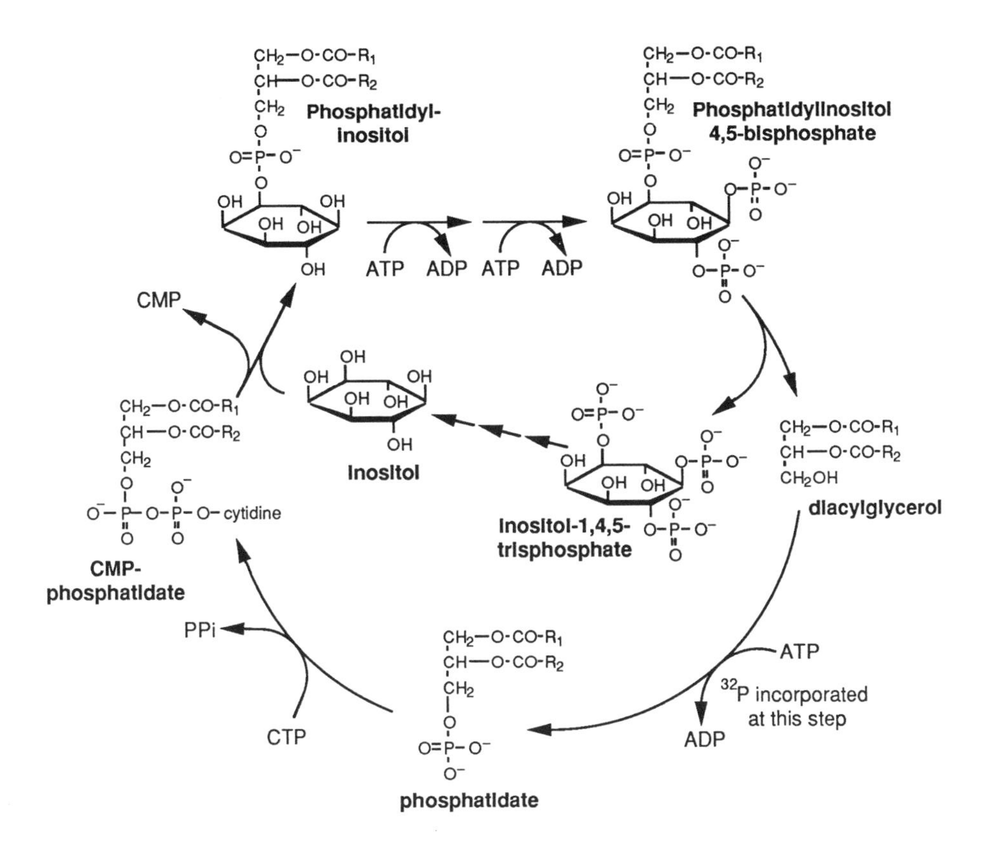

Fig. 7.17 The phosphatidylinositol lipid cycle. Although first messengers activate the phosphoinositidase C step, the response was originally detected by the secondary effect of increased incorporation of radioactivity from extracellular [^{32}P]phosphate into phosphatidylinositol.

of muscarinic acetylcholine receptors. The muscarinic agonist carbamyl choline caused measurable release of Ca^{2+} in cells in which the plasma membrane had been made leaky by prior incubation with Ca^{2+}-free medium. It was found that inositol-1,4,5-trisphosphate (IP_3), but not several related analogues such as inositol-1,4-bisphosphate (IP_2) or inositol-1-phosphate (IP_1), would mimic the effect of carbamylcholine (Figure 7.18). Similar results were rapidly reported in many other cell types. To overcome the problem of the lack of permeability of inositol phosphates to the plasma membrane, IP_3 was either microinjected, or cells were made permeable with digitonin or saponin, detergents which cause the formation of holes in the plasma membrane by binding to cholesterol (endoplasmic reticulum membranes contain much less cholesterol and under the right conditions remain intact).

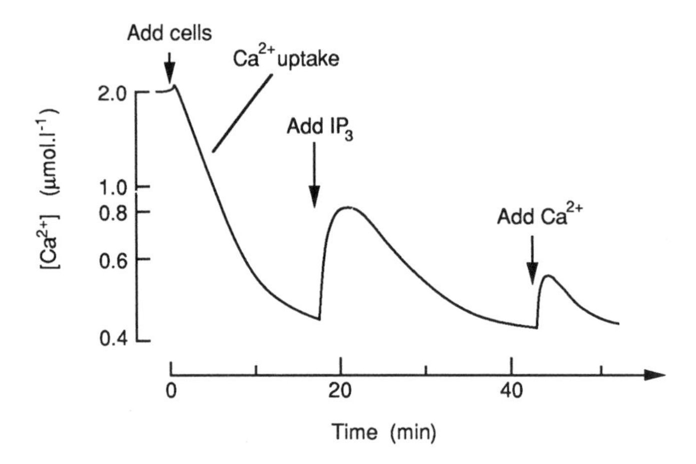

Fig. 7.18 Release of Ca^{2+} from intracellular stores by inositol-1,4,5-trisphosphate (IP_3) in pancreatic acinar cells. The cells were made permeable to IP_3 by pre-incubation in Ca^{2+}-free medium. On addition of cells to the incubation, they take up Ca^{2+} into intracellular stores, with external Ca^{2+} concentration reaching a steady state value of ~0.4 $\mu mol.l^{-1}$. Addition of IP_3 at this point causes a transient release of Ca^{2+}, followed by re-uptake to the steady state value. The final addition of Ca^{2+} was to calibrate the system. External Ca^{2+} concentration was monitored using a Ca^{2+} electrode: note that the scale is logarithmic. Redrawn from Streb *et al.* (1983) Nature **306**, 67-69.

Although our picture is still incomplete, a variety of evidence suggest that there are two distinct sub-fractions of endoplasmic reticulum which can store and release Ca^{2+}:

1) An IP_3-sensitive pool, containing generally between 30 and 50% of the total releasable Ca^{2+}, which appears to have a perinuclear location. As discussed in Section 4.1.2, cytoplasmic Ca^{2+} concentrations can be imaged within individual cells using the indicator fura-2 and a fluorescence microscope, and in adrenal chromaffin cells IP_3-generating agonists appear to release the cation into the perinuclear region of the cytoplasm. In the same cells, a monoclonal antibody raised against the Ca^{2+}-transporting ATPase of muscle sarcoplasmic reticulum recognizes a 140 kDa polypeptide which appears to co-localize with the IP_3-sensitive Ca^{2+} release (Figure 7.19).
2) An IP_3-insensitive pool, which is located in the peripheral region of the cell away from the nucleus. This sub-fraction appears to have several similarities with the sarcoplasmic reticulum of skeletal muscle (Section 8.1.3):
   a) its Ca^{2+} can be released by the drug caffeine, which also activates the Ca^{2+}-release channels of sarcoplasmic reticulum (**ryanodine receptors**). In

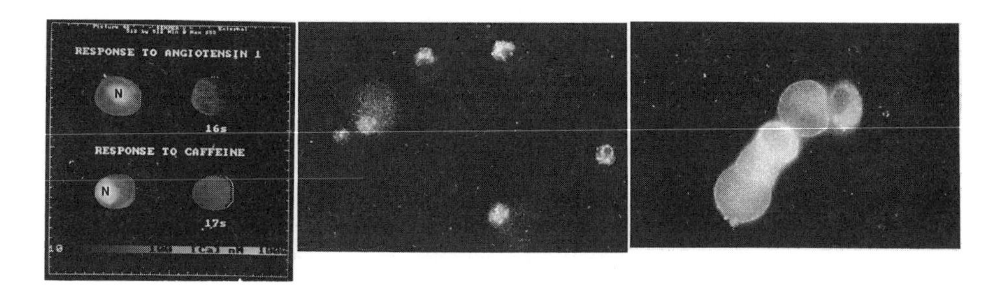

Fig. 7.19 Fluorescence micrographs of adrenal chromaffin cells. Left panel: cells loaded with ethidium bromide (left) to label the nucleus (N), or with fura-2 (right), to monitor Ca^{2+}. The pale areas in the fura-2 images show that an IP_3-releasing agonist (angiotensin I) releases Ca^{2+} from the perinuclear region, while caffeine releases it throughout the cytosol. The other two panels show results using two different fluorescent-labelled antibodies raised against Ca^{2+}-transporting ATPase, indicating two distinct pools of transporter which seem to mirror the location of the two releasable pools of Ca^{2+}. Reproduced from Burgoyne *et al.* (1989) Nature **342**, 72–74, courtesy of Robert Burgoyne.

adrenal chromaffin cells caffeine releases Ca^{2+} into the peripheral region of the cell (Figure 7.19).

b) a different monoclonal antibody raised against the Ca^{2+}-transporting ATPase of sarcoplasmic reticulum recognizes a 100 kDa polypeptide that has a peripheral distribution in the cell (Figure 7.19).

c) peripheral vesicles lying close to the plasma membrane are also recognized by antibodies against skeletal muscle calsequestrin, a Ca^{2+}-binding protein that is thought to act as a reservoir of the cation within the sarcoplasmic reticulum. The vesicles containing this calsequestrin-like protein in non-muscle cells have been termed **calciosomes**.

How Ca^{2+} release is normally triggered from the IP_3-insensitive vesicles is not yet clear, but an attractive possibility is that it is induced by Ca^{2+} itself. This could explain the slowly propagating waves of Ca^{2+} release which have been observed in many cells by fura-2 imaging.

Thus inositol-1,4,5-trisphosphate appears to be a second messenger which triggers the release of Ca^{2+} from a sub-fraction of the endoplasmic reticulum. A prediction of this model is that the endoplasmic reticulum membranes should contain a receptor for IP_3 which is coupled to the opening of Ca^{2+} channels. An IP_3-binding protein has been purified from rat cerebellum, and reconstitution of the purified protein into artificial membranes shows that binding of IP_3 causes opening of Ca^{2+} channels. Thus the IP_3 receptor appears to be a ligand-gated ion channel.

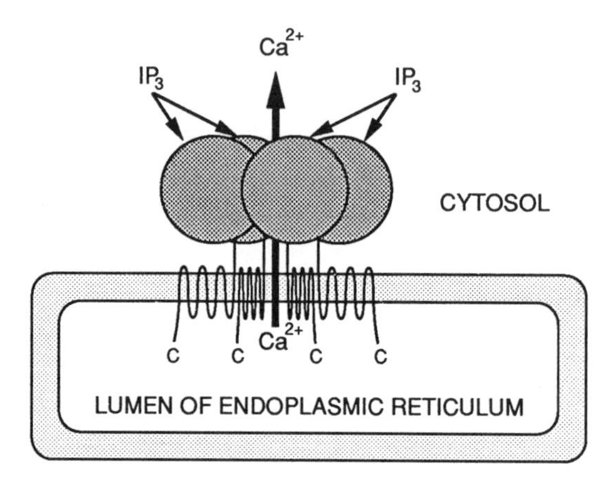

Fig. 7.20 Model for the inositol-1,4,5-trisphosphate receptor, based on the cDNA sequence and analogies with the skeletal muscle sarcoplasmic reticulum Ca^{2+} release channel (ryanodine receptor).

cDNA encoding the receptor has been cloned and sequenced, and reveals that it is a very large polypeptide containing over 2700 amino acids, which shows some similarity in sequence with the Ca^{2+}-releasing channels of muscle sarcoplasmic reticulum (the **ryanodine** receptor – Section 8.1.3). Current evidence favours a model similar to that for the ryanodine receptor (Figure 7.20) in which there is a very large N-terminal cytoplasmic region which presumably contains the IP_3-binding site, a transmembrane region which must form the Ca^{2+} channel, and a rather short C-terminal tail in the lumen of the vesicle. The native receptor appears to be a tetramer, and a high degree of cooperativity (see Box 7.1) suggests that at least three subunits need to bind IP_3 before the channel opens. Immunohistochemical studies using antibody against the receptor show that it is highly enriched in the Purkinje cells of the cerebellum, and one study at the electron microscope level suggested that it is localized within these cells in the outer membrane of the nuclear envelope, and in some, but not all, cisternae of rough and smooth endoplasmic reticulum (Figure 7.21). This would be consistent with the apparent perinuclear location of the IP_3-released Ca^{2+}.

7.3.2 *Could other inositol phosphates be second messengers?*

If IP_3 is to be a rapidly acting second messenger, there must be mechanisms for its removal. This is particularly important for a second messenger which releases Ca^{2+}, since a prolonged increase of cytosol Ca^{2+} can be toxic to cells, and first messengers usually only produce a transient increase. Recent work on the metabolism of IP_3 has revealed that it is remarkably complex (Figure 7.22). One route of degradation is via three specific phosphatases which hydrolyse in turn the

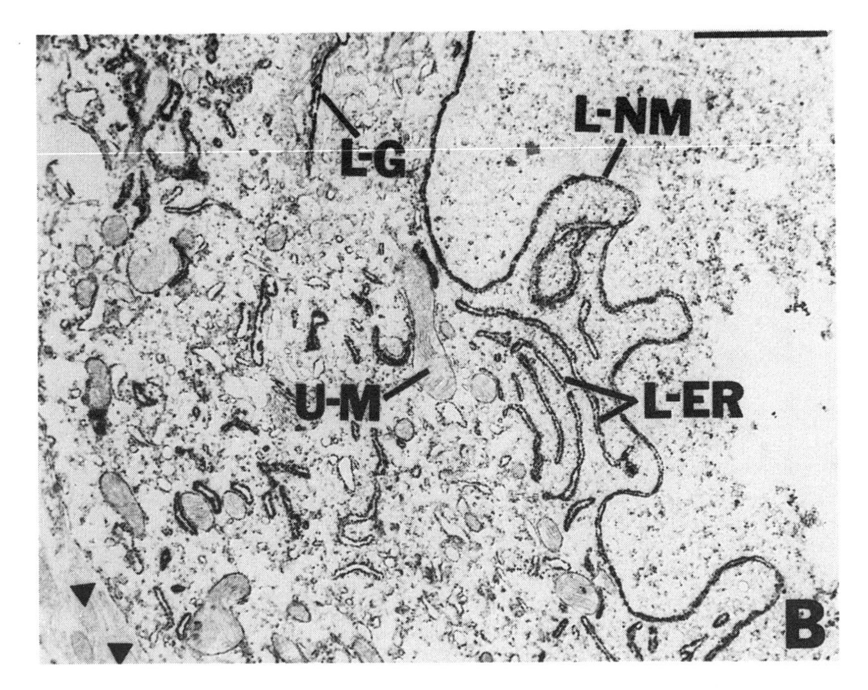

Fig. 7.21 Immunohistochemical localization of the inositol-1,4,5-trisphosphate receptor in the perinuclear region. Dark staining corresponds to the location of anti-receptor antibodies. Key: L-NM, labelled nuclear membrane; L-ER, labelled endoplasmic reticulum; L-G, labelled Golgi apparatus; U-M, unlabelled mitochondrion. Reproduced from Ross *et al* (1989) Nature **339**, 468–470, courtesy of Solomon Snyder.

5-, 1-, and 4-phosphates to release inositol. Intriguingly, the latter two phosphatases are inhibited by the therapeutic agent lithium (Box 7.4).

More recently, an alternative pathway has been found in which IP_3 is phosphorylated by a specific IP_3 kinase to inositol-1,3,4,5-tetrakisphosphate (IP_4; Figure 7.22). The kinase responsible for this is stimulated by Ca^{2+}, so that release of Ca^{2+} by IP_3 accelerates its own removal. IP_4 is subsequently broken down to inositol-1,3,4-trisphosphate, an IP_3 isomer which is not active in Ca^{2+} release but

Lithium has been used for many years in the treatment of manic depression. It has been found that lithium ions potently inhibit the two phosphatases which degrade inositol-1,4-bisphosphate and inositol-1-phosphate to free inositol. Lithium therefore prevents inositol from recycling into phosphoinositides. Since many peptide neurotransmitters activate phosphoinositide turnover it is possible, although not proven, that this explains the therapeutic effect of the compound.

Box 7.4 Lithium ions and manic depression

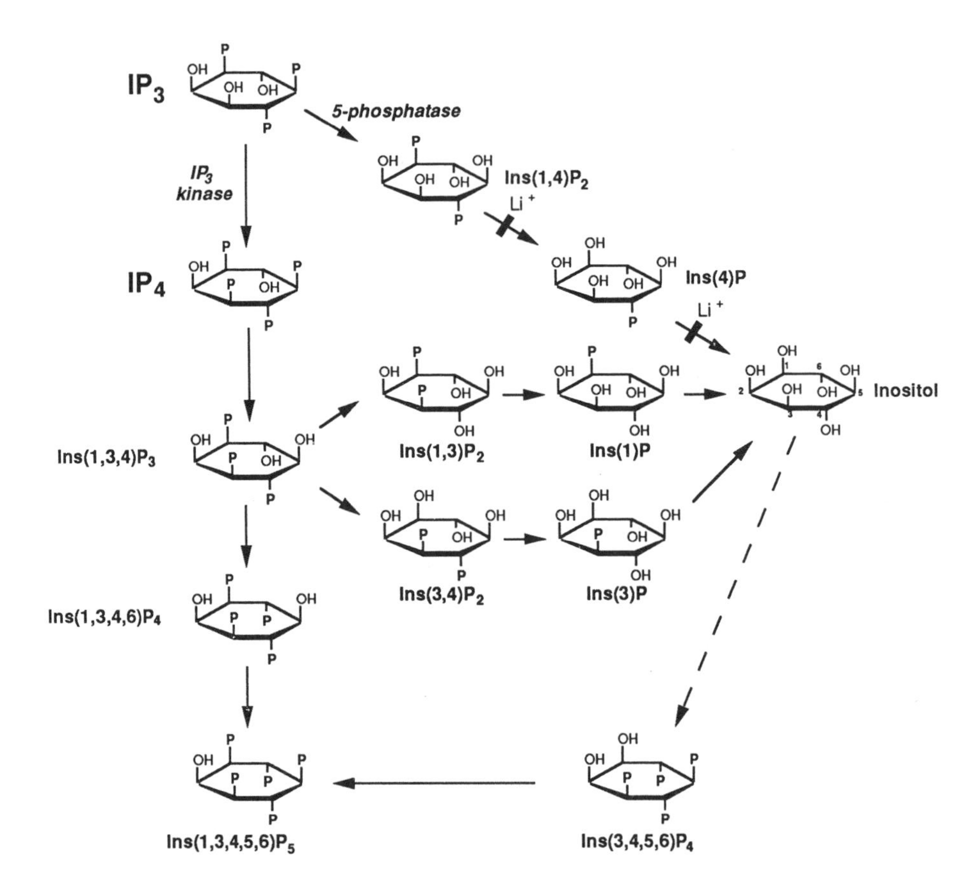

Fig. 7.22 Summary of the known metabolic pathways for inositol phosphates. *Abbreviations:* Ins, inositol; P_2, P_3, P_4, P_5, bisphosphate, trisphosphate, tetrakisphosphate, pentakisphosphate. The dashed line between inositol and Ins(3,4,5,6)P_4 indicates that the intermediate steps have not been elucidated.

which accumulates to much higher levels than the 1,4,5-isomer after receptor stimulation. Inositol-1,3,4-trisphosphate is then normally degraded by various phosphatases back to inositol. However, kinases also exist which can phosphorylate inositol-1,3,4-trisphosphate to inositol-1,3,4,6-tetrakisphosphate and even inositol-1,3,4,5,6-pentakisphosphate. The latter may also be formed directly from inositol, although the pathway has not been completely elucidated (Figure 7.22).

Could any of these other inositol phosphates have second messenger roles in addition to the established role of IP_3? None of the other naturally occurring inositol phosphates release Ca^{2+} from endoplasmic reticulum. However, there is

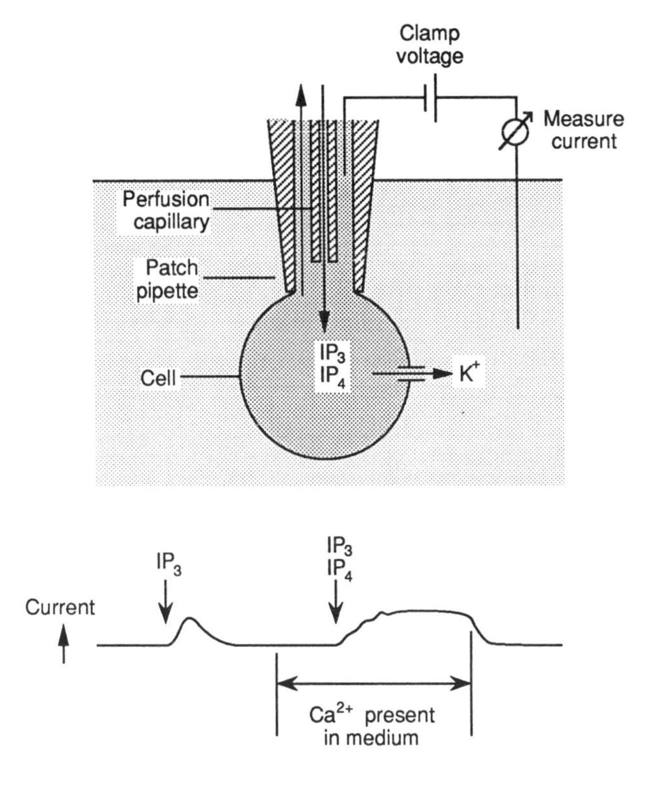

Fig. 7.23 Whole cell patch clamp experiment suggesting that IP_4, in the presence of IP_3, promotes entry of external Ca^{2+}. The patch of membrane attached to the pipette (see Figure 4.14) is sucked away, leaving the remainder of the cell attached. The cell interior is then perfused with medium through the central capillary. The trace at the bottom shows that IP_3 causes a transient opening of K^+ channels in the absence of external Ca^{2+}. IP_3 plus IP_4 cause a more prolonged effect, but it ceases as soon as external Ca^{2+} is removed. Redrawn from Petersen, Cell Calcium (1989) **10**, 375–383.

some evidence for a distinct role for IP_4 in promoting Ca^{2+} entry from outside the cell. The most convincing evidence for this comes from experiments in which whole lachrymal (tear gland) acinar cells attached to a patch pipette (see Section 4.2.1) are perfused internally with inositol phosphates (Figure 7.23). These cells contain Ca^{2+}-activated K^+ channels which can be monitored by voltage clamping. IP_3 alone causes a transient activation of the channels, whereas IP_3 and IP_4 together give a prolonged response which more closely mimics that given by the extracellular first messenger, acetylcholine. The prolonged response to IP_3 plus IP_4, but not the transient response to IP_3 alone, is dependent on the presence of external Ca^{2+} (Figure 7.23). These results suggest that IP_4 promotes Ca^{2+} entry, but this does not appear to be a direct effect (e.g. via an IP_4-gated Ca^{2+} channel) because IP_4 alone has no effect. One suggestion is that IP_4 somehow promotes the refilling from external Ca^{2+} of the internal pools when they are depleted by IP_3.

7.3.3 *Yet another second messenger – diacylglycerol*

The other product of messenger-stimulated breakdown of phosphatidylinositol-4,5-bisphosphate is diacylglycerol, a lipophilic compound that remains associated with the membrane. The possibility that this might act as a second messenger arose in 1979 when Nishizuka and coworkers were studying a protein kinase, protein kinase C, which was originally thought to be only active after proteolysis, especially by a Ca^{2+}-activated proteinase. However, they discovered that the kinase could also be activated reversibly (and probably more physiologically) by Ca^{2+} in the presence of crude membrane phospholipids. Sonicated suspensions of the acidic phospholipid, phosphatidylserine, would also activate the kinase, but only at unphysiologically high Ca^{2+} concentrations. However, if traces of diacylglycerol were included in the phosphatidylserine suspension, activation occurred at concentrations of 10^{-7} to 10^{-6} mol.l^{-1} Ca^{2+} (see Figure 8.37), exactly the range over which cytosol Ca^{2+} varies in response to IP_3 releasing messengers.

Protein kinase C should, strictly speaking, now be called the Ca^{2+}- and phospholipid-dependent protein kinase, although the shorter original name is still in widespread use. The consequences for the cell of activation of this kinase are discussed further in Chapter 8.

7.3.4 *Coupling of receptors to phosphoinositidase C also involves G proteins*

We have already discussed in Chapter 6 the "seven-pass" family of receptors of which numerous examples have now been sequenced. Among the original members of this family were the β-adrenergic receptors and rhodopsin, both of which are coupled to their effectors (adenylate cyclase and cyclic GMP phosphodiesterase) by G proteins. However, the family includes numerous examples of receptors which are coupled to phosphoinositidase C, including the α_1-adrenergic receptor and muscarinic acetylcholine receptors. The similarities between these receptors immediately suggest that the phosphoinositidase systems may also be coupled via G proteins. Although these putative "Gp" proteins have not yet been identified, there is considerable evidence that they must exist:

1) in several cell-free membrane preparations in which agonists activate phosphoinositide breakdown, non-hydrolysable analogues of GTP such as GTP-γ-S stimulate the process;
2) in some systems, hormones coupled to phosphoinositidase C have been

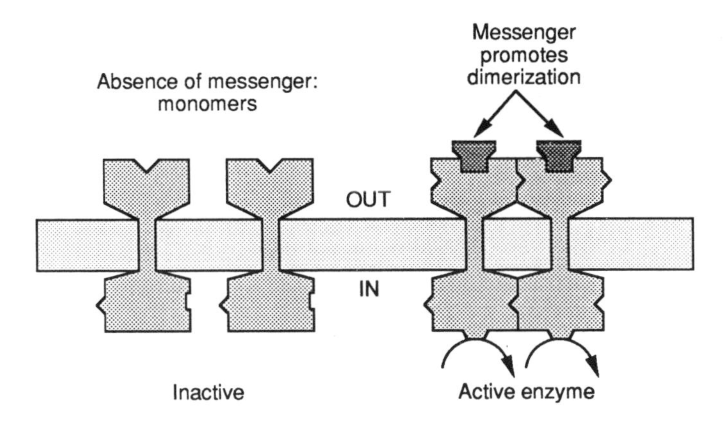

Fig. 7.24 Dimerization model for signal transduction by "single pass" receptors exhibiting cytoplasmic enzyme activity.

reported to stimulate GTPase activity in isolated membranes;

3) in some cases, pertussis toxin blocks the effects of messengers on phosphoinositidase C. This does not happen in all cell types, suggesting there may be multiple Gp proteins, only some of which are ADP-ribosylated by the toxin.

At least five isoforms of phosphoinositidase C have now been purified from different sources, and cDNA encoding four of these (α, β, γ, and δ) has been cloned and sequenced. It is not yet clear which, if any, of these is coupled to receptors via G proteins. However, there is evidence that the γ isoform is phosphorylated on tyrosine and activated by the platelet-derived growth factor (Section 7.4.1). This appears to be an alternative method of coupling receptors to phosphoinositidase breakdown which does not involve a G protein.

7.4 *Receptors with integral enzyme activity*

As discussed in Chapter 6, at least two classes of receptors have been found in which binding of the messenger to the ectoplasmic domain activates enzyme activities in the cytoplasmic domain. These are the protein (tyrosine) kinase (Section 6.3.3) and guanylate cyclase (Section 6.3.4) receptors. In one sense, these discoveries solved at a stroke the problem of how the signal was transduced across the membrane. However, it is not easy to visualize how a conformational change in the ectoplasmic domains could be transmitted through a single transmembrane helix to the cytoplasmic domains. One possible model is shown in Figure 7.24. Binding of messenger to the ectoplasmic domain causes a conformational change leading to dimerization. This juxtaposes the intracellular

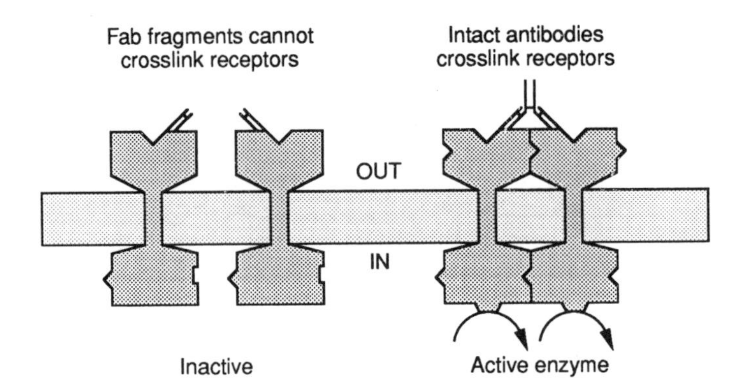

Fig. 7.25 Model for the activation of insulin receptor by divalent, but not monovalent, monoclonal antibodies.

domains and produces a conformational change and/or a covalent modification (e.g. phosphorylation in the case of the kinase receptors). With the insulin receptor there is some evidence for this model, since some antibodies which bind to the ectoplasmic domain outside of the insulin-binding site, mimic the effects of insulin. However, this effect only works with intact, divalent antibodies and not with Fab fragments, which would be unable to crosslink two receptors (Figure 7.25). Many receptors also cluster in coated pits after activation, but as discussed in Section 6.5 this may be concerned more with messenger degradation rather than signal transduction.

All of the protein (tyrosine) kinase receptors autophosphorylate on tyrosine residues in response to their messenger, and in some cases there is evidence that this activates the receptor. For the insulin receptor, it has been shown that the protein (tyrosine) kinase remains active after autophosphorylation, even when insulin is removed. This suggested an interesting model for receptor activation. If the ectoplasmic domains dimerized on insulin binding, this would juxtapose the cytoplasmic domains (Figure 7.24) which could then phosphorylate and activate each other. However there is little direct evidence to support this model, and it clearly could not account for signal transduction by guanylate cyclase receptors.

7.4.1 *Role of protein (tyrosine) kinase activity*

Whatever the mechanism for activation of the cytoplasmic domain, production of cyclic GMP by the guanylate cyclase receptors does solve the problem of signal transduction, since the nucleotide is known to serve as a second messenger. Since protein (serine/threonine) phosphorylation is an important general downstream effect of messengers acting on cell surface receptors (Chapter 8), it seemed very likely that the effects of the protein (tyrosine) kinase receptors would be mediated

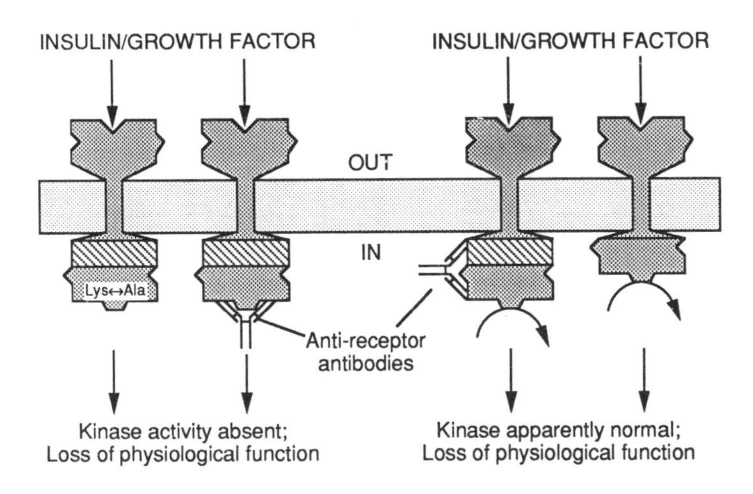

Fig. 7.26 Summary of evidence for and against obligatory involvement of protein (tyrosine) kinase activity in signal transduction. Evidence in favour is shown on the left: mutations in the kinase active site, or antibodies which block the active site, also block the function of the receptor in signal transduction. Evidence against is shown on the right: antibodies which do not block the kinase active site can still lead to loss of function, and deletion of the "kinase insert" in receptors of the PDGF family leads to loss of physiological function, but not kinase activity.

via their kinase activities. However, there has been apparently conflicting evidence as to whether the growth factor-activated protein kinase activity is essential to signal transduction (Figure 7.26). Evidence in favour of this view may be summarized as follows:

1) Site-directed mutagenesis has revealed residues involved in the protein kinase active site, such as a lysine residue which is conserved in all protein kinases, and which is believed to be involved in binding of the phosphate groups of the substrate, ATP. If the codon for this lysine is changed in insulin receptor cDNA, and the mutant DNA is transfected into cells, an insulin-binding receptor is expressed but its signal transducing functions are abolished. Similar results are obtained if the autophosphorylation sites are mutated.
2) Some antibodies which inhibit the protein (tyrosine) kinase activity of the receptors interfere with signal transduction when microinjected into cells.

Evidence against an involvement of protein (tyrosine) kinase activity (Figure 7.26) includes the following:

1) Some monoclonal antibodies against the cytoplasmic domain of the insulin receptor abolish downstream effects of insulin when microinjected into cells, without affecting protein (tyrosine) phosphorylation of the receptor or of other proteins.
2) In the class of receptor which has an "insert" in the kinase domain (e.g. the PDGF receptor, Section 6.3.3 and Figure 6.27) deletion of the insert does not affect the protein (tyrosine) kinase activity of the receptor on model substrates. However, transfection of these mutant receptors shows that although many of the rapid responses to PDGF are maintained, the longer term effects on DNA synthesis and cell proliferation are abolished.

How can these apparently conflicting observations be reconciled? This can be achieved by making two propositions:

1) Many of the intracellular protein (tyrosine) phosphorylation events seen in response to these growth factors are irrelevant to their action, with the key substrates being minor components. This is a reasonable postulate, because most purified protein (tyrosine) kinases seem to be rather non-specific, and will slowly phosphorylate almost any protein if it has an exposed tyrosine residue. Also, some of the proteins found to be phosphorylated on tyrosine in response to growth factors are modified to only a very small extent.
2) While the protein (tyrosine) kinase activity of these receptors is essential to their function, the cytoplasmic domains of these receptors also contain regions (e.g. the "insert" in the PDGF receptor) which are not directly involved in kinase activity, but which are nevertheless essential for the complete response to the growth factor. One possibility is that these regions are required for protein–protein interactions, and that the key protein substrates for the protein (tyrosine) kinase form stable aggregates with the receptor.

In the case of the PDGF receptor there is now some evidence that the latter hypothesis is correct. It has been found that several polypeptides co-precipitate when solubilized membranes containing the PDGF receptor are incubated with anti-receptor antibody, suggesting that they are associated by stable protein-protein interactions. A key point is that this **only occurs in the presence of PDGF**. Four of these proteins, all of which are phosphorylated on tyrosine residues by the PDGF receptor, have recently been identified, and this has suggested several intriguing new possibilities regarding signal transduction mechanisms. These four proteins are:

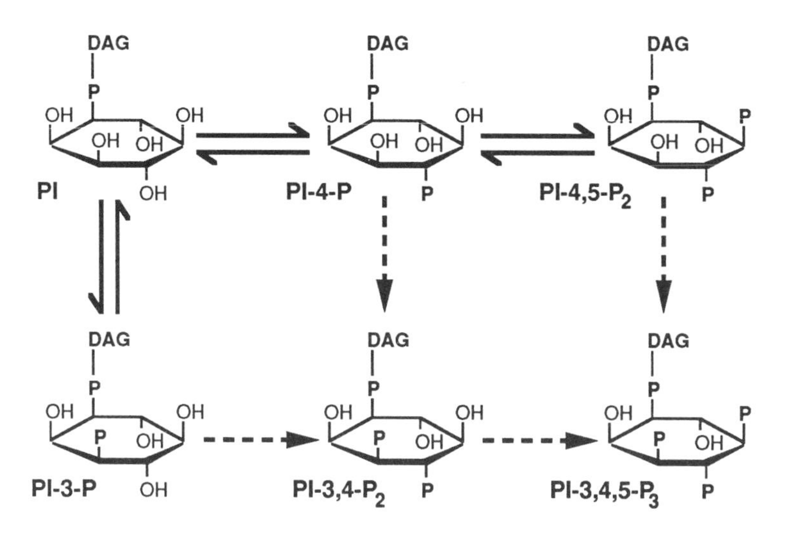

Fig. 7.27 Structure of phosphatidylinositol-3-phosphate (PI-3-P), and possible metabolic routes for the formation of other 3-phosphorylated inositol lipids. Some reactions are shown with dashed arrows because it is not clear which metabolic routes are followed *in vivo*. Phosphate groups are represented by "P" and diacylglycerol by "DAG": see Figure 7.17 for complete structures for phosphatidylinositol and its 4- and 4-, 5- phosphorylated forms (PI, PI-4-P, PI-4,5-P_2).

1) The γ isoform of phosphoinositidase C. PDGF activates phosphatidylinositol-4,5-bisphosphate breakdown in intact cells, and it is tempting to speculate that this is due to phosphorylation of phosphoinositidase Cγ. It has also been shown that antibodies against phosphoinositidase Cγ can co-precipitate both the PDGF receptor and the phosphatidylinositol kinase mentioned below, suggesting that these proteins can bind simultaneously to the PDGF receptor at distinct sites.
2) A novel **phosphatidylinositol-3-kinase** which gives rise to PI-3-phosphate, PI-3,4-phosphate and PI-3,4,5-phosphate *in vitro* (Figure 7.27). Although it has not been proved that tyrosine phosphorylation activates the kinase, treatment of cells with PDGF does increase the amount of phosphoinositides phosphorylated at the 3-position. A particularly intriguing finding is that co-precipitation of the PI kinase with the receptor does not occur in mutants lacking the "kinase insert", which suggests that the latter could be involved in interaction with the PI kinase. Since the kinase insert is also necessary for the proliferative response, it is possible that these 3-phosphorylated lipids perform some novel signalling function concerned with cell growth and division.

```
GAP 1:    WYHGKLDRTIAEERLR-QAGKSGSYLIRESDRRPGSFVLSF..FSSLSDLIGYYSH
          |||* | |* ||  *|  *   |**|*|     | |* *||  | || |||*||
PIC-γ:    WYHASLTRAQAE-HMR-VPRD-GAFLVRKRNE-PNSYAISF..FDSLVDLISYYEK
          |*|* ***  |   *  |*    *||||     |  |** |  * |* |*|  | |
GAP 2:    WFHGKISKQEAYNLLMTVGQAC-SFLVRPSDNTPGDYSLYF..YNSIGDIIDHYRK
```

Fig. 7.28 Amino acid sequence comparisons of two regions of GTPase activator protein (GAP 1 and GAP 2)) and a region of phospholipase C γ (PIC–γ). The single letter code is used (see Appendix). Identities are marked by vertical lines and conservative replacements by asterisks.

3) ***Raf-1***, a serine/threonine kinase of unknown function. There is evidence that the PDGF-dependent phosphorylation on tyrosine activates the enzyme, an exciting finding that is discussed further in Chapter 8.
4) **GTPase activator protein** (GAP), which interacts with the *ras* gene product, which in turn is implicated in control of cell proliferation (Section 9.3.3). Intriguingly, the sequence of GAP contains two repeated segments which are closely related to a region found in phosphoinositidase Cγ (Figure 7.28). It is tempting to speculate that these are involved in interactions with the PDGF receptor.

While much still remains to be learned about signal transduction by the protein (tyrosine) kinase receptors, some intriguing clues are therefore beginning to emerge. It seems likely that the key target proteins which are phosphorylated on tyrosine must be in the correct subcellular location, where they may form stable interactions with their substrate (Figure 7.29):

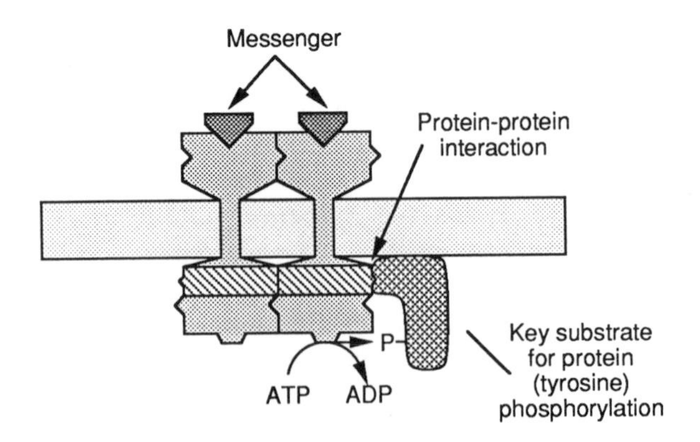

Fig. 7.29 Revised hypothesis for mechanism of signal transduction by protein (tyrosine) kinase receptors. Key substrates need to form a stable protein–protein interaction with the receptor: this may be one function of the "kinase insert" region in the PDGF receptor family.

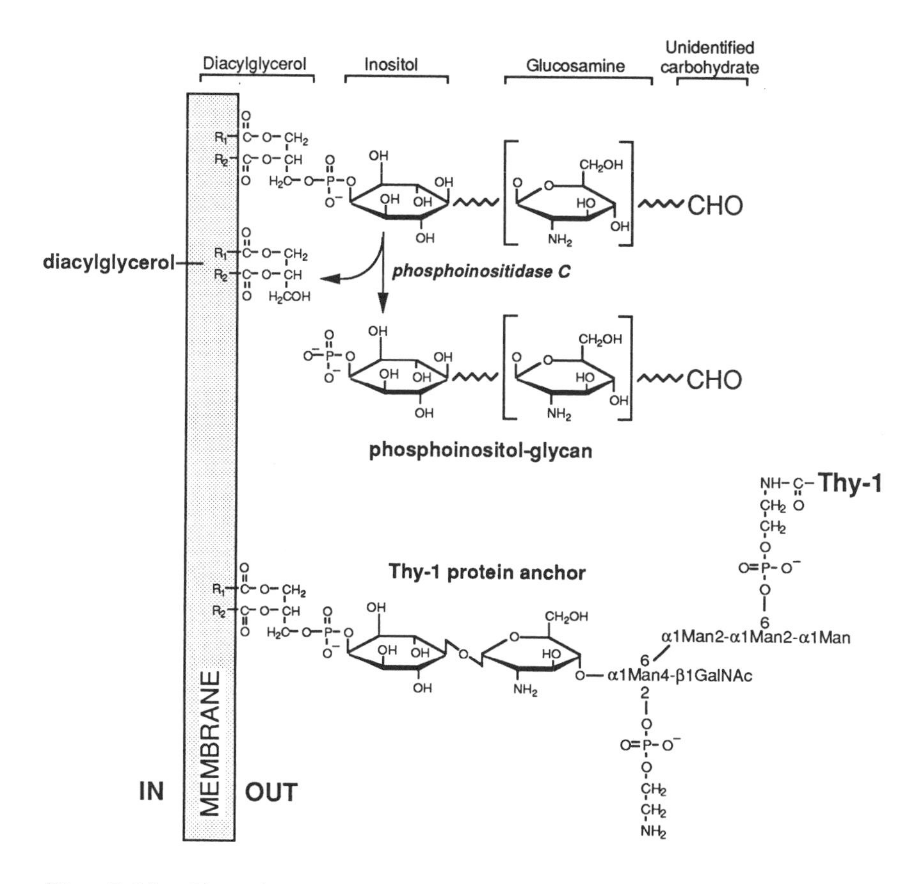

Fig. 7.30 Tentative structure for the insulin-mimicking phosphoinositol glycan, and its generation by phosphoinositidase C. Also shown for comparison is the structure of the phosphatidylinositol glycan which is the membrane anchor for the Thy-1 protein (a surface antigen on cells of the immune system). The Thy-1 protein is attached through its C-terminus.

7.4.2 *Second messengers for insulin?*

Although the mechanisms discussed in the previous section would obviate the need for a second messenger in the case of a protein (tyrosine) kinase receptor, over the years numerous compounds have been put forward as candidates for second messengers which mediate the action of insulin. These suggestions have ranged from peptides to polyamines, and even hydrogen peroxide, but none of them have stood the test of time. An interesting recent candidate for this role is a phosphoinositol-linked glycan which can be released from membranes using a bacterial phosphoinositidase C (Figure 7.30). The exact structure of the compound is not known, but it appears to be related to the phosphatidylinositol-glycan structures which anchor certain proteins to the external surface of the plasma membrane. It was originally reported that the compound mimicked effects of insulin in cell-free systems, but some of these effects have turned out to be artefacts. There is no doubt that the phosphoinositol glycan does mimic some effects of insulin when given to **intact** cells, and evidence has been presented that it can be released from membranes by insulin as well as using phosphoinositidase C. However it now appears that this compound is released on the **outside** of the cell, so if anything it is a local mediator (i.e. a first messenger) which modulates the response of cells to insulin, rather than an intracellular second messenger. Further progress awaits the determination of the structure of the compound.

SUMMARY

Signal transduction at cell-surface receptors appears to work by one of three basic mechanisms: (1) opening of ion channels; (2) second messenger mechanisms; (3) receptors with integral ligand-activated enzyme activity.

The major ion channel mechanism is the ligand-gated ion channel, in which the channel is an integral part of the receptor. The classical case is the nicotinic acetylcholine receptor which is an excitatory cation channel, but two receptors for inhibitory transmitters ($GABA_A$ and glycine) now appear to be anion channels closely related to the nicotinic receptor. All of these receptors have multiple, related subunits (five in the case of the nicotinic acetylcholine receptor) which associate in pseudosymmetrical fashion and line a central ion channel. Each subunit contains four transmembrane α-helices, and site-directed mutagenesis is currently defining the functions of the different subunits and domains. Another mechanism for regulation of ion channels involves indirect coupling to the receptor via GTP-binding (G) proteins.

The classical example of a second messenger is cyclic AMP, produced in response to β-adrenergic agonists and many other first messengers. Second messengers are produced catalytically in response to the extracellular (first) messenger, and thus amplify the signal. The enzyme producing cyclic AMP (adenylate cyclase) is coupled indirectly to receptor(s) via GTP-binding (G) proteins. The receptors themselves are invariably members of the "seven-pass" family containing seven transmembrane α-helices. This system shows remarkable analogies to the system coupling light absorption by rhodopsin to the activation of cyclic GMP phosphodiesterase in the eye. Some receptors inhibit adenylate cyclase via distinct, inhibitory G proteins, which may also be involved in coupling to ion channels.

Many first messengers cause a transient elevation of cytosolic Ca^{2+}, initially by releasing the cation from endoplasmic reticulum. The second messenger in this case is now known to be inositol-1,4,5-trisphosphate, released from *phosphatidylinositol-4,5-bisphosphate by the action of phosphoinositidase C. The latter exists as several isoforms, some of which appear to be coupled to receptors via G proteins. Another second messenger released by these systems is diacylglycerol, which activates protein kinase C. Numerous other inositol phosphates, some of which may have second messenger effects, are now known to exist.*

Receptors containing integral enzyme activity include the insulin and growth factor class, which exhibit ligand activated protein (tyrosine) kinase activity, and the resact and atrial natriuretic peptide receptors, which directly produce the second messenger cyclic GMP. Signal transduction by the protein (tyrosine) kinase class of receptor is not completely understood, but recent experiments suggest that the protein (tyrosine) kinase activity is necessary, but not sufficient, to explain the downstream effects. The ability of the receptor to form stable protein–protein interactions with effector proteins may also be crucial. These effector proteins are currently being identified. The protein (tyrosine) kinase receptors may not require a second messenger to mediate their effects, although various candidates have been put forward as second messengers for insulin.

FURTHER READING

1. "The molecular basis of communication within the cell" – introductory review on signal transduction. Berridge, M.J. (1985), Scientific American, October, pp. 124–134.
2. "Rings of negatively charged amino acids determine the acetylcholine receptor channel conductance" – site-directed mutagenesis of nicotinic receptors and expression in *Xenopus* oocytes. Imoto, K., Busch, C., Sakmann, B., Mishina, M., Konno, T., Nakai, J., Bujo, H., Mori, Y., Fukuda, K. and Numa, S. (1988) Nature **335**, 645–648.
3. "Sequence and functional expression of the $GABA_A$ receptor shows a ligand-gated receptor super-family" – expression of GABA-activated Cl^- channels after reconstitution of cloned receptor in *Xenopus* oocytes. Schofield, P.R., Darlison, M.G., Fujita, N., Burt, D.R., Stephenson, F.A., Rodriguez, H., Rhee, L.M., Ramachandran, J., Reale, V., Glencorse, T.A., Seeburg, P.H. and Barnard, E.A. (1987) Nature **328**, 221–227.
4. "Intracellular cyclic AMP production enhanced by a hormone receptor transferred from a different cell. β-Adrenergic responses in cultured cells conferred by fusion with turkey erythrocytes". Schulster, D., Orly, J., Seidel, G. and Schramm, M. (1978) J. Biol. Chem. **253**, 1201–1206.
5. "G proteins: transducers of receptor-generated signals" – review. Gilman, A.G. (1987) Ann. Rev. Biochem. **56**, 615–649.
6. "The molecules of visual excitation" – review. Stryer, L. (1987) Scientific American, July, pp.32–40.

7. "Single cyclic GMP-activated channel activity in excised patches of rod outer segment membrane". Haynes, L.W., Kay, A.R. and Yau, K.W. (1986) Nature **321**, 66–70.
8. "G protein involvement in receptor–effector coupling" – minireview. Casey, P.J. and Gilman, A.G. (1988) J. Biol. Chem. **263**, 2577–2580.
9. "Mechanism of guanine nucleotide regulatory protein-mediated inhibition of adenylate cyclase. Studies with isolated subunits of transducin in a reconstituted system". Cerione, R.A., Staniszewski, C., Gierschik, P., Codina, J., Somers, R.L., Birnbaumer, L., Spiegel, A.M., Caron, M.G. and Lefkowitz, R.J. (1986) J. Biol. Chem. **261**, 9514–9520.
10. "Occurrence of an inhibitory guanine nucleotide-binding regulatory component of the adenylate cyclase system in cyc– variants of S49 lymphoma cells". Jakobs, K.H., Gehring, U., Gaugler, B., Pfeuffer, T. and Schulz, G. (1983) Eur. J. Biochem. **130**, 605–611.
11. "The G protein-gated atrial K^+ channel is stimulated by three distinct Gi α subunits". Yatani, A., Mattera, R., Codina, J., Graf, R., Okabe, K., Padrell, E., Iyengar, R., Brown, A.M. and Birnbaumer, L. (1988) Nature **336**, 680–682.
12. "The guanylate cyclase/receptor family of proteins" – review. Schulz, S., Chinkers, M. and Garbers, D.L. (1989) FASEB J. **3**, 2026–2035.
13. "Release of Ca^{2+} from a nonmitochondrial intracellular store in pancreatic acinar cells by inositol-1,4,5-trisphosphate" – key paper in discovery of second messenger role of IP3. Streb, H., Irvine, R.F., Berridge, M.J. and Schulz, I. (1983) Nature **306**, 67–69.
14. "Inositol trisphosphate, a novel second messenger in cellular signal transduction" – review. Berridge, M.J. and Irvine, R.F. (1984) Nature **312**, 315–321.
15. "Inositol 1,4,5-trisphosphate receptor localized to endoplasmic reticulum in cerebellar Purkinje neurons". Ross, C.A., Meldolesi, J., Milner, T.A., Satoh, T., Supattapone, S. and Snyder, S.H. (1989) Nature **339**, 468–470.
16. "Primary structure and functional expression of the inositol 1,4,5-trisphosphate-binding protein P400". Furuichi, T., Yoshikawa, S., Miyawaki, A., Wada, K., Maeda, N. and Mikoshiba, K. (1989) Nature **342**, 32–38.
17. "Distribution of two distinct Ca^{2+}-ATPase-like proteins and their relationships to the agonist-sensitive calcium store in adrenal chromaffin cells". Burgoyne, R.D., Cheek, T.R., Morgan, A., O'Sullivan, A.J., Moreton, R.B., Berridge, M.J., Mata, A.M., Colyer, J., Lee, A.G. and East, J.M. (1989) Nature **342**, 72–74.
18. "Synergism of inositol trisphosphate and tetrakisphosphate in activating Ca^{2+}-dependent K^+ channels". Morris, A.P., Gallacher, D.V., Irvine, R.F. and Petersen, O.H. (1987) Nature **330**, 653–655.
19. "Studies of inositol phospholipid-specific phospholipase C" – review. Rhee, S.G., Suh, P.G., Ryu, S.H. and Lee, S.Y. (1989) Science **244**, 546–550.

20. "After insulin binds" – review. Rosen, O.M. (1987) Science **237**, 1452–1458.
21. "Acute insulin action requires insulin receptor kinase activity: introduction of an inhibitory monoclonal antibody into mammalian cells blocks the rapid effects of insulin". Morgan, D.O. and Roth, R.A. (1987) Proc. Natl. Acad. Sci. USA **84**, 41–45.
22. "Signal transduction by the platelet-derived growth factor receptor" – review. Williams, L.T. (1989) Science **243**, 1564–1570.
23. "Platelet-derived growth factor (PDGF) binding promotes physical association of PDGF receptor with phospholipase C". Kumjian, D.A., Wahl, M.I., Rhee, S.G. and Daniel, T.O. (1989) Proc. Natl. Acad. Sci. USA **86**, 8232–8236.
24. "Signal transduction from membrane to cytoplasm: growth factors and membrane-bound oncogene products increase Raf-1 phosphorylation and associated protein kinase activity". Morrison, D.K., Kaplan, D.R., Rapp, U. and Roberts, T.M. (1988) Proc. Natl. Acad. Sci. USA **85**, 8855–8859.
25. "PDGF induction of tyrosine phosphorylation of GTPase activating protein". Molloy, C.J., Bottaro, D.P., Fleming, T.P., Marshall, M.S., Gibbs, J.B. and Aaronson, S.A. (1989) Nature **342**, 711–714.
26. "Direct activation of the serine/threonine kinase activity of Raf-1 through tyrosine phosphorylation by the PDGF β-receptor". Morrison, D.K., Kaplan, D.R., Escobedo, J.A., Rapp, U.R., Roberts, T.M. and Williams, L.T. (1989) Cell **58**, 649–657.

EIGHT

CELL SURFACE RECEPTORS - PROTEIN PHOSPHORYLATION AND OTHER INTRACELLULAR EVENTS

The signal transduction mechanisms discussed in Chapter 7 result in the generation of three types of signal within the cell (Figure 8.1):

1) depolarization induced by a ligand-gated ion channel;
2) production of a second messenger, i.e. cyclic AMP, cyclic GMP, inositol-1,4,5-trisphosphate (and hence Ca^{2+}) or diacylglycerol;
3) activation of a protein (tyrosine) kinase activity.

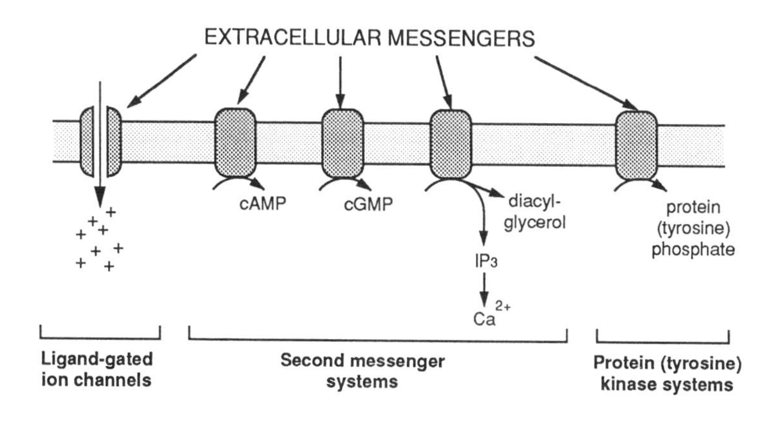

Fig. 8.1 The three basic types of signal transduction mechanisms at cell surface receptors.

In different cell types a huge variety of responses result from these relatively few signalling systems. However, with a few exceptions discussed below, a common element in all of these responses is that they are mediated by modulation of phosphorylation of target proteins on serine/threonine residues. A discussion of protein (serine/threonine) kinases and phosphatases and their targets will therefore form the bulk of this chapter. We begin by considering the three types of signal in turn.

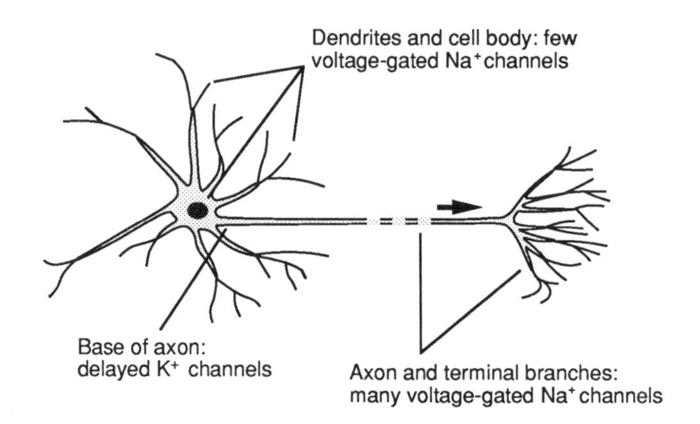

Fig. 8.2 Diagram of a typical neurone, showing the regions where voltage-gated Na^+ and delayed K^+ channels are found.

8.1 *Responses to depolarization by ligand-gated ion channels*

The response to depolarization depends on the nature of the target cell. We will consider three cases, i.e. neurones, adrenal chromaffin cells, and skeletal muscle.

8.1.1 *Neurones*

Neurones are, of course, excitable because they contain voltage-gated ion channels, especially Na^+ channels (Chapter 4). A simplistic view is that, if the depolarization induced by the excitatory receptor exceeds a certain threshold, the Na^+ channels will open and an action potential will proceed along the target neurone. In practice the situation is more complicated: a neurone may receive inputs from as many as several thousand nerve terminals, some excitatory and some inhibitory, and has to sum these many inputs into a particular frequency of action potential down its own axon.

Each input causes a small transient depolarization or hyperpolarization of the post-synaptic membrane, which can be quite variable in magnitude. The dendrites where these inputs are generally received, and the cell body, contain rather few voltage-gated ion channels (Figure 8.2), so these individual voltage changes are carried by simple ionic conduction to the start of the axon, their magnitude falling with distance from the point of input. The net voltage change at this point will be a summation of all the individual changes: clearly a train of rapid, transient

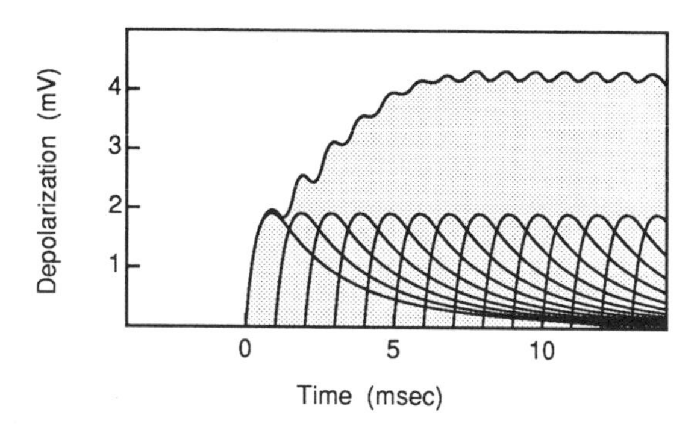

Fig. 8.3 Diagram showing the summation of many transient depolarization events into a continuous depolarization.

depolarization events could give rise to a continuous depolarization (Figure 8.3). A problem arises here because, as discussed in Chapter 4, voltage-gated Na^+ channels only open transiently in response to depolarization and, once closed, cannot reopen until normal membrane potential is recovered. So at first sight a continuous depolarization could only ever give rise to one action potential.

The solution to this paradox lies in the fact that other types of voltage-gated ion channels exist in the axon, particularly at the point where it leaves the cell body (Figure 8.2). One type is the **delayed K^+ channel**, which opens in response to depolarization, but with a short time delay compared with the Na^+ channels. After the latter have triggered the action potential, the K^+ channels open and restore the normal K^+ equilibrium potential, repolarizing the membrane and allowing the Na^+ channels to convert to their "ready" state. As soon as repolarization is achieved, the K^+ channels (being voltage-gated) will of course close again, and the depolarizing stimulus can now trigger a second action potential. Thus a continuous depolarization is converted into a train of action potentials. Other channel types, discussion of which is outside the scope of this book, ensure that the frequency of these potentials reflects the size of the depolarizing stimulus.

8.1.2 *Adrenal chromaffin cells*

Adrenal chromaffin cells represent a model for secretory cells which are controlled by the autonomic nervous system. They contain nicotinic acetylcholine receptors which cause a local depolarization when the cell is stimulated by the sympathetic nerves (Figure 8.4). The immediate effect of this is the opening of voltage-gated Ca^{2+} channels in the plasma membrane. Because of the 10,000-fold higher concentration of the cation outside the cell than in the cytosol, Ca^{2+} ions flood in and cause a >10-fold rise in concentration close to the plasma membrane. This rise

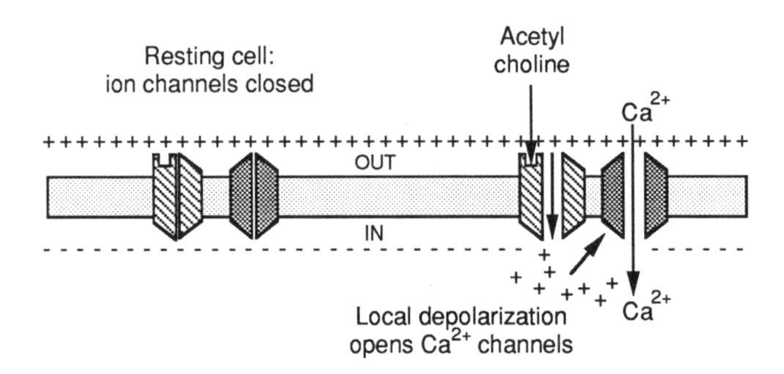

Fig. 8.4 Mechanism of Ca^{2+} elevation in response to acetylcholine in an adrenal chromaffin cell. Depolarization in response to opening of nicotinic acetylcholine receptor channels causes opening of voltage-gated Ca^{2+} channels.

in Ca^{2+} may also trigger release of Ca^{2+} from IP_3-insensitive stores in endoplasmic reticulum (Section 7.3.1). The elevated Ca^{2+} then triggers exocytosis of chromaffin granules and release of adrenaline (epinephrine) and peptide messengers, as discussed in Section 5.2.

8.1.3 *Skeletal and cardiac muscle*

Like secretion in chromaffin cells, contraction in **striated** (i.e. skeletal and cardiac) muscles is triggered by a rise in Ca^{2+}. The cation causes contraction by binding to the muscle protein **troponin C** (Section 8.2.1). However, due to the very high concentration of troponin C in muscle, it seems that the required amounts of external Ca^{2+} cannot enter rapidly enough, and muscle contains large stores of the cation in a specialized form of endoplasmic reticulum called **sarcoplasmic reticulum** (Figure 8.5). The very abundant Ca^{2+} pump in sarcoplasmic reticulum has already been described (Section 4.1.2): another abundant protein called **calsequestrin** sequesters the cation inside the sarcoplasmic reticulum. To release Ca^{2+} for contraction, the sarcoplasmic reticulum also contains Ca^{2+} channels which are activated pharmacologically by caffeine, and by the plant alkaloid **ryanodine**. Since ryanodine binding is the assay used for this protein during purification, it is often referred to as the ryanodine receptor.

When a motor nerve leading to a skeletal muscle is stimulated, acetylcholine is released at the synapse between it and the muscle cell: the **neuromuscular junction**. The plasma membrane of the muscle cell (**sarcolemma**) contains nicotinic acetylcholine receptors at this point, and their stimulation leads to depolarization. The sarcolemma is an excitable membrane containing voltage-gated Na^+ channels. An action potential therefore spreads across the membrane and is carried into the cell at long tubular invaginations known as **transverse tubules**.

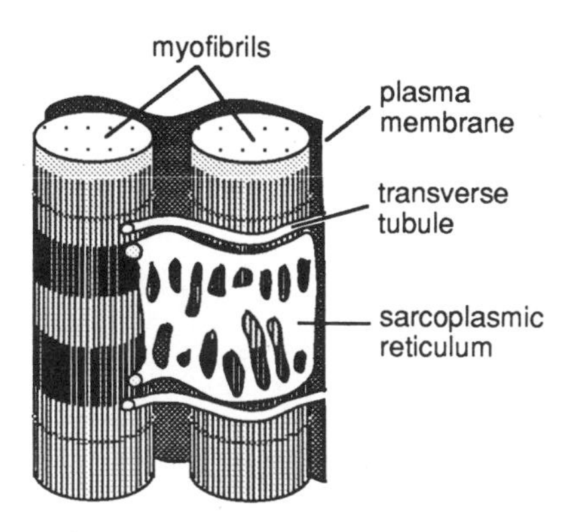

Fig. 8.5 Cut-away diagram of a skeletal muscle cell. Only a portion of the sarcoplasmic reticulum and transverse tubules is shown, in order to reveal the myofibrils underneath.

One edge of the sarcoplasmic reticulum lies closely apposed to the transverse tubules, separated by a gap of 15 nm. How then is the effect of depolarization of the tubule transmitted across this gap? The details are still not clear, but the best guess at present is that there is an interaction between the ryanodine receptors, which line up on the sarcoplasmic reticulum opposite the tubule, and Ca^{2+} channels on the tubule. These latter channels are termed variously L channels, "slow" Ca^{2+} channels, or, because they are blocked by dihydropyridine drugs (e.g. verapamil), dihydropyridine receptors.

cDNAs coding for both the ryanodine and the dihydropyridine receptors have recently been cloned and sequenced from skeletal muscle. The ryanodine receptors are very large molecules composed of four identical subunits, each with >5000 amino acids. Hydropathy plots suggest four transmembrane domains at the extreme C-terminus of each subunit: together these may form the Ca^{2+} channel. The remaining N-terminal 4500 amino acids are thought to lie on the cytoplasmic side of the membrane (Figure 8.6) consistent with electron micrographs in which structures called "feet" project about 12 nm into the gap between the sarcoplasmic reticulum and the transverse tubules. Here they may associate with the dihydropyridine receptors. The predicted amino acid sequences of the latter show that they are remarkably similar to voltage-gated Na^+ channels (Section 4.2.1), with four internal repeats showing sequence similarity, each containing six putative transmembrane helices. As in the Na^+ channels, the fourth putative transmembrane helix in each repeat has regularly spaced positively charged residues which may represent the voltage sensor (Figure 8.7).

The Ca^{2+} release channels of isolated sarcoplasmic reticulum vesicles are activated by Ca^{2+}, so an obvious mechanism for the coupling would be that entry of the cation through the dihydropyridine receptor channels triggers release from

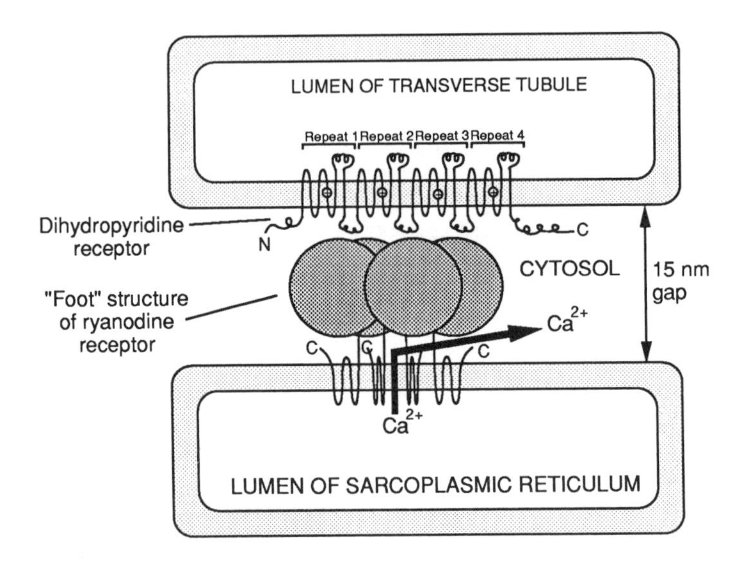

Fig. 8.6 Model for the membrane topology of the ryanodine and dihydropyridine receptors, showing how the cytoplasmic "foot" of the former may form an interaction with the latter. The fourth transmembrane helix in each repeat of the dihydropyridine receptor sequence, believed to be the voltage sensor, is marked with a '+'.

DHP receptor Repeat 1:	V-**K**-A-L-**R**-A-F-**R**-V-L-**R**-P-L-**R**-L-V-S-G-V
Na^+ channel Repeat 1:	V-S-A-L-**R**-T-F-**R**-V-L-**R**-A-L-**K**-T-I-S-V-I
DHP receptor Repeat 2:	I-S-V-L-**R**-C-I-**R**-L-L-**R**-L-F-**K**-I-T-**K**-Y-W
Na^+ channel Repeat 2:	L-S-V-L-**R**-S-F-**R**-L-L-**R**-V-F-**K**-L-A-**K**-S-W
DHP receptor Repeat 3:	V-**K**-I-L-**R**-V-L-**R**-V-L-**R**-P-L-**R**-A-I-N-**R**-A
Na^+ channel Repeat 3:	I-**K**-S-L-**R**-T-L-**R**-A-L-**R**-P-L-**R**-A-L-S-**R**-F
DHP receptor Repeat 4:	S-A-F-F-**R**-L-F-**R**-V-M-**R**-L-I-**K**-L-L-S-**R**-A
Na^+ channel Repeat 4:	I-**R**-L-A-**R**-I-G-**R**-I-L-**R**-L-I-**K**-G-A-**K**-G-I

Fig. 8.7 Amino acid sequences of the fourth transmembrane helix in each fourfold repeat of the sequence of the dihydropyridine (DHP) receptor and the voltage-gated sodium channel. The one-letter code is used (R = arginine, K = lysine; see Appendix for the remainder).

sarcoplasmic reticulum via the ryanodine receptor channels. However there is a paradox, in that extracellular Ca^{2+} is not required for muscle contraction. Possibly the dihydropyridine receptor acts primarily as a voltage sensor, transmitting a conformational change by direct protein–protein interaction to the ryanodine receptor, whose Ca^{2+} channel then opens to allow Ca^{2+} to flow out from the sarcoplasmic reticulum. The much slower entry of external Ca^{2+} through the dihydropyridine receptor may merely replace Ca^{2+} lost to the exterior due to the action of the plasma membrane Ca^{2+} pumps.

8.2 *Responses to calcium and second messengers*

The effects of ligand-gated ion channels discussed in Section 8.1 result either in stimulation of a target neurone or, in the case of muscle cells and secretory cells, in a rise in cytosolic Ca^{2+}. The latter can also occur, of course, in response to inositol-1,4,5-trisphosphate release (Section 7.3.1) by receptors coupled to phosphoinositidase C.

8.2.1 *Responses to Ca^{2+} – direct activation of enzymes by troponin C and calmodulin*

Ca^{2+} exerts most of its intracellular effects through binding to a family of small heat-stable proteins which includes troponin C and calmodulin. Exceptions to this rule are proteins which may be involved in Ca^{2+} regulation of secretion (e.g. the annexin family, Section 5.3.1) and at least one protein kinase (protein kinase C, Section 8.3.3) which are discussed elsewhere.

Structure of troponin C and calmodulin

Troponin C is found in striated muscles (i.e. skeletal and cardiac muscle) and is responsible for activating contraction in response to Ca^{2+}, whereas calmodulin is an abundant protein of ubiquitous distribution, which in the presence of Ca^{2+} activates many different enzymes, including several protein kinases (Section 8.2.2). They both contain about 150 amino acids, and dot matrix comparisons show that they are closely related to each other (Figure 8.8). The presence of additional diagonals in the dot matrix plot arises from the fact that both proteins contain a sequence which is repeated four times with variations: this is clearly seen when the sequence of calmodulin is compared with itself (Figure 8.8). These four repeats correspond to the four Ca^{2+}-binding sites on troponin C and calmodulin. The structures of both proteins have now been elucidated by X-ray crystallography: troponin contains a pair of Ca^{2+}-binding sites at each end of the polypeptide chain, separated by a long central α-helix (Figure 8.9). The structure of calmodulin is very similar. Each Ca^{2+}-binding site comprises two α-helices approximately at

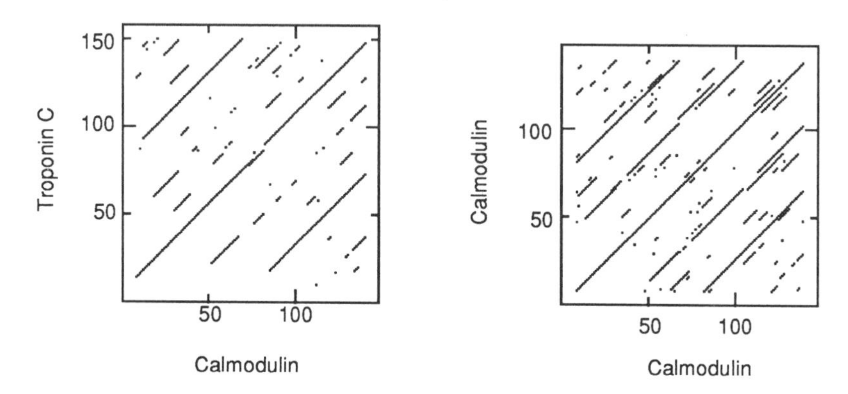

Fig. 8.8 Dot matrix comparison of calmodulin with troponin C (left) and with itself (right). The multiple diagonal lines evident (see Figure 3.15) are due to the fact that both proteins contain four internal sequence repeats.

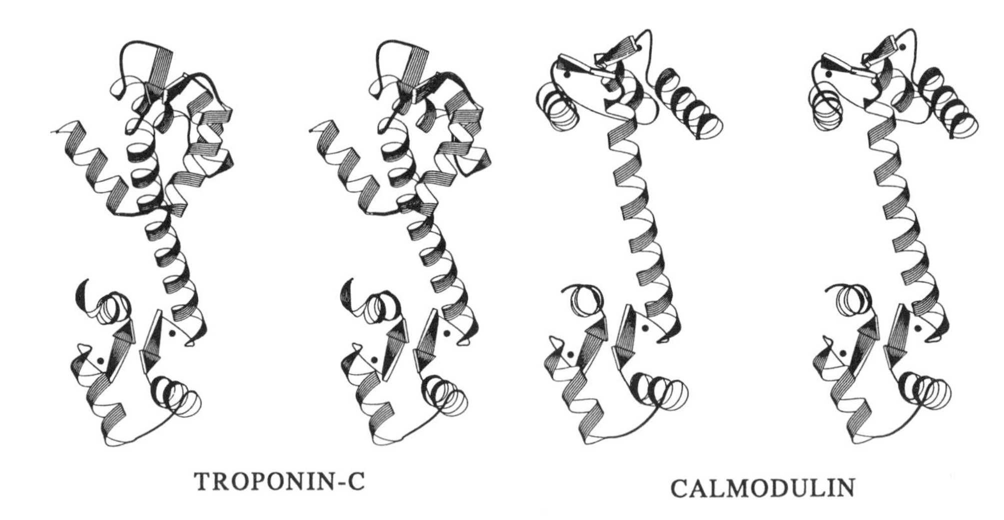

Fig. 8.9 Ribbon and tape models for the structure of troponin C (left) and calmodulin (right). Ca^{2+} atoms are shown as small black spheres. These are stereo diagrams: by converging the eyes until the two paired images fuse, it is possible to obtain a three-dimensional view. Courtesy of Michael James.

right angles to each other, connected by a loop that wraps around the Ca^{2+} (Figure 8.10). In this loop are several acidic residues (Figure 8.11): together with a main chain carbonyl group these form an octahedral coordination sphere around the Ca^{2+} ion. Thus each binding site on calmodulin or troponin C binds the cation by a similar mechanism to the well-known Ca^{2+} chelator EGTA, which has four

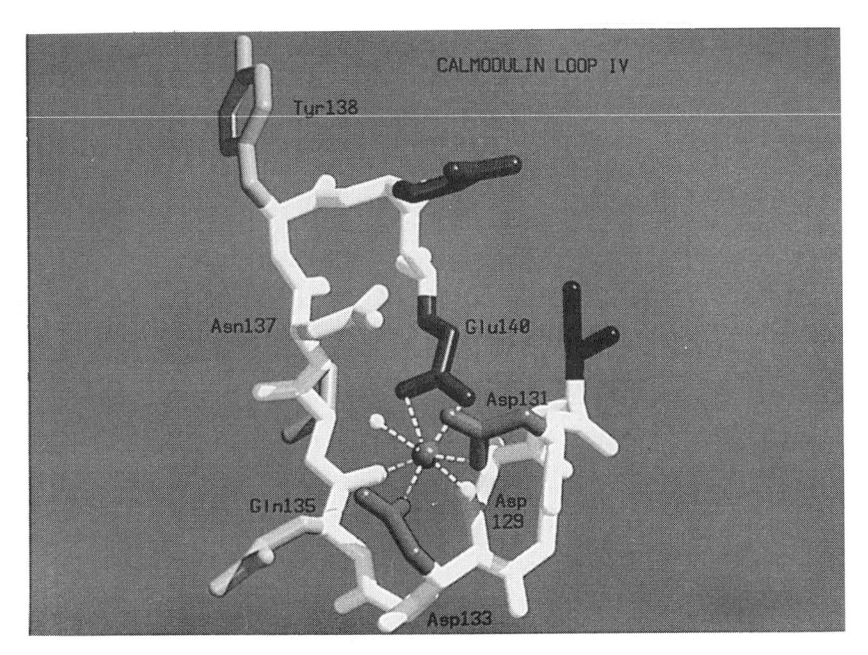

Fig. 8.10 Close-up of a model of the fourth Ca^{2+}-binding loop in calmodulin, showing the atoms which form the octahedral coordination sphere around the Ca^{2+}. Courtesy of Michael James.

	* * * * * *
Troponin I:	asp-ala-asp-gly-gly-gly-asp-ile-ser-val-lys-glu
Troponin II:	asp-glu-asp-gly-ser-gly-thr-ile-asp-phe-glu-glu
Troponin III:	asp-arg-asn-ala-asp-gly-tyr-ile-asp-ala-glu-glu
Troponin IV:	asp-lys-asn-asn-asp-gly-arg-ile-asp-phe-asp-glu
Calmodulin I:	asp-lys-asp-gly-asp-gly-thr-ile-thr-thr-lys-glu
Calmodulin II:	asp-ala-asp-gly-asn-gly-thr-ile-asp-phe-pro-glu
Calmodulin III:	asp-lys-asp-gly-asn-gly-tyr-ile-ser-ala-ala-glu
Calmodulin IV:	asp-ale-asp-gly-asp-gly-gln-val-asn-tyr-glu-glu

Fig. 8.11 Sequences of the four Ca^{2+}-binding loops in troponin C and calmodulin. Residues thought to be involved in Ca^{2+} binding are marked by asterisks; residues with carboxylic acid side chains are underlined.

carboxyl groups which wrap around the ion.

For both troponin C and calmodulin, the adjacent sites in each half of the molecule appear to bind Ca^{2+} with strong positive cooperativity, and the N-terminal pair of binding sites bind Ca^{2+} with lower affinity than the C-terminal pair. In the crystals of troponin C used for crystallography (Figure 8.9), the N-terminal sites are unoccupied, and although all of the helices are still present, the Ca^{2+}-binding loops are in a more open conformation. Although the exact site(s) at which other proteins interact with troponin C and calmodulin are not yet clear, it is thought to involve hydrophobic residues in the helical regions, which are exposed by conformational changes induced by the tight folding of the loops around Ca^{2+}. The sensitivity of the various calmodulin-activated proteins to Ca^{2+} appears to differ, and one explanation for this is that they may interact selectively with either the N-terminal or the C-terminal domain, or both.

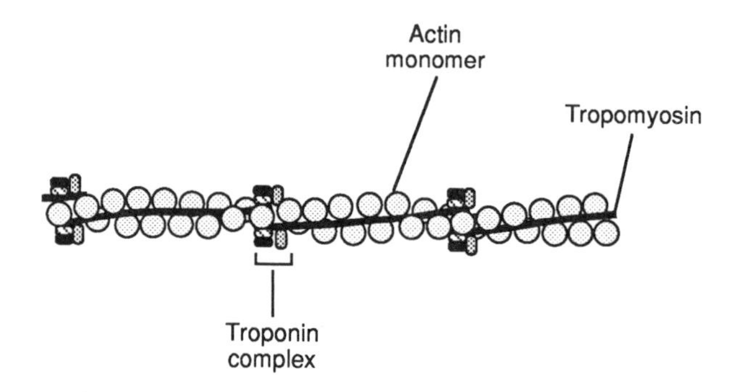

Fig. 8.12 Schematic structure of a thin filament from striated muscle, showing the tight helical array of actin monomers, with associated tropomyosin/troponin complexes.

Role of troponin C in contraction of striated muscles

Skeletal and cardiac muscle fibres are known as **striated** muscles because of the regular arrangement of the thin and thick filaments, giving a striped appearance. Contraction occurs when the **thick filaments**, composed mainly of myosin, slide past the **thin filaments**, which contain actin. This sliding is due to conformational changes of the cross-bridges formed by the myosin heads (see Figure 8.14), and is driven by the hydrolysis of ATP catalysed by the myosin heads, a process activated by the proximity of actin. A fundamental difference between smooth and striated muscle is the mechanism by which this actin–myosin interaction is prevented in the resting state. The major components of the thin filaments are monomers of actin packed into a helical polymer (Figure 8.12). In striated muscles they also contain two accessory proteins, **tropomyosin** and

troponin. Tropomyosin is a long rod-shaped protein which lies in the groove of the actin helix. Troponin molecules occur at regular intervals (one for every seven actins) and consist of three subunits:

1) troponin T, which interacts with tropomyosin;
2) troponin I, which binds to actin and, in association with troponin T and tropomyosin prevents it from interacting with myosin;
3) troponin C, which relieves inhibition by troponin I in the presence of Ca^{2+}.

Troponin C can perhaps be thought of as a specialized isoform of calmodulin which is permanently associated with its protein target, troponin I. As will become clear below, calmodulin also appears to act by relieving the effects of inhibitory sequences within its protein targets.

Activation of the plasma membrane Ca^{2+} pump by calmodulin

As discussed in Section 4.1.2, the plasma membranes of all cells contain a Ca^{2+}-transporting ATPase which helps to maintain the cytosolic Ca^{2+} concentration at a value 10,000 times lower than that outside. This pump has a high affinity but a relatively low capacity, and appears to control the "fine-tuning" of Ca^{2+} concentration. The Ca^{2+}-transporting ATPase from many different cells can be activated by Ca^{2+}/calmodulin: this makes physiological sense because any rise of Ca^{2+} above the basal level of 10^{-7} mol.l^{-1} would activate the pump, which would then exert a feedback effect by increased export of the cation from the cell. Since Ca^{2+} is also a substrate for the enzyme, a similar effect could be achieved by ensuring that the pump was not saturated with Ca^{2+}, but activation by Ca^{2+}/calmodulin would make the system much more sensitive to small rises in the cation.

Like most calmodulin-activated proteins, the Ca^{2+}-ATPase can be purified by Ca^{2+}-mediated binding to calmodulin-Sepharose columns. The enzyme purified from red blood cells is activated 10-fold by Ca^{2+}/calmodulin. It can also be activated by proteolysis, which appears to remove an inhibitory region whose effect is normally relieved by binding of calmodulin.

Calmodulin-stimulated adenylate cyclases

Adenylate cyclase is the enzyme which produces the second messenger cyclic AMP (Section 7.2). The exact number of isoforms of adenylate cyclase in mammalian systems is not yet clear, but at least one form is activated up to 10-fold by calmodulin in the presence of Ca^{2+}. The purified enzyme is a single polypeptide of 120 kDa, and it is clear that it can be activated independently by hormone receptors coupled via Gs (Section 7.2.3) and by Ca^{2+}/calmodulin. Recently cDNA encoding this polypeptide has been cloned and sequenced, and tentative modelling suggests a structure in which three cytoplasmic domains are separated by two clusters of six

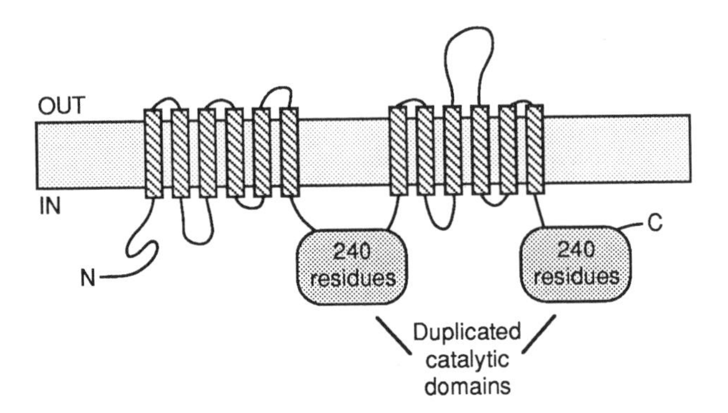

Fig. 8.13 Model for the membrane topology and domain structure of calmodulin-sensitive adenylate cyclase.

transmembrane helices (Figure 8.13). The central and C-terminal cytoplasmic domains are related to each other, and to both soluble guanylate cyclases and the cytoplasmic domains of the atrial natriuretic peptide and resact receptors (Section 6.3.4), and presumably represent duplicated catalytic domains. The site at which calmodulin interacts is not yet clear.

Calmodulin-activated phosphodiesterases

Calmodulin-activated adenylate cyclase provides one of many systems by which first messengers acting through Ca^{2+} and cyclic AMP can interact. Calmodulin activation of cyclic nucleotide phosphodiesterase provides a second. At first sight it might appear paradoxical that Ca^{2+}/calmodulin should activate the enzymes which both synthesize and break down cyclic AMP. However, it is not certain that these two enzymes are expressed in the same cell type. In addition, the calmodulin-activated phosphodiesterase has higher activity with cyclic GMP rather than cyclic AMP as substrate, and it is possible that its true function *in vivo* is the hydrolysis of cyclic GMP.

Calmodulin was discovered by virtue of its ability to activate cyclic nucleotide phosphodiesterase from brain. Cyclic nucleotide phosphodiesterases exist as numerous isoforms, and calmodulin-stimulated types seem to be present in all tissues, although their activity in crude extracts of some tissues is masked by calmodulin-insensitive enzymes. Calmodulin-stimulated phosphodiesterases are generally found to be dimers of subunits of around 60 kDa, which need to bind 2 molecules of calmodulin per dimer for full activity. Like the Ca^{2+}-transporting ATPase, they can also be activated by proteolysis, which generates a constitutively active fragment of about 40 kDa, presumably due to loss of the inhibitory calmodulin-binding region.

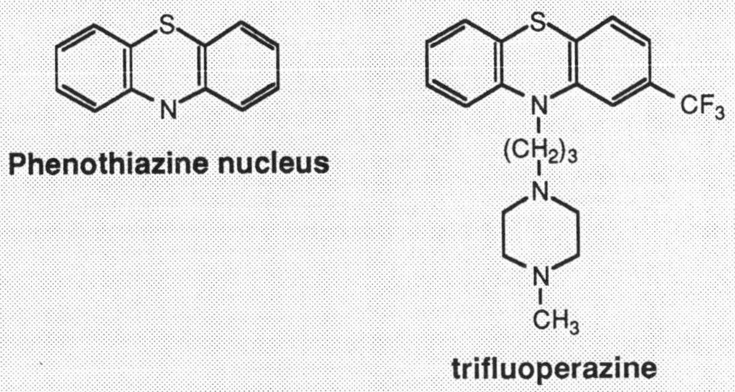

Phenothiazine nucleus

trifluoperazine

The phenothiazine drugs such as trifluoperazine are used to treat various psychotic conditions such as schizophrenia. Trifluoperazine potently antagonizes the activation of calmodulin-dependent enzymes by calmodulin, although it is not clear that this underlies all of the pharmacological effects of the drug.

Box 8.1 Trifluoperazine, a calmodulin antagonist

8.2.2 *Responses to Ca^{2+}: calmodulin-activated protein kinases*

The troponin C- and calmodulin-activated enzymes discussed in the previous section were all regulated by direct binding of Ca^{2+}-binding proteins. However, most effects of Ca^{2+} are mediated more indirectly, via protein phosphorylation catalysed by a family of calmodulin-activated protein kinases, which will be discussed next.

Contraction of smooth muscle – myosin light chain kinase

Smooth muscle is involved in involuntary actions such as peristalsis in the gut, contraction of airways and blood vessels, and changes of size of the iris in the eye. Compared with striated muscle, contraction of smooth muscle occurs slowly, and there is now good evidence that it is triggered by a protein phosphorylation event catalysed by a calmodulin-activated **myosin light chain kinase**. Thus, for example, the toxin **okadaic acid** (see Box 8.4), which inhibits protein phosphatases, triggers contraction of smooth muscle, while calmodulin antagonists such as trifluoperazine (Box 8.1) inhibit it.

The mechanism of contraction in smooth muscle appears to be broadly similar to that in striated muscle (Section 8.2.1), although the organization of the actin and myosin filaments is very different. Myosin in all types of muscle consists of a dimer of "heavy chain" polypeptides which forms a long rod-like tail and a globular head containing the ATP-hydrolysing activity (Figure 8.14). Associated with each head domain are two "light chains" and one of these (the "P" light chain) can be

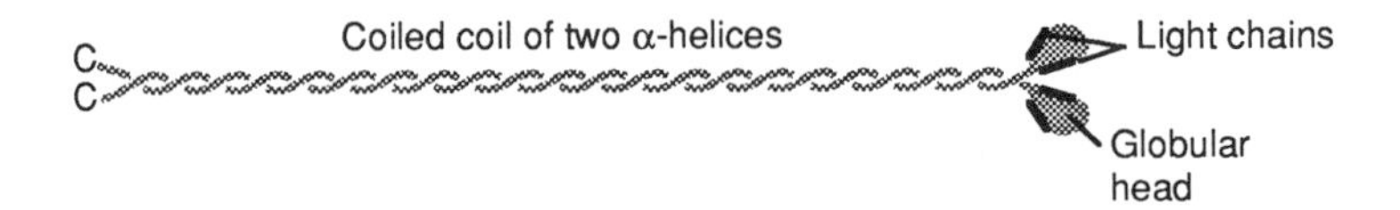

Fig. 8.14 Domain structure of muscle myosin.

phosphorylated by myosin light chain kinase. Experiments in purified, reconstituted systems from smooth muscle show that this phosphorylation is essential before actin will activate the ATPase activity of myosin. There is also an excellent correlation between the phosphorylation of myosin light chain and **initiation** of contraction in smooth muscle, although it appears that phosphorylation is not necessary to maintain the contraction.

The amino acid sequences of smooth muscle myosin light chain kinases suggest that they contain at least three regions (Figure 8.15):

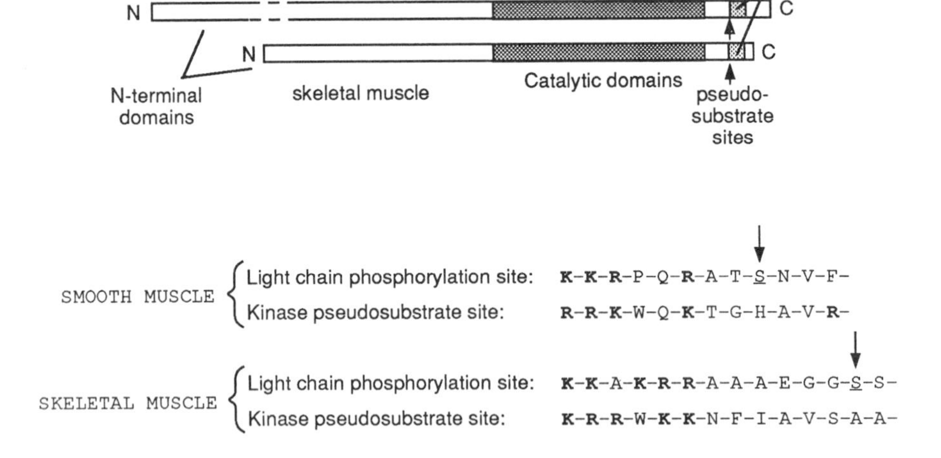

Fig. 8.15 Domain structure of smooth and skeletal muscle myosin light chain kinases (top), and comparison of the sequences around the phosphorylated sites (arrowed) on the myosin light chains, and the "pseudosubstrate sites" on the protein kinases. The one-letter code is used to represent amino acids (R = arginine, K = lysine, S = serine; see Appendix for the remainder). The N-terminal domain of the smooth muscle enzyme is very large and is not shown to scale.

1) a large N-terminal region whose functions are unknown;
2) a central domain which is homologous with the catalytic domains of other protein kinases;
3) a C-terminal regulatory domain containing the calmodulin-binding site. A 27 residue hydrophobic peptide derived from this region competitively inhibits binding of the kinase to Ca^{2+}/calmodulin.

Unlike many other protein kinases, smooth muscle myosin light chain kinase is extremely specific for its substrate. The single serine residue phosphorylated on the light chain has a cluster of basic residues (lys and arg) on the N-terminal side, and experiments using synthetic peptide substrates based on this sequence have shown that these are essential for recognition. Intriguingly, just N-terminal to, and overlapping, the calmodulin-binding region, myosin light chain kinase contains a sequence with the same cluster of basic residues, except that the serine is replaced by a non-phosphorylatable histidine residue (Figure 8.15). It has been suggested that this is a "pseudosubstrate site" which competitively inhibits the catalytic site in the absence of calmodulin. Binding of calmodulin adjacent to this site would displace this inhibitory region, and expose the catalytic site for phosphorylation of light chains. The evidence for this model, which has general applicability for second-messenger stimulated protein kinases, is now rather strong:

1) a synthetic peptide based on the pseudosubstrate sequence is a potent competitive inhibitor of the kinase;
2) a 64 kDa fragment generated from the kinase by trypsin, which has lost most of the calmodulin-binding site but retains the pseudosubstrate sequence (Figure 8.15), is **inactive** in the presence or absence of calmodulin;
3) a 61 kDa tryptic fragment which has also lost the pseudosubstrate site (Figure 8.15) is **active** in the presence or absence of calmodulin.

Skeletal muscle contains a distinct isoform of myosin light chain kinase with a shorter N-terminal region (Figure 8.15), which appears to regulate the **strength** of contraction, although it is not concerned with initiation of contraction, which is controlled by troponin C. The sequences phosphorylated on skeletal and smooth muscle light chains by the two kinases are somewhat different, and the distribution of basic residues at the pseudosubstrate sites reflect this difference (Figure 8.15). This provides excellent support for the pseudosubstrate concept.

Regulation of glycogen breakdown – phosphorylase kinase

Phosphorylase kinase occupies a special place in that it was the first protein kinase to be discovered, by Krebs and Fischer in 1956. Skeletal muscle, particularly in the **fast twitch** fibres used during vigorous exercise such as sprinting, relies heavily on breakdown of stored glycogen and anaerobic glycolysis to provide ATP for contraction. The enzyme phosphorylase, which catalyses the initial breakdown

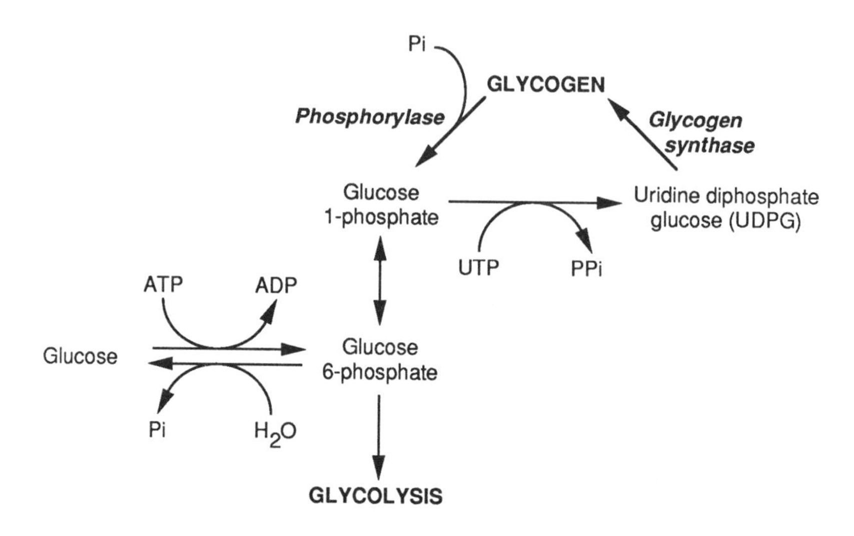

Fig. 8.16 Pathways of glycogen metabolism.

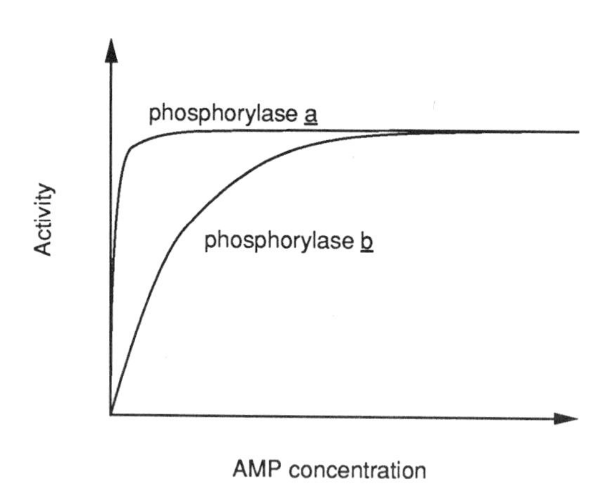

Fig. 8.17 Regulation of phosphorylase a and b by AMP.

of glycogen (Figure 8.16) can be converted from the b form, which requires high concentrations of the allosteric activator AMP for activity, to the a form, which is much less dependent on AMP (Figure 8.17). This conversion occurs when the muscle is stimulated, either via the motor nerve, or via adrenaline (epinephrine) (which would cause rises in cytosolic Ca^{2+} or cyclic AMP respectively). AMP tends to rise during sustained contraction due to the action of adenylate kinase (Box 8.2). However conversion of phosphorylase to the a form makes it active at very low AMP, and thus glycogen can be broken down in the muscle before ATP levels

1) The enzyme adenylate kinase, which is particularly active in skeletal muscle, maintains the concentrations of ATP, ADP and AMP at close to their equilibrium ratio of about 1:

$$2ADP \leftrightarrow ATP + AMP$$

$$K_{eq} = \frac{[AMP][ATP]}{[ADP]^2} \cong 1$$

$$\therefore\ [AMP] \cong \frac{[ADP]^2}{[ATP]}$$

2) In fully energized cells, the respiratory chain maintains the concentrations of ATP and ADP at about 4 mM and 0.4 mM (or 10:1, which is about 10 orders of magnitude away from the equilibrium ratio of the reaction ATP $\leftrightarrow$ ADP + Pi). If the adenylate kinase reaction is at equilibrium, [AMP] will therefore be approximately equal to $0.4^2 \div 4 = 0.04$ mM.

3) If ATP was to be broken down to ADP due to muscle contraction under anaerobic conditions, such that [ATP] = [ADP] $\cong$ 2 mM, and adenylate kinase maintained its equilibrium ratio, then [AMP] would become $2^2 \div 2 = 2$ mM. Hence a 2-fold drop in ATP would produce a *50-fold* rise in [AMP].

Box. 8.2 AMP is a sensitive indicator of cellular energy status

start to drop and AMP to rise. By this mechanism the cell is effectively **anticipating** the drop in ATP that would be produced by vigorous exercise.

Krebs and Fischer observed that the phosphorylase b to a conversion occurred in a crude muscle extract when it was passed through filter paper, a somewhat bizarre finding which was traced to the presence of Ca^{2+} in the filter paper. They went on to show that the conversion involved phosphorylation of the protein, and were able to detect a Ca^{2+}-dependent enzyme, **phosphorylase kinase**, that catalysed the transfer of phosphate from ATP to phosphorylase. Krebs and Fischer had thus discovered a common control element, Ca^{2+} ions, which could activate glycogen breakdown as well as muscle contraction.

Phosphorylase kinase is now known to be a large and complex enzyme containing four different subunits, with the native enzyme being $\alpha_4\beta_4\gamma_4\delta_4$. The γ subunit (45 kDa) has been isolated and shown to be catalytically active in a constitutive manner (i.e. in the presence or absence of Ca^{2+}). The δ subunit proved to be identical with calmodulin, and confers the Ca^{2+} dependence on the enzyme (Figure 8.18). Phosphorylase kinase is unusual among calmodulin-regulated enzymes in that the Ca^{2+}-binding protein remains bound stoichiometrically even in the absence of Ca^{2+}. The α (140 kDa) and β (130 kDa) subunits are homologous with one another: their functions are not completely clear,

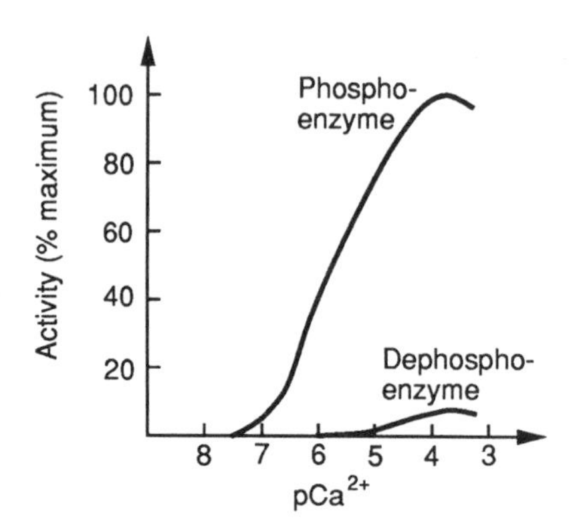

Fig. 8.18 Ca^{2+} dependence of phosphorylase kinase, before and after phosphorylation by cyclic AMP-dependent protein kinase. pCa is analogous to pH, and is equal to $-\log_{10}[Ca^{2+}]$. A pCa of 6 is therefore a Ca^{2+} concentration of 1 $\mu mol.l^{-1}$.

but one or both subunits presumably inhibit the catalytic activity of the γ subunit unless Ca^{2+} is bound to the δ subunit. Both α and β are phosphorylated at one major site by cyclic AMP-dependent protein kinase (Section 8.2.2), an event which occurs in response to adrenaline (epinephrine) in intact muscle. This dramatically stimulates phosphorylase kinase (Figure 8.18), and by this mechanism there would be a very rapid breakdown of glycogen when the muscle contracted – a classical component of the "fight or flight" adrenaline response.

In addition to activation by the endogenous calmodulin (δ subunit), the dephosphorylated form of phosphorylase kinase is also activated by exogenous calmodulin (Figure 8.19). At higher concentrations troponin C produces the same effect, and due to the high abundance of this protein in skeletal muscle, this may be the natural activator. Heart muscle and **slow twitch** skeletal muscle fibres contain an isoform of phosphorylase kinase with a slightly smaller α subunit (α'), and these isoforms are not activated by exogenous troponin C or calmodulin. Possibly these more slowly contracting muscle types, which utilize aerobic rather than anaerobic metabolism, do not require such rapid glycogen breakdown to satisfy their demands for ATP.

Phosphorylase kinase has a very restricted substrate specificity, although in cell-free assays it does phosphorylate glycogen synthase, the enzyme which regulates glycogen synthesis, at a site which causes inactivation of the enzyme (see Figure 8.26). This effect would simultaneously switch off glycogen synthesis while phosphorylation of phosphorylase activated breakdown. However it has not yet been demonstrated that phosphorylation of glycogen synthase occurs during muscle contraction *in vivo*.

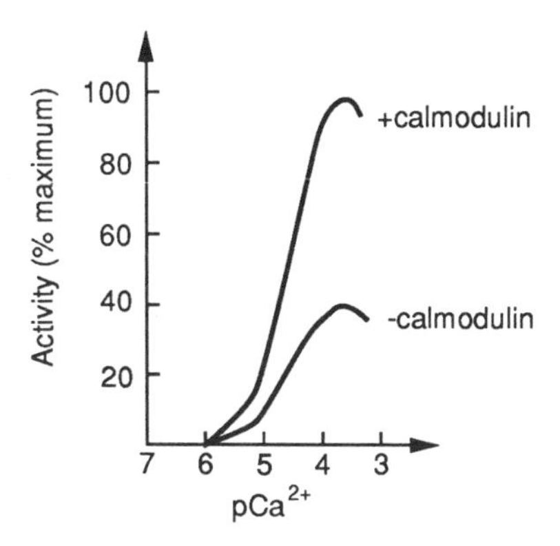

Fig. 8.19 Activation of the dephosphorylated form of phosphorylase kinase by exogenous calmodulin.

Elongation factor-2 kinase

This protein kinase was originally described in brain and termed calmodulin-dependent protein kinase III. It was defined by its ability to phosphorylate a 100 kDa polypeptide in a Ca^{2+}/calmodulin-dependent manner. Subsequently it was realized that both the kinase and the substrate had a ubiquitous distribution, and the substrate was identified as elongation factor-2, a protein involved in the elongation reactions of protein synthesis. Phosphorylation of elongation factor-2 completely abolishes protein synthesis, and dephosphorylation reverses this. The physiological significance of this intriguing effect is not clear, but it has been shown that Ca^{2+}-mobilizing hormones do cause a temporary halt in protein synthesis. Possibly this conserves ATP for other energy-requiring processes that are activated by the increase in Ca^{2+}, or enables the cell to switch to translation of new mRNAs induced by the extracellular signal.

A multisubstrate Ca^{2+}-dependent protein kinase – calmodulin-dependent multiprotein kinase

All of the other calmodulin-dependent protein kinases described in this section are highly specific for their protein substrate. However, there is at least one broad specificity calmodulin-dependent protein kinase which appears to occupy an analogous place in cellular regulation, with respect to Ca^{2+}, as does the cyclic AMP-dependent protein kinase (Section 8.2.4) with respect to cyclic AMP. The kinase was discovered simultaneously by a number of laboratories using different substrates, and consequently received many different names. It is now most widely known as calmodulin-dependent protein kinase II, but the more informative name of calmodulin-dependent multiprotein kinase is preferred here.

The kinase is apparently ubiquitous in all animal cells from *Drosophila* to

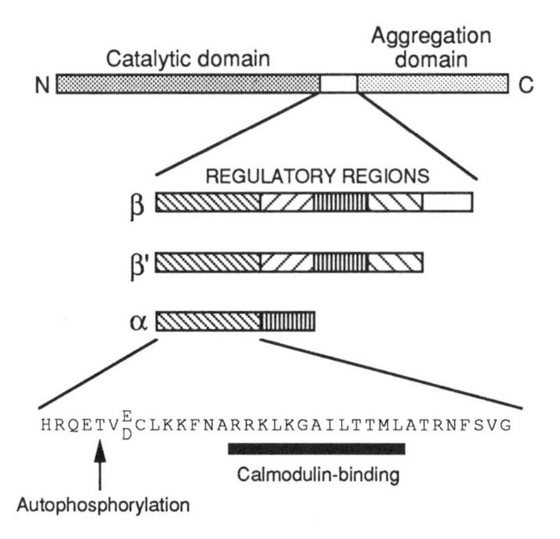

Fig. 8.20 Domain structure of the calmodulin-dependent multiprotein kinase. Only the regulatory regions are shown in detail: the β/β' isoforms contain two regions (light diagonal hatching) not present in the α isoform, while β contains an additional segment compared with β' due to alternative splicing. The sequences around the calmodulin-binding site and the primary auto-phosphorylation site are identical (apart from one conservative replacement) in all three isoforms. See Appendix for the single letter amino acid code.

humans, but is particularly enriched in neural tissue, representing no less than 2% of total protein in the hippocampus of rat brain. The kinase protein comprises 12 identical subunits, probably arranged as a stack of two hexameric discs, with a molecular mass of 600–700 kDa. In mammalian brain and other tissues the subunits exist as three isoforms, i.e. α (50 kDa) and β/β' (58/60 kDa), with the latter two being produced by alternative splicing from the same gene. Sequence analysis and proteolytic studies suggest a three domain structure for the subunit (Figure 8.20):

1) The N-terminal region is homologous with catalytic regions of other protein kinases and is clearly the catalytic domain.
2) The central region has the regulatory functions and is the most variable between the three isoforms. It contains a calmodulin-binding site and, just N-terminal to it, a region which is homologous with the sequences around several sites phosphorylated by the kinase (Figure 8.21). This is reminiscent of the inhibitory pseudosubstrate sequences on the myosin light chain kinases, except that in this case the threonine residue is actually phosphorylated, i.e. it is an **autophosphorylation site** rather than a pseudosubstrate site. Intriguingly, autophosphorylation of this site appears to relieve the inhibition, converting the

	↓
Synapsin-1 (site 2):	-gln-ala-thr-arg-gln-ala-**ser**-ile-ser-gly-
Synapsin-1 (site 3):	-gly-pro-ile-arg-gln-ala-**ser**-gln-ala-gly-
Glycogen synthase:	-pro-leu-ser-arg-thr-leu-**ser**-val-ser-ser-
Tyrosine hydroxylase:	-gly-phe-arg-arg-ala-val-**ser**-glu-gln-asp-
Cardiac phospholamban:	-ala-ile-arg-arg-ala-ser-**thr**-ile-glu-met-
β autophosphorylation site:	-met-met-his-arg-gln-glu-**thr**-val-glu-cys-
α autophosphorylation site:	-cys-met-his-arg-gln-glu-**thr**-val-glu-cys-

Fig. 8.21 Comparison of the sequences around the primary autophosphorylation site on the calmodulin-dependent multiprotein kinase, and of sites phosphorylated on other protein substrates. The phosphorylated serine/threonine is marked with an arrow, while arginine residues are underlined.

enzyme into a calmodulin-independent form. This implies that the enzyme could remain active even after the Ca^{2+} transient had returned to basal levels. However the situation is complicated, because autophosphorylation at additional sites, catalysed by the calmodulin-independent form, leads to subsequent inactivation of the enzyme.

3) The C-terminal region appears to be necessary for aggregation into the dodecamer, in that its removal by proteolysis converts the enzyme into a monomeric form which still displays calmodulin-dependent activity. The different isoforms appear to be able to form hybrid dodecamers.

As its name suggests, the calmodulin-dependent multiprotein kinase appears to have a broad specificity and phosphorylates many proteins in purified cell-free systems. Studies with synthetic peptide substrates suggest that the minimal consensus sequence is -arg-X-X-ser/thr-, where ser/thr represents the site phosphorylated (Figure 8.21). Full discussion of all substrate proteins is beyond the scope of this book, but I will briefly discuss some examples where there is some evidence that the phosphorylation is a physiologically important event:

1) *Neurotransmitter release*

One of the best substrates for the kinase in cell-free assays is synapsin-1. It is a neurone-specific protein, now known to comprise two isoforms, which is found to be associated with the special synaptic vesicles containing "fast acting neurotransmitters" (Section 5.3.3). Synapsin-1 is phosphorylated at two sites (sites 2 and 3) in the C-terminal region by calmodulin-dependent multiprotein kinase, and this has been shown to cause dissociation from the synaptic vesicles. The initial release of neurotransmitter when the nerve is stimulated appears to be too rapid (<10 msec) for phosphorylation of synapsin-1 to be a causal event. One possibility is that synapsin-1 mediates binding to the cytoskeleton of a back-up reservoir of secretory vesicles, and that Ca^{2+}-mediated phosphorylation causes release of these vesicles so that they can be exocytosed. It would thus regulate the amount of neurotransmitter released in response to **prolonged** stimulation. In support of this, microinjection of dephospho-synapsin-1 into nerve terminals of squid giant axons decreases, while injection of the kinase increases, the rate of rise and amplitude of the post-synaptic potentials. There is also good evidence that synapsin-1 is phosphorylated in response to stimulation both in isolated nerves and *in vivo*.

2) *Hormone and neurotransmitter biosynthesis*
The regulatory steps in the synthesis of the catecholamines and serotonin (Sections 3.1.2 and 3.1.3) are believed to be catalysed by tyrosine and tryptophan hydroxylases respectively. Both enzymes are phosphorylated in cell-free assays by the calmodulin-dependent multiprotein kinase, and this is associated with activation, although only in the presence of an acidic "activator protein". It has been known for some time that depolarization of relevant neurones or chromaffin cells, which increases cytosolic Ca^{2+} via voltage-gated Ca^{2+} channels, not only causes release of catecholamines or serotonin, but stimulates their synthesis. In the case of tyrosine hydroxylase there is now good evidence that depolarization induces phosphorylation at the major site (Figure 8.21) phosphorylated by the calmodulin-dependent multiprotein kinase.

3) *Ca^{2+} sequestration in heart muscle*
The kinase also phosphorylates phospholamban, which regulates Ca^{2+} uptake into sarcoplasmic reticulum in heart muscle. This protein is also a substrate for cyclic AMP-dependent protein kinase, and is discussed in more detail in Section 8.2.4.

8.2.3 *Effect of elevated Ca^{2+} on mitochondrial enzymes*

As discussed in Section 4.1.2, mitochondria take up Ca^{2+} by an electrogenic uniport system which is driven by the proton electrochemical gradient, and this is opposed by an electroneutral efflux system. There are at least three mitochondrial matrix enzymes which are activated by Ca^{2+} in the range 1 to 10 $\mu mol.l^{-1}$ (Figure 8.22). These are:

1) Pyruvate dehydrogenase phosphatase. This enzyme antagonizes the phosphorylation and inactivation of the pyruvate dehydrogenase complex by a tightly bound protein kinase, and thus converts pyruvate dehydrogenase into its active form.
2) Isocitrate dehydrogenase and 2-oxoglutarate dehydrogenase, two of the NAD-linked dehydrogenases of the tricarboxylic acid cycle.

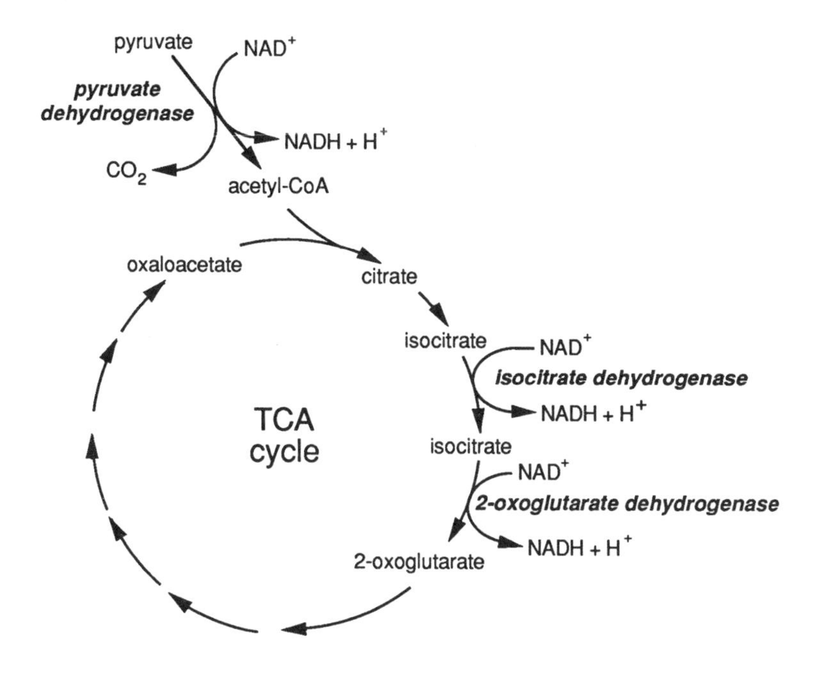

Fig. 8.22 Metabolic roles of Ca^{2+}-activated enzymes in mitochondria.

The molecular mechanism of these effects of Ca^{2+} is not known, although it does not involve calmodulin. There is now good evidence that changes in cytosolic Ca^{2+} within the physiological range (0.1 to 1 $\mu mol.l^{-1}$) do produce changes in intramitochondrial Ca^{2+} in the slightly higher range which activates these three enzymes. Many of the situations in which Ca^{2+} is elevated by first messengers (e.g. the effect of adrenaline on heart muscle, discussed in the next Section) are associated with increased demand for ATP, and the activation of these three dehydrogenases could help to meet that demand by stimulating the TCA cycle.

8.2.4 *Responses to cyclic nucleotides – cyclic AMP- and cyclic GMP-dependent protein kinases*

In the late 1960s Krebs and coworkers noticed that preparations of phosphorylase kinase from skeletal muscle were activated to various extents by addition of cyclic AMP. Eventually it was realized that this was due to binding of cyclic AMP to a contaminating activity, now termed **cyclic AMP-dependent protein kinase**, which phosphorylated and activated phosphorylase kinase. The key to this was the discovery by Walsh of a specific protein inhibitor of cyclic AMP-dependent protein kinase, which blocked the cyclic AMP stimulation of phosphorylase kinase, but not

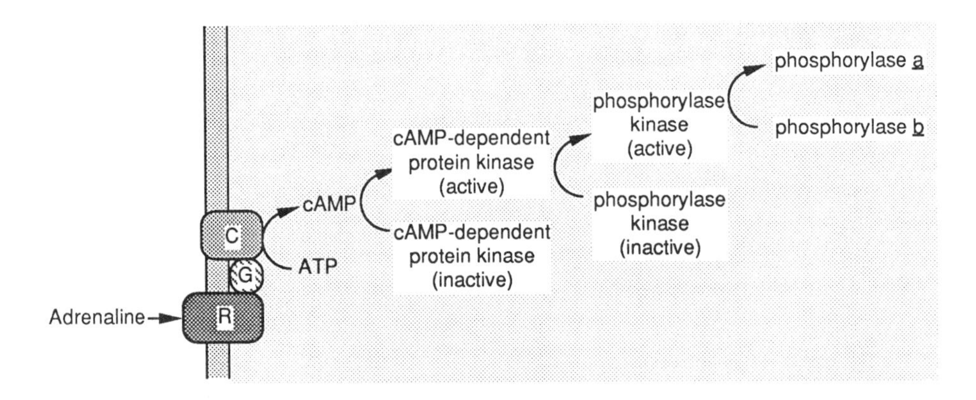

Fig. 8.23 The cascade of events leading from the β-adrenergic receptor to activation of glycogen breakdown in skeletal muscle. Key: R, the receptor; G, the stimulatory G protein; C, the catalytic unit of adenylate cyclase.

the basal activity. Thus the missing link in the chain of events from binding of adrenaline on the muscle membrane to activation of glycogen breakdown had been found (Figure 8.23). Since the discovery of the cyclic AMP-dependent protein kinase over 20 years ago, no other high affinity binding proteins for cyclic AMP (other than the phosphodiesterases which hydrolyse it) have been found in eukaryotic cells. This is an important point because it suggests that all of the multitude of effects of cyclic AMP are mediated by protein phosphorylation.

Models for the domain structures of cyclic AMP-dependent protein kinases, based on the amino acid sequence and other biochemical studies, are shown in Figure 8.24. The enzyme exists as two major isoforms, Type I and Type II, each of which consists of apparently identical catalytic C subunits complexed with different regulatory subunits (RI and RII). As so often happens, cDNA cloning has now shown that this is an oversimplification, with at least two isoforms being reported for each of RI, RII and C. In all cases the native holoenzymes are inactive R_2C_2 complexes, but binding of cyclic AMP to the regulatory subunits causes dissociation to release active catalytic subunits. The regulatory subunits contain four regions or domains:

1) An N-terminal "dimerization domain" that seems to provide the main contacts holding the R_2 dimer together.
2) A "hinge region" which interacts with and inhibits the catalytic subunit in the absence of cyclic AMP. Figure 8.24 shows that in the case of RI this region contains a pseudosubstrate site (analogous to myosin light chain kinase), whereas in RII it is an autophosphorylation site (analogous to calmodulin-dependent multiprotein kinase). Autophosphorylation of the RII subunit inhibits reassociation with the catalytic subunit, which could represent a

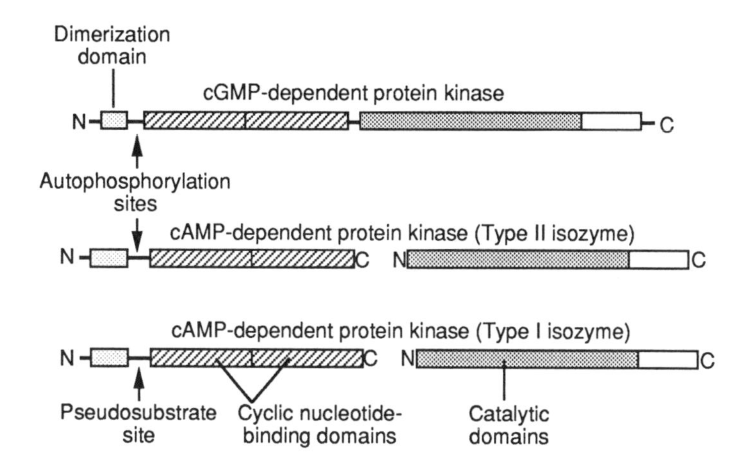

Fig. 8.24 Models for the domain structures of cyclic AMP- and cyclic GMP-dependent protein kinases. Homologous regions are shown by similar shading or hatching.

mechanism for prolonging the response even after cyclic AMP has been removed. This is analogous with the initial effect of autophosphorylation on calmodulin-dependent multiprotein kinase.

3) Two tandem cyclic AMP-binding domains, each of which is homologous with the catabolite gene activator protein of *Escherichia coli* (a transcriptional activator protein that mediates responses to cyclic AMP in that bacterium). Cyclic AMP binds to the two sites on the R subunits of the kinase with a high degree of positive cooperativity (see Box 7.1). This means that a large change in kinase activity will result from a small change in cyclic AMP: just the response that is needed if the system is to act as a molecular "switch".

Cyclic GMP also appears to act almost exclusively through a protein kinase, although there are cyclic GMP-gated ion channels in the retina (Section 7.2.4). Purification of cyclic GMP-dependent protein kinase showed that it was a dimer of identical subunits. At first sight this is very different from its cyclic AMP-dependent cousin, but amino acid sequencing revealed that the two kinases were in fact highly homologous, and that the single subunit of the cyclic GMP-dependent enzyme had probably arisen by fusion of genes encoding separate catalytic and regulatory subunits (Figure 8.24).

Cyclic AMP-dependent protein kinase has been shown to phosphorylate many sites on numerous different proteins in cell-free assays. Based on sequencing of these sites and studies with synthetic peptide substrates, the minimal requirement for significant phosphorylation by the kinase appears to be a pair of basic residues on the N-terminal side of the phosphorylated residue (Figure 8.25). In one case

	↓
Phosphorylase kinase α subunit:	-val-glu-phe-arg-arg-leu-**ser**-ile-
Phosphorylase kinase β subunit:	-arg-thr-lys-arg-ser-asn-**ser**-val-
Glycogen synthase site 1A:	-gln-trp-pro-arg-arg-ala-**ser**-cys-
Glycogen synthase site 1b:	-gly-ser-lys-arg-ser-asn-**ser**-val-
Glycogen synthase site 2:	-pro-leu-ser-arg-thr-leu-**ser**-val-
Pyruvate kinase (L type):	-gly-tyr-leu-arg-arg-ala-**ser**-val-
6-Phosphofructo-2-kinase:	-leu-gln-arg-arg-arg-gly-**ser**-ser-
Hormone-sensitive lipase:	-pro-met-arg-arg-**ser**-val-
Tyrosine hydroxylase:	-phe-ile-gly-arg-arg-gln-**ser**-leu-
Cardiac phospholamban:	-ala-ile-arg-arg-ala-**ser**-thr-
Protein phosphatase inhibitor-1:	-ile-arg-arg-arg-arg-pro-**thr**-pro-
RI subunit pseudosubstrate site:	-gly-arg-arg-arg-arg-gly-**ala**-ile-
RII subunit autophosphorylation site:	-arg-phe-asp-arg-arg-val-**ser**-val-

Fig. 8.25 Comparison of the sequences around the pseudosubstrate and autophosphorylation sites on the regulatory subunits of Type I and Type II cyclic AMP-dependent protein kinase, and of sites phosphorylated on other protein substrates. Phosphorylated residues and pseudosubstrate sites are shown in bold and marked with an arrow. Basic residues are underlined.

(protein phosphatase inhibitor-1) the kinase phosphorylates a threonine residue. The substrate specificity of cyclic GMP-dependent protein kinase appears to be qualitatively similar to that of the cyclic AMP-dependent enzyme, although there are quantitative differences. Cyclic GMP-dependent protein kinase has a much more restricted tissue distribution than the cyclic AMP-dependent enzyme, and it seems likely that it will play specialized roles in the cell types where it is found, such as smooth muscle (Section 7.2.7) and certain cell types in the brain. As yet there is no case in which a physiological substrate of cyclic GMP-dependent protein kinase with a known function has been established. By contrast, a very large number of physiological substrates for cyclic AMP-dependent protein kinase have been characterized. An exhaustive discussion of these is beyond the scope of this book, but I will discuss several examples which should illustrate the diverse nature of cyclic AMP-mediated responses:

1) *Glycogen metabolism*

The cascade from cyclic AMP-dependent protein kinase to phosphorylase kinase to phosphorylase was the first protein phosphorylation network to be elucidated, and has already been discussed (Section 8.2.2). Phosphorylation of phosphorylase

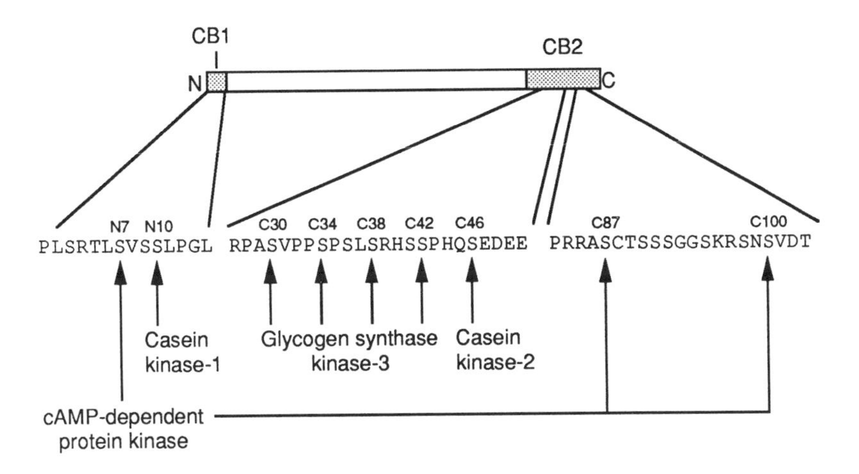

Fig. 8.26 Location of phosphorylation sites in skeletal muscle glycogen synthase. The phosphorylation sites are numbered according to their locations in the N-terminal or C-terminal cyanogen bromide peptides (CB1 and CB2). The one-letter code is used to represent amino acids (S = serine; see Appendix for the remainder).

kinase by cyclic AMP-dependent protein kinase greatly stimulates the maximal activity as well as increasing the sensitivity of the enzyme to Ca^{2+} (Figure 8.18). This means that a partial conversion of phosphorylase b to a can occur even in resting muscle, and would also ensure that a very rapid conversion to the a form occurred as soon as muscle contraction was initiated. In this way stimulation by adrenaline can anticipate the demand for ATP and prepare the muscle for rapid action.

Cyclic AMP-dependent protein kinase also phosphorylates skeletal muscle glycogen synthase, the regulatory enzyme of glycogen synthesis, at three sites (Figure 8.26). Phosphorylation at one of these sites (N7) increases the dependence of the enzyme on the allosteric activator, glucose-6-phosphate (Figure 8.27), such that it will only synthesize glycogen in the presence of high concentrations of the metabolite. At first sight this would appear to provide an explanation for the inhibition of glycogen synthesis (simultaneous with activation of glycogen breakdown) by adrenaline (epinephrine). However, studies of the sites phosphorylated on glycogen synthase *in vivo* show that although N7 is phosphorylated, paradoxically this hardly increases in response to adrenaline. The major effect of adrenaline is to increase phosphorylation at other sites, i.e. N10 phosphorylated by casein kinase-1, and C30, C34, C38 and C42 phosphorylated by glycogen synthase kinase-3 (Figure 8.26). Phosphorylation at these latter sites produces a much more dramatic inactivation of the enzyme (Figure 8.27). The mechanism for adrenaline-induced phosphorylation at these other sites involves

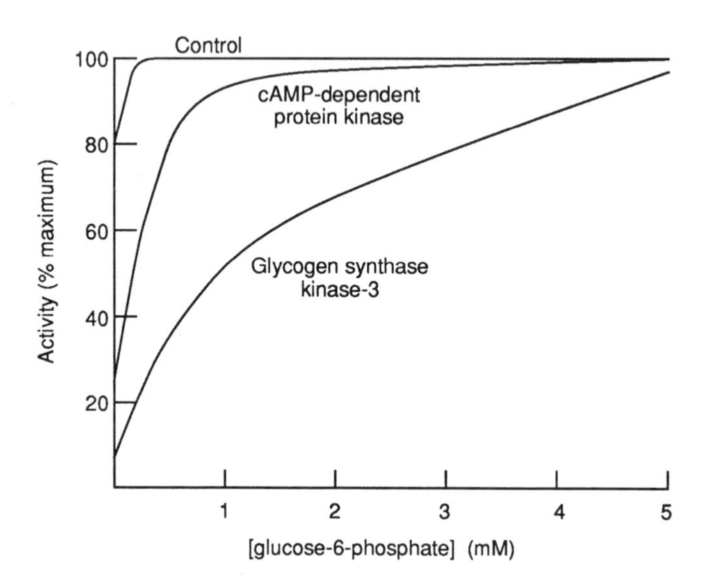

Fig. 8.27 Kinetic properties of dephosphorylated glycogen synthase (control), and after phosphorylation by cyclic AMP-dependent protein kinase or glycogen synthase kinase-3.

regulation of protein phosphatase-1 by cyclic AMP-dependent protein kinase (Section 8.4.1).

Muscle glycogen is used as a store of energy for muscle contraction, and when it is broken down the glucose-6-phosphate produced is channelled into glycolysis for production of ATP. By contrast, liver glycogen is used as a store of glucose for the rest of the body and is mobilized during fasting. Although different isoforms of phosphorylase, phosphorylase kinase and glycogen synthase exist in liver, the regulation of glycogen metabolism is broadly similar to that described above. Glucagon released during starvation increases cyclic AMP in liver (as can adrenaline if β-receptors are present), while adrenaline can also act through α_1-adrenergic receptors to release inositol trisphosphate, and hence Ca^{2+}. Both effects would of course switch on phosphorylase kinase and activate glycogen breakdown. A important difference between muscle and liver is that the latter contains the enzyme glucose-6-phosphatase (Figure 8.28), so that glucose-6-phosphate derived from glycogen breakdown is channelled to blood glucose rather than into glycolysis.

2) *Regulation of glycolysis and gluconeogenesis*

As well as releasing glucose from glycogen during fasting, the liver synthesizes it from non-carbohydrate precursors (lactate and some amino acids) by the pathway

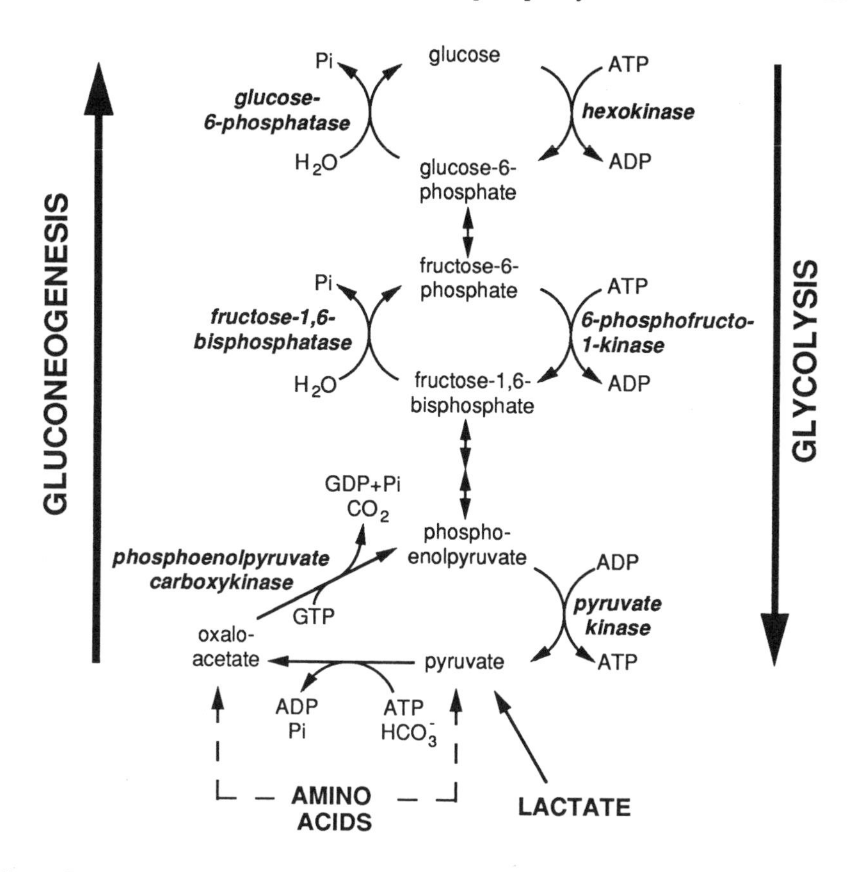

Fig. 8.28 Pathways of glycolysis and gluconeogenesis, showing sites of regulation.

of gluconeogenesis (Figure 8.28). Since this is essentially the reverse of glycolysis, it is obviously essential that liver glycolysis is switched off during starvation. The two important regulatory enzymes in glycolysis are 6-phosphofructo-1-kinase and pyruvate kinase, which catalyse steps involving different enzymes in the gluconeogenetic direction (Figure 8.28). Glucagon released during starvation activates cyclic AMP-dependent protein kinase, which phosphorylates and inactivates pyruvate kinase at a site near the N-terminus (see Figure 8.25). This does not change the V_{max} of pyruvate kinase, but increases the K_m for the substrate, phosphoenolpyruvate, makes it more sensitive to allosteric inhibitors (ATP and alanine), and less sensitive to the allosteric activator (fructose-1,6-bisphosphate).

Cyclic AMP-dependent protein kinase also inactivates 6-phosphofructo-1-kinase, but not by a simple phosphorylation as might have been expected. Instead the kinase phosphorylates and inactivates 6-phosphofructo-2-kinase, an enzyme

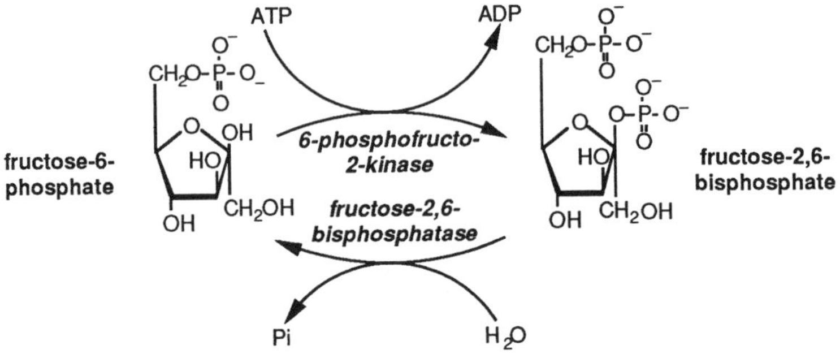

Fig. 8.29 Fructose-2,6-bisphosphate, and mechanism of its formation and breakdown.

catalysing a related reaction, the synthesis of fructose-2,6-bisphosphate from fructose-6-phosphate (Figure 8.29). Intriguingly, 6-phosphofructo-2-kinase is a bifunctional enzyme which carries a second activity, fructose-2,6-bisphosphatase, catalyzing the reverse reaction. Phosphorylation of this bifunctional enzyme activates its phosphatase activity and inactivates its kinase activity, leading to a dramatic drop in the concentration of fructose-2,6-bisphosphate.

Fructose-2,6-bisphosphate is not an ordinary metabolite in the sense that its only fate is to be broken down to fructose-6-phosphate again (Figure 8.29). It appears to be a pure signal molecule (or, perhaps, a third messenger!), being a potent activator of the glycolytic enzyme 6-phosphofructo-1-kinase, and an inhibitor of the enzyme which catalyses the reverse step in gluconeogenesis, fructose-1,6-bisphosphatase. The overall effect of these intriguing mechanisms is that glucagon, released during starvation, causes a dramatic drop in fructose-2,6-bisphosphate, and the liver switches from net glycolysis to net gluconeogenesis.

Since glycolysis needs, if anything, to be elevated in skeletal muscle in response to adrenaline (epinephrine), it would make no sense if the same mechanisms operated in muscle. Both pyruvate kinase and the bifunctional kinase/phosphatase exist in muscle, but as different isoforms from which the cyclic AMP-dependent protein kinase sites appear to have been deleted.

3) *Regulation of triacylglycerol/cholesterol ester storage*

Liver glycogen only provides a sufficient reserve of glucose for 1–2 days of starvation. A major long-term energy reserve in the body is in the form of the fatty acids in triacylglycerol, stored mainly in adipose tissue. Adipose tissue contains a **hormone-sensitive lipase** which breaks down triacylglycerols to free fatty acids, which can be oxidized for ATP production by most tissues. The lipase is almost inactive unless it is phosphorylated at a single site (Figure 8.25) by cyclic AMP-dependent protein kinase. Stimulation of β-adrenergic receptors by catecholamines, and/or stimulation of glucagon receptors, is the mechanism by which fatty acids are mobilized from adipose tissue during starvation. It has recently been demonstrated that hormone-sensitive lipase is also the major intracellular cholesterol esterase in most cells. Thus elevation of cyclic AMP leads to breakdown of stored cholesterol esters to release free cholesterol. This happens, for example, when adrenal cortex cells are stimulated by adrenocorticotropin. Since cholesterol is the precursor for steroid hormone synthesis (Section 3.2.2), this explains, at least in part, how adrenocorticotropin stimulates the latter pathway.

4) *Amino acid breakdown*

During prolonged starvation, amino acids derived from muscle protein breakdown become an important source of energy. Phenylalanine breakdown occurs in liver, and the enzyme which regulates this pathway, phenylalanine hydroxylase, is phosphorylated and activated by cyclic AMP-dependent protein kinase.

5) *Catecholamine biosynthesis*

I have already discussed how calmodulin-dependent multiprotein kinase can activate tyrosine hydroxylase, the enzyme which regulates catecholamine biosynthesis in neurones and the chromaffin cells of the adrenal medulla. Cyclic AMP-dependent protein kinase phosphorylates and activates the enzyme at a different site (Figure 8.25). This causes a dramatic activation, and can explain how certain neuropeptides (e.g. vasoactive intestinal peptide) which are coupled to adenylate cyclase can stimulate catecholamine biosynthesis.

6) *Contractility of the heart*

The effect of adrenaline (mediated by β_1-receptors) in increasing both the strength and rate of contraction of heart muscle is well-known, to the extent that the effect of proximity of the opposite sex on "heart beat" is a common theme in popular songs! Although these effects are not completely understood, they probably arise from at least three phosphorylation events:

(i) Cyclic AMP-dependent protein kinase phosphorylates cardiac troponin I. This accelerates the dissociation of Ca^{2+} from troponin C, leading to a faster rate of recovery before the next beat.

(ii) Cyclic AMP increases both the probability of opening, and the number, of "slow Ca^{2+} channels" (**dihydropyridine receptors**, Section 8.1.3) in the plasma membrane, these being the major route for entry of Ca^{2+} into the cell. Dihydropyridine receptors are substrates for cyclic AMP-dependent protein kinase *in vitro*, and it seems likely that this plays a part in the activation of the Ca^{2+} channels.

(iii) As already mentioned (see calmodulin-dependent multiprotein kinase), cyclic AMP-dependent protein kinase phosphorylates phospholamban, a small

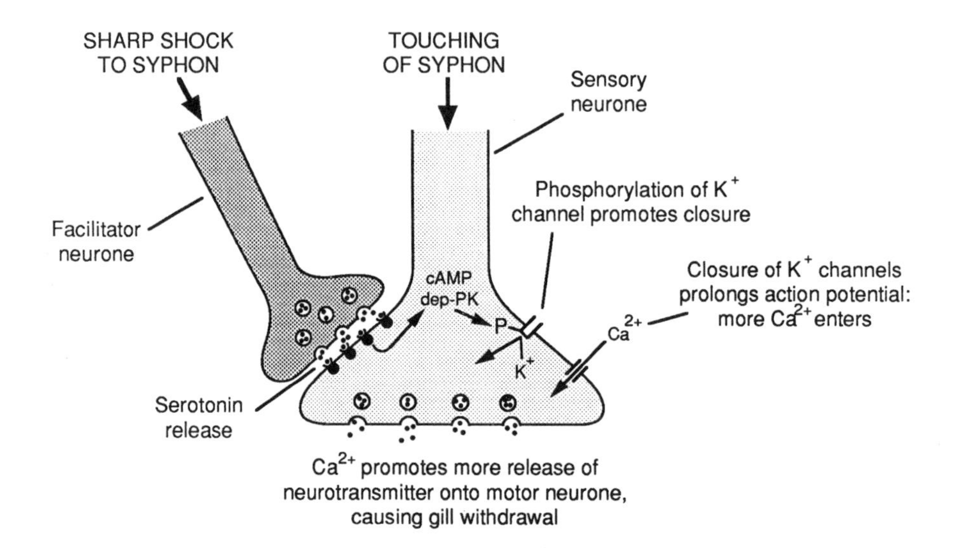

Fig. 8.30 The gill withdrawal mechanism in *Aplysia punctata*.

protein which inhibits the Ca^{2+}-transporting ATPase of cardiac sarcoplasmic reticulum. Phospholamban is a pentamer of 6 kDa subunits, each thought to consist of an N-terminal cytoplasmic domain and a C-terminal membrane domain. Phosphorylation of serine-16 in the N-terminal domain by cyclic AMP-dependent protein kinase relieves the inhibitory effect of phospholamban on the ATPase and thus speeds the uptake of Ca^{2+} into the sarcoplasmic reticulum. This allows a faster recovery from each heart beat. In addition, the increase of cytosolic Ca^{2+} due to mechanisms (i) and (ii) above leads to phosphorylation at threonine-17 by the calmodulin-dependent multiprotein kinase, which causes further stimulation of Ca^{2+} uptake.

7) *Phosphorylation of ion channels – short-term memory in sea snails*

In the sea snail *Aplysia punctata*, touching of the siphon leads to a reflex withdrawal of the gills. If this is repeated many times, the animal becomes **habituated** and ceases to respond. However, a single hard blow or an electrical shock reverses habituation, and this effect can persist for hours – a form of short term memory called **sensitization**. The sea snail has a particularly simple nervous system with large neurones, which makes it a good model system for study. The sensory neurones from the siphon release excitatory neurotransmitter onto the motor neurones which initiate contraction of the muscles responsible for gill withdrawal (Figure 8.30).

During habituation, increased influx of K^+ through "S type" voltage-gated K^+ channels in the nerve terminal shortens the action potential (Section 8.1.1), less

Ca^{2+} enters through voltage-gated Ca^{2+} channels, and less neurotransmitter is released. Conversely, patch clamping experiments shows that after sensitization, far fewer K^+ channels open, so that the action potential is prolonged and more neurotransmitter is released. This is caused by release of serotonin from **facilitator** neurones which form synapses on the nerve terminal of the sensory neurone (Figure 8.30). The effects of serotonin are mediated by cyclic AMP, and microinjection of cyclic AMP-dependent protein kinase also reduces the opening of the S type channels. Conversely, microinjection of the heat-stable protein inhibitor of cyclic AMP-dependent protein kinase blocks the effect of serotonin.

The S type channels of *Aplysia* are not well characterized at the biochemical level. It is therefore not yet clear that these effects are due to direct phosphorylation of the channel proteins. However, a variety of other ion channels have been shown to be substrates for cyclic AMP-dependent protein kinase in cell-free assays. These include voltage-gated Na^+ channels (Section 4.2.1), the "ryanodine receptor" Ca^{2+} channels of sarcoplasmic reticulum (Section 8.1.3) and the inositol trisphosphate receptor (Section 7.3.1). In most cases the physiological roles of these phosphorylation events are not yet clear. Phosphorylation of the nicotinic acetylcholine receptor was discussed in Chapter 6.

8) *Regulation of gene expression*

Although many of the effects exerted by cyclic AMP are acute effects occurring within seconds, the second messenger also has more long-term effects by switching on transcription of specific genes. An example is the gene for phosphoenolpyruvate carboxykinase, a key enzyme in gluconeogenesis (Figure 8.28) which is elevated in liver in response to glucagon released during starvation. Experiments involving microinjection of the heat-stable protein inhibitor (which does not interact with the regulatory subunits), or the catalytic subunit of the kinase, indicate that it is the latter, rather than the regulatory subunit, which is responsible for modulating gene expression. In support of this, immunohistochemical evidence clearly shows that the catalytic subunit is found in the nucleus after activation of the cell with cyclic AMP-elevating messengers (Figure 8.31).

The promoter regions of the gene for phosphoenolpyruvate carboxykinase and other cloned cyclic AMP-responsive genes contain short sequences which are almost identical (Figure 8.32) and which are thought to be the binding sites for a regulatory DNA-binding protein. This protein has been termed cyclic AMP-response element binding-protein or CREB. Recently CREB has been shown to be a substrate for cyclic AMP-dependent protein kinase in cell-free assays, and abolition of the phosphorylation site by site-directed mutagenesis prevents activation of cyclic AMP-responsive genes. These results suggest that, like all other effects of cyclic AMP, the effects on gene expression may also be mediated by protein phosphorylation.

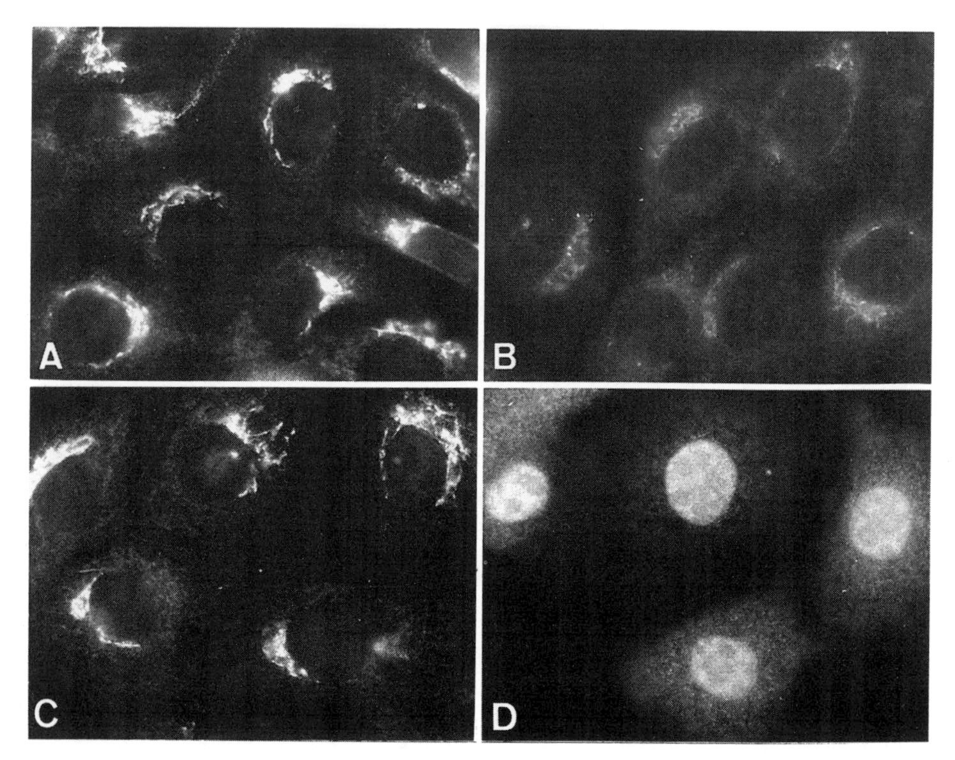

Fig. 8.31 Localization, using fluorescently labelled antibodies, of the RII regulatory subunit (A,C; left) and catalytic subunit (B,D; right) of cyclic AMP-dependent protein kinase in unstimulated bovine kidney cells (A,B; top) and after stimulation by forskolin, which activates adenylate cyclase and elevates cyclic AMP levels (C,D; bottom). The nuclear fluorescence in the lower right panel indicates that the catalytic subunit has entered the nucleus in response to forskolin. Reproduced from Nigg, E.A., Hilz, H., Eppenberger, H.M. and Dutly, F. (1985) EMBO J. **4**, 2801–2806, courtesy of Erich Nigg.

8.2.5 *Response to diacylglycerol – protein kinase C*

Diacylglycerol is one of the two products of cleavage of phosphatidylinositol-4,5-bisphosphate by messengers coupled to phosphoinositidase C (see Figure 7.18). In addition it can be produced from any other phospholipid by the action of phospholipase C. In 1979 Nishizuka and coworkers discovered that a protein kinase, previously only known to be active after proteolysis, could be activated reversibly by Ca^{2+} and a mixture of crude phospholipids. The activity received at that time the name of Ca^{2+}- and phospholipid-dependent protein kinase, although the earlier name of protein kinase C is still commonly used. Experiments with

Somatostatin:	TGACGTCA
Phosphoenolpyruvate carboxykinase:	TTACGTCA
Vasoactive intestinal peptide:	TGACGTCT
Proenkephalin:	TGCGTCAG
c-fos :	TGACGTAG

```
Consensus:  5'TGACGTCA 3'
              ||||||||
            3'ACTGCAGT 5'
```

Fig. 8.32 Nucleotide sequences of cyclic AMP response elements in cyclic AMP-regulated genes. Using the most common base at each position one can derive an ideal or consensus sequence. Note that this is palindromic, i.e. the complementary sequences on the two strands are identical when both are read in the 5'→3' direction.

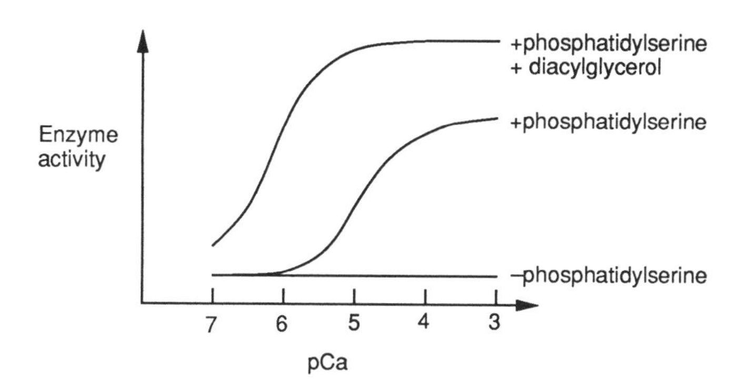

Fig. 8.33 Activation of protein kinase C by Ca^{2+} and phospholipids. $pCa = -\log_{10}[Ca^{2+}]$.

purified phospholipids showed that sonicated vesicles of phosphatidylserine would activate the enzyme, but only in the presence of unphysiologically high levels of Ca^{2+}. However if small quantities of diacylglycerol were included in the vesicles, activation occurred at lower, physiological Ca^{2+} concentrations (Figure 8.33).

These results suggested that the kinase might mediate some of the actions of messengers which are coupled to phosphoinositidase C. In support of this, such messengers cause a translocation of the kinase from the soluble fraction of the cell to membrane fractions, presumably because the generation of diacylglycerol causes it to bind to phospholipids. Another key discovery was the finding that phorbol esters (Box 8.3) would substitute for diacylglycerol in activation of the kinase. Naturally occurring diacylglycerols are insoluble and rapidly metabolized, and do

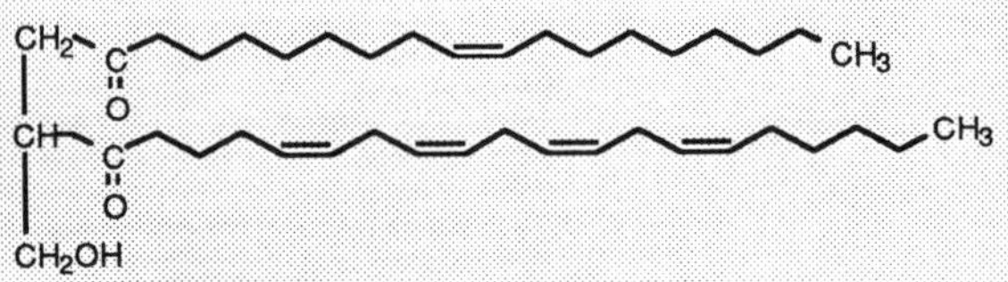

1-oleoyl 2-arachidonyl glycerol

1-oleoyl 2-acetyl glycerol (OAG)

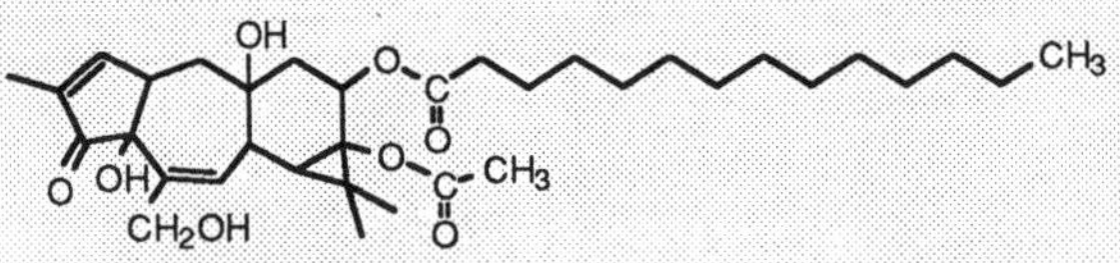

phorbol myristoyl acetate (PMA)

(also known as tetradecanoyl phorbol acetate, TPA)

The physiological regulators of protein kinase C are diacylglycerols which tend to have an unsaturated fatty acid at the 2-position, such as 1-oleoyl 2-arachidonyl glycerol (top). However these are very insoluble, and do not activate protein kinase C when presented to intact cells, unlike the synthetic diacylglycerol OAG (centre). Phorbol esters such as PMA/TPA (bottom) are products of the *Euphorbaciae* family of higher plants, and are tumour promoters, i.e. they greatly potentiate the action of carcinogens. Although the mechanism of tumour promotion remains unknown, it is now clear that phorbol esters act by mimicking the effect of diacylglycerols on protein kinase C. Being metabolized much more slowly than diacylglycerols, they are very useful experimental tools for activating protein kinase C in intact cells.

Box 8.3 Diacylglycerols, phorbol esters and protein kinase C.

not activate the enzyme when presented to intact cells. Some success was achieved with synthetic diacylglycerols such as 1-oleoyl-2-acetyl glycerol (Box 8.3), but the phorbol esters, probably because they are more slowly metabolized, represent a more reliable method for activating the kinase in intact cells. This discovery led to literally hundreds of papers in which it was shown that phorbol esters would mimic some, but not all, actions of messengers which are coupled to phosphoinositidase C.

Three isoforms (α, β and γ) of protein kinase C have now been separated from

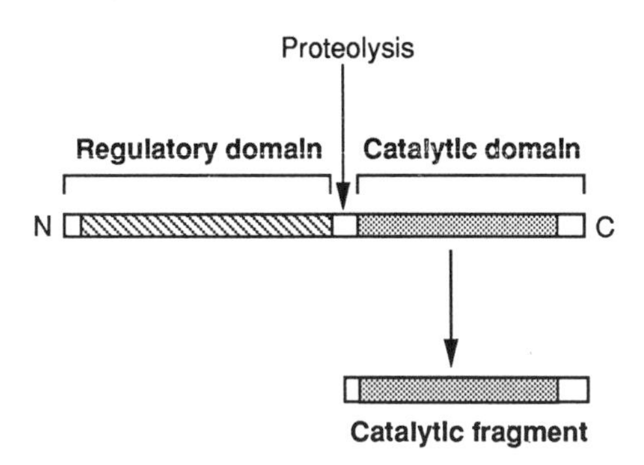

Fig. 8.34 Generalized domain structure for protein kinase C isoforms, and proteolytic generation of catalytic fragment.

brain tissue. The β isoform is a mixture of two products (βI and βII), probably produced by alternative splicing of transcripts from a single gene. In addition, three more forms (δ, ε and ζ) have been characterized by cDNA cloning. The reasons for this heterogeneity are not yet clear, although expression of isoforms seems to be tissue-specific, with γ only being found in nervous tissue. As well as being activated by diacylglycerol and phospholipids in the presence of Ca^{2+}, α and γ can be activated by free fatty acids such as arachidonic acid, and for γ this does not require Ca^{2+}. This is interesting because arachidonic acid, the precursor of eicosanoids, can be released in response to some first messengers (Section 3.2.1).

The cDNA sequences of the various isoforms show that their monomeric 70–80 kDa subunits contain domains at the C-termini which are homologous with the catalytic regions of other protein kinases. Catalytically active fragments, which are no longer regulated by lipids, can be derived from this C-terminal domain by treatment with proteinases (Figure 8.34). Although this proteolytic activation probably has no physiological significance, it was the phenomenon through which the kinase was originally discovered. The N-terminal domain contains the binding sites for Ca^{2+} and lipids, although these have not yet been clearly defined.

Studies with physiological substrates for protein kinase C such as the EGF receptor (see below) show that it preferentially phosphorylates serine or threonine residues which have clusters of basic residues on both the N- and C-terminal sides (Figure 8.35). This requirement has been confirmed by studies of phosphorylation of synthetic peptides. All seven isoforms contain, close to the N-terminus, a sequence which has similar clusters of basic residues, but with alanine in place of serine/threonine. These represent potential inhibitory pseudosubstrate sites and, as expected, synthetic peptides based on these sequences are potent competitive inhibitors of the kinase.

To be classed as a physiological target for protein kinase C, at least two criteria should be met for any potential substrate protein:

	↓
MARCKS (1):	-lys-lys-lys-arg-phe-**ser**-phe-lys-lys-ser-phe-
MARCKS (2):	-phe-ser-phe-lys-lys-**ser**-phe-lys-leu-ser-gly-
MARCKS (3):	-lys-ser-phe-lys-leu-**ser**-gly-phe-ser-phe-lys-
MARCKS (4):	-lys-leu-ser-gly-phe-**ser**-phe-lys-lys-asn-lys-
EGF receptor:	-ile-val-arg-lys-arg-**thr**-leu-arg-arg-leu-leu-
pp60src:	-lys-pro-lys-asp-pro-**ser**-glu-arg-arg-arg-ser-
interLeukin-2 receptor:	-arg-arg-glu-arg-lys-**ser**-arg-arg-thr-ile-
transferrin receptor:	-tyr-thr-arg-phe-**ser**-leu-ala-arg-
α/βI/βII isoform:	-phe-ala-arg-lys-gly-**ala**-leu-arg-glu-lys-asn-
γ isoform:	-phe-cys-arg-lys-gly-**ala**-leu-arg-glu-lys-val-
δ isoform:	-met-asn-arg-arg-gly-**ala**-ile-lys-glu-ala-lys-
ε isoform:	-arg-lys-arg-glu-gly-**ala**-val-arg-arg-arg-val-
ζ isoform:	-ile-tyr-arg-arg-gly-**ala**-arg-arg-trp-arg-lys-

Fig. 8.35 Comparison of the sequences around sites (arrowed, bold type) phosphorylated by protein kinase C, and the putative pseudosubstrate sites on protein kinase C isozymes. MARCKS (an eponym for "myristoylated, alanine-rich C-kinase substrate") is a widely distributed, soluble protein of unknown function that is phosphorylated on multiple sites by the kinase. pp60src is a protein (tyrosine) kinase (Section 9.2.1). Basic residues are double underlined.

1) the protein should be phosphorylated rapidly by purified protein kinase C in cell-free assays;
2) the protein should be phosphorylated **at the same sites** in response to phorbol esters in intact cells.

Although a large number of proteins have met one or other of these criteria, to date very few have fulfilled both, particularly with regard to the identical sites of phosphorylation in the two circumstances. Of those that have met both criteria, several are receptor proteins, e.g. the receptors for EGF, interleukin-2, and transferrin (the latter involved with iron uptake rather than signal transduction). The only case in which the function of this phosphorylation is understood is the EGF receptor, which is phosphorylated on a threonine (thr-654) nine residues from the transmembrane domain on the cytoplasmic side (Figure 8.36). Treatment of intact cells with phorbol ester leads to phosphorylation of thr-654, and to both a reduction in the affinity of the receptor for EGF and its protein (tyrosine) kinase activity. Conversion of thr-654 to an alanine residue by site-directed mutagenesis demonstrates that the reduction in kinase activity, although not the reduction in affinity, is accounted for by this phosphorylation. Phosphorylation of the purified

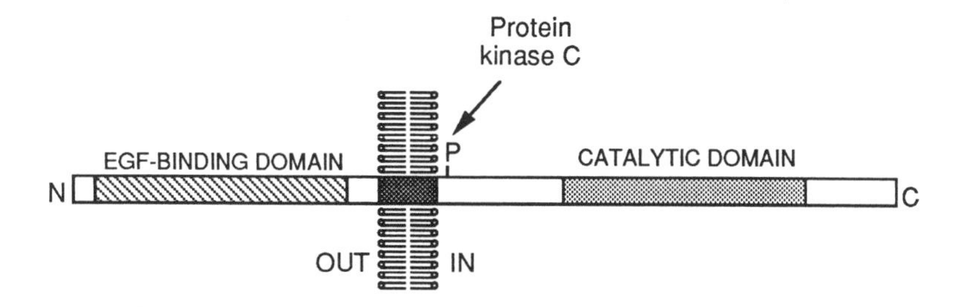

Fig. 8.36 Location of site of phosphorylation by protein kinase C on the epidermal growth factor receptor.

receptor by protein kinase C also reduces its protein (tyrosine) kinase activity. Since EGF promotes diacylglycerol release, at least in some cell types, this represents a potential feedback mechanism.

Since the active form of protein kinase C is complexed with lipids, it seems most likely that physiological substrates will be mainly associated with membranes. Although the number of well-established physiological substrates is still small, the diverse effects of phorbol esters suggest that protein kinase C will turn out to have numerous targets, as in the case of cyclic AMP-dependent protein kinase. Some processes which are affected by phorbol esters are:

1) tumour formation in mouse skin in response to chemical carcinogens, which is potentiated: this tumour promotion activity was the original biological assay for phorbol esters (Box 3.3);
2) differentiation or cell proliferation, with phorbol esters often causing a switch from one to the other;
4) expression of specific genes: a phorbol ester response element, analogous to the cyclic AMP response element, has been defined in the promoters of susceptible genes (Section 9.3.4);
5) secretion, effects of Ca^{2+} ionophores and phorbol esters often being additive;
6) movement of ions through transporters and channels, several of which are good candidates as substrates for the kinase. For example, treatment with phorbol esters activates the Na^+/H^+ antiporter discussed in Section 4.1.3, which regulates intracellular pH. Several first messengers, particularly growth factors, which activate phosphoinositidase C and hence, protein kinase C, stimulate the antiporter and cause an intracellular pH increase of around 0.2 units. Whether this is achieved via a direct phosphorylation of the Na^+/H^+ antiporter by protein kinase C is not yet clear. Experiments with mutant cells lacking the Na^+/H^+ antiporter suggest that the rise in pH may be necessary for cell proliferation, although it does not trigger proliferation *per se*. It may be that activation of the transporter is necessary to deal with an increased acid load generated by increased metabolism.

Table 8.1 Proteins modified on serine/threonine residues in response to insulin.

Protein	Pathway	Change in Phosphorylation	Effect
Glycogen synthase	Glycogen synthesis	Decrease	Activation
Acetyl-CoA carboxylase	Fatty acid synthesis	Increase	None*
ATP-citrate lyase	Fatty acid synthesis	Increase	None*
Pyruvate dehydrogenase	Fatty acid synthesis	Decrease	Activation
Ribosomal protein S6	Protein synthesis	Increase	None*
S6 kinase	Protein synthesis?	Increase	Activation

* In these cases no convincing effect has been demonstrated on the function of the purified proteins, although flux through the pathways (fatty acid or protein synthesis) is increased by insulin.

8.3 *Responses to protein (tyrosine) kinase receptors*

When it was discovered that the protein (tyrosine) kinase activity of the receptors for insulin, and growth factors such as epidermal growth factor (EGF), were essential for signal transduction (Section 7.4), it might have appeared that their mechanism of action had been solved. It would merely be a question of identifying the protein targets which are phosphorylated on tyrosine residues.

Insulin and growth factors such as EGF do indeed increase the phosphorylation of many intracellular proteins, but somewhat paradoxically this occurs predominantly on serine and threonine, rather than tyrosine, residues. Some proteins whose phosphorylation is modified on serine/threonine residues in response to insulin are listed in Table 8.1. In some, but not all cases, these changes are associated with measurable changes in the function of the target protein. It will immediately be noticed that both phosphorylation and dephosphorylation events occur. Attention has therefore focused on the protein (serine/threonine) kinases and phosphatases which mediate these events: could these be the elusive critical targets for modification by the protein (tyrosine) kinases?

A number of protein (serine) kinases have been shown to be activated in response to insulin and growth factors such as EGF and PDGF. The best characterized of these are those which phosphorylate ribosomal protein S6. The designation S6 indicates that this is (somewhat arbitrarily) protein 6 of the small ribosomal subunit, and it is thought to be involved in binding of mRNA to the ribosome. Protein synthesis is activated by insulin and growth factors, and there is a good correlation with phosphorylation of S6, although it is still unproven that

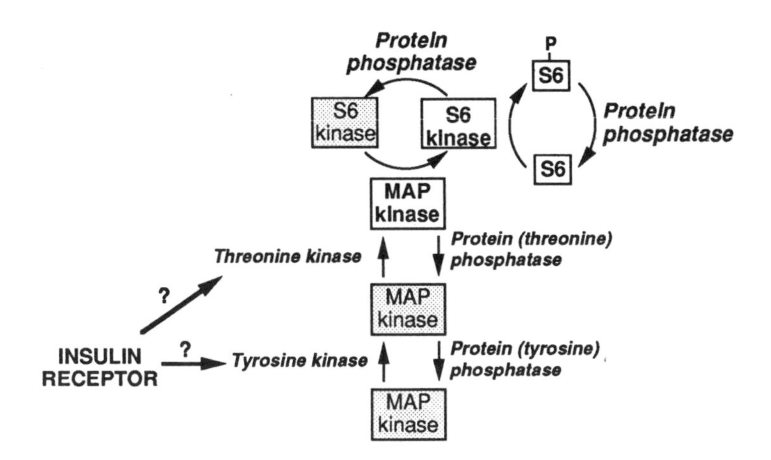

Fig. 8.37 Speculative model to account for phosphorylation of ribosomal protein S6 in response to insulin and/or growth factors. Inactive forms of proteins are shown in lighter type in shaded boxes.

there is a causal connection. Nevertheless, this event is an excellent marker for growth factor action. At least two distinct S6 kinases, one from rat liver and one from *Xenopus* oocytes, have been shown to be activated in response to insulin and growth factors. The activation is stable during purification in the presence of protein phosphatase inhibitors, and can be reversed by protein phosphatases. However it is now clear that in both cases the activation is produced by phosphorylation of **serine/threonine** residues. The research has therefore merely taken one step back up the phosphorylation cascade, and if anything is activated by tyrosine phosphorylation it must be the **S6 kinase kinase**.

Given the difficulties of purification of the S6 kinases in reasonable amounts, characterization of the kinase kinase(s) is a daunting task. However, it has been shown that a protein kinase purified originally by its ability to phosphorylate a different protein (microtubule-associated protein-2, MAP2) can phosphorylate and activate the *Xenopus* S6 kinase, at least *in vitro*. Intriguingly, this kinase, which is commonly referred to as **MAP kinase**, is itself regulated by phosphorylation. In several cell types it is phosphorylated on both threonine and tyrosine residues, and dephosphorylation of either type of residue by specific phosphatases leads to inactivation. Thus it must be phosphorylated on **both** threonine and tyrosine residues to be active (Figure 8.37). The kinases which phosphorylate these sites (which could be termed kinase kinase kinases!) have not been identified, although disappointingly the insulin receptor does not appear to be one of them.

As discussed in Section 7.4.1, several proteins which form stable complexes with, and are phosphorylated by, the PDGF receptor have now been identified. One of these, *raf-1,* is a protein (serine/threonine) kinase, and therefore provides

Table 8.2 Classification of protein (serine/threonine) phosphatases

	PP-1	PP-2A	PP-2B	PP-2C
Preference for α or β subunit of phosphorylase kinase	β	α	α	α
Inhibition by inhibitor-1 and -2	Yes	No	No	No
Inhibition by okadaic acid	Potent	Very potent	Weak	No
Divalent cation-dependent	No	No	Ca^{2+}	Mg^{2+}
Calmodulin-dependent	No	No	Yes	No
Inhibition by trifluoperazine	No	No	Yes	No

another possible connection between tyrosine and serine/threonine phosphorylation systems. Recent results suggest that phosphorylation of *raf-1* on tyrosine activates the protein kinase activity by 5- to 6-fold. Although the targets for the *raf-1* protein kinase are not known, this exciting result would, if confirmed, be the first case in which a protein (tyrosine) kinase has been shown to directly activate a protein (serine/threonine) kinase. To end on a more optimistic note, it is therefore perhaps only a matter of time before the complex web of protein phosphorylation events in insulin and growth factor action are deciphered. Since this may also involve regulation of protein phosphatases, it is now appropriate to discuss what is known of these activities.

8.4 *Protein phosphatases*

Counting all known isoforms, at least seventy protein kinases have now been described. At present the number of distinct protein phosphatases would appear to be smaller, although with the application of cDNA cloning the number of known isoforms is increasing all the time. As is the case for the protein kinases, protein phosphatases are specific for either serine/threonine or tyrosine residues, and the two types will be described separately.

8.4.1 *Protein (serine/threonine) phosphatases*

For many years the protein phosphatase field was very confused because various

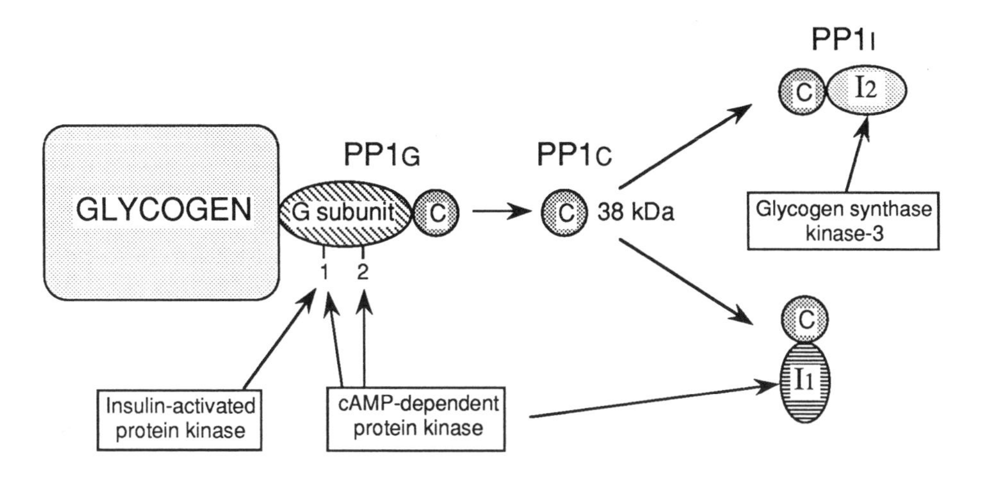

Fig. 8.38 Multiple forms and regulation of protein phosphatase-1 (PP1). Key: G, glycogen-binding subunit; C, catalytic subunit; I1, I2, inhibitor-1, -2.

laboratories were using different substrates to characterize their activities, and consequently a profusion of apparently distinct enzymes was described. The relationships between these have now been clarified, and it appears that all of the known extramitochondrial protein (serine/threonine) phosphatase activities can be ascribed to only four enzymes, although a number of isoforms of each of these are being defined. In addition there are at least two protein phosphatases in the mitochondrion (pyruvate dehydrogenase phosphatase and branched chain ketoacid dehydrogenase phosphatase). The extramitochondrial phosphatases may be classified using their specificity for the α and β subunits of phosphorylase kinase, their sensitivity to inhibitors and inhibitor proteins, and their regulation by divalent cations (Table 8.2).

Protein phosphatase-1

Protein phosphatase-1 (PP1) contains a 36–38 kDa catalytic subunit, complexed with various other regulatory subunits. There are at least two isoforms of the catalytic subunit (PP1α and PP1β) encoded by different genes. In skeletal muscle a large proportion of the activity is found to be associated with glycogen particles. Since phosphorylase and glycogen synthase are also found to be bound to these particles, this indicates that protein phosphatase-1 plays a particularly important role in regulation of glycogen metabolism. The glycogen-bound form ($PP1_G$) comprises a single catalytic subunit complexed to a 160 kDa glycogen-binding (G) subunit. The latter can be phosphorylated at site 2 by cyclic AMP-dependent protein kinase in response to adrenaline (epinephrine) (Figure 8.38), causing the catalytic subunit to dissociate from the glycogen particles. Thus activation of cyclic AMP-dependent protein kinase not only increases the activity of the kinases acting

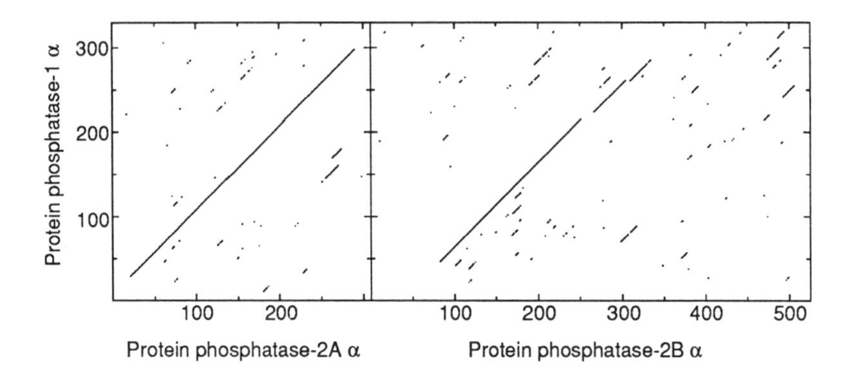

Fig. 8.39 Dot matrix comparisons of the catalytic subunits (α isoforms) of protein phosphatase-1, -2A and -2B. The obvious diagonals (see Figure 3.15) reveal that PP-1 and -2A are related except at their N- and C-termini, while PP-2B is related to PP-1 over residues 80–350 approx.

on the glycogen-metabolizing enzymes, but also decreases the activity of the protein phosphatase which reverses their effects. This mechanism can explain how adrenaline increases phosphorylation of glycogen synthase at sites which are not phosphorylated by cyclic AMP-dependent protein kinase, and causes a large conversion of phosphorylase b to a in resting muscle despite the fact that Ca^{2+} concentrations, and hence phosphorylase kinase activity, are very low (Section 8.2.4).

The G subunit of $PP1_G$ is phosphorylated at an additional site, site 1, by both cyclic AMP-dependent protein kinase and an insulin-stimulated protein (serine) kinase that is not yet well characterized. This increases the glycogen synthase phosphatase activity of $PP1_G$, an exciting observation that may explain how one case of insulin-stimulated **dephosphorylation** operates. Since site 1 is dephosphorylated more slowly than site 2, this mechanism would also allow a more rapid resynthesis of glycogen when the activity of cyclic AMP-dependent protein kinase returns to basal levels after adrenaline stimulation.

The catalytic subunit that is released from the glycogen particles in response to adrenaline is probably not active, because the cytosol contains a heat-stable inhibitor protein called inhibitor-1. When phosphorylated on a threonine residue by cyclic AMP-dependent protein kinase, inhibitor-1 becomes an extremely potent inhibitor of the catalytic subunit (Figure 8.38). This probably ensures that release of catalytic subunit does not lead to dephosphorylation of cytosolic proteins. An important feature of regulation by inhibitor-1 is that protein phosphatase-1 cannot dephosphorylate inhibitor-1 itself. A distinct isoform of inhibitor-1 called DARPP (dopamine and cyclic AMP-regulated phosphoprotein) is expressed at high levels, along with inhibitor-1, in neurones containing dopamine receptors coupled to

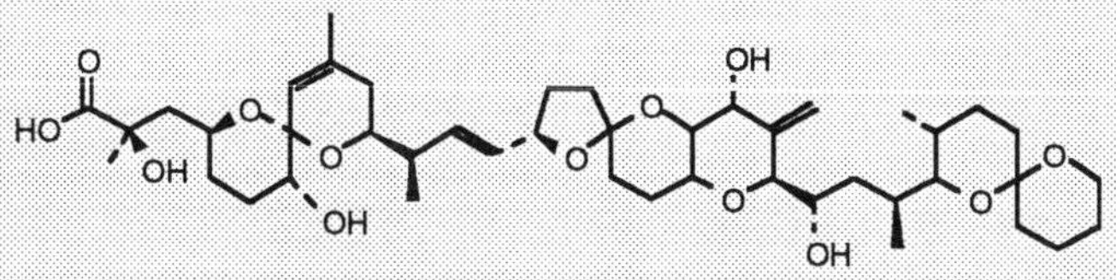

Okadaic acid is a toxic polyether derivative of a long chain fatty acid. It is synthesized by marine dinoflagellates, which are ingested by filter feeders such as sponges and mussels, and is believed to be a major cause of diarrhetic shellfish poisoning in humans. The toxin is an extremely potent inhibitor of protein phosphatases-1 and -2A: with PP2A, only one molecule of toxin is required to inhibit one molecule of phosphatase. Being a specific, cell-permeable protein phosphatase inhibitor, okadaic acid is a valuable experimental tool.

Box 8.4 Okadaic acid, a protein phosphatase inhibitor

adenylate cyclase.

An additional form of protein phosphatase-1 ($PP1_I$), found in the cytosol, is a complex between the catalytic subunit and another heat-stable inhibitor protein, inhibitor-2. In contrast to inhibitor-1, inhibitor-2 is a potent inhibitor in the **dephosphorylated** form, but the inactive complex can be reactivated by phosphorylation of a threonine residue by glycogen synthase kinase-3 (Figure 8.38). Since this requires MgATP, $PP1_I$ was originally termed the MgATP-dependent protein phosphatase. The function of $PP1_I$ is not yet understood.

Although PP1 has been discussed in the context of its established role in glycogen metabolism, it is clear that it is involved in regulating many other processes. For example, mutations in PP1-like catalytic subunits in the fungi *Saccharomyces cerevisiae* and *Aspergillus nidulans*, and the insect *Drosophila melanogaster*, are defective in various aspects of the mitotic cell cycle.

Protein phosphatase-2A and other related phosphatases

Protein phosphatase-2A consists of a catalytic (C) subunit of about 36 kDa complexed with larger A and B subunits in various combinations. The C subunit exists as at least 2 isoforms (α and β) which are distinct gene products, and which are both closely related to the catalytic subunits of protein phosphatase-1 (Figure 8.39). There are at least three forms of the B subunit.

Protein phosphatase-2A has a very broad specificity, and for most substrates is the major protein phosphatase found in the soluble fraction derived from homogenates. With some substrates, dephosphorylation is dramatically activated by polyamines such as spermine, and an alternative name for the enzyme is the polycation-stimulated protein phosphatase. Since the levels of polyamines present in all cells are normally in the range which would cause maximal activation

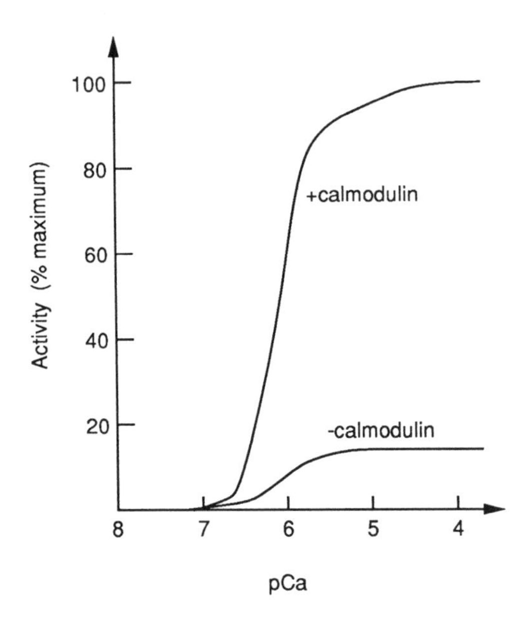

Fig. 8.40 Regulation of protein phosphatase-2B by Ca^{2+} and calmodulin. $pCa = -\log_{10}[Ca^{2+}]$.

the physiological importance of this is not clear. However as yet no other mechanism of regulation of protein phosphatase-2A has been found, and the functions of the A and B subunits are not clear. Treatment of cells with okadaic acid, a potent inhibitor of protein phosphatase-1 and -2A (Box 8.4), leads to dramatic increases in phosphorylation of many phosphoproteins. This suggests that protein phosphatases-1 and/or -2A have wide-ranging and important roles *in vivo*.

The C subunits of protein phosphatases-1 and -2A are among the most highly conserved sequences of any enzymes known, being >90% identical between the rabbit and the fruit fly *Drosophila melanogaster*. This may reflect the fact that they have to interact with a very large number of other proteins, so that any mutation is likely to be harmful. Recently at least three additional putative protein phosphatase catalytic subunits have been found by screening of cDNA libraries with PP1 or PP2A probes. Their sequences appear to be intermediate between those of PP1 and PP2A, but their functions are not yet clear.

Protein phosphatase-2B

Protein phosphatase-2B is a Ca^{2+}-dependent enzyme. It comprises a 61 kDa catalytic (A) subunit complexed with a 19 kDa (B) subunit. The catalytic subunit contains an N-terminal region which is related to the catalytic subunits of protein phosphatase-1 and -2A (Figure 8.39). The B subunit is related to calmodulin, and confers the Ca^{2+} dependence on the enzyme activity. In addition, calmodulin itself can bind at a distinct site, and activate the phosphatase a further 10-fold (Figure 8.40). In this respect, protein phosphatase-2B is similar to phosphorylase kinase, which has the tightly-bound δ subunit as well as a site for further activation by calmodulin. Protein phosphatase-2B is particularly abundant in brain tissue, and before it was discovered to be a protein phosphatase, it had been described as a major calmodulin-binding protein, and termed **calcineurin**.

Protein phosphatase-2B is much more specific than protein phosphatase-1 and –2A, and does not dephosphorylate the metabolic enzymes which are substrates for the other phosphatases. The best substrates found to date are regulatory subunits of protein kinases and phosphatases which are phosphorylated by cyclic AMP-dependent protein kinase. Thus for example, it dephosphorylates inhibitor-1 and its neural isoform, DARPP, the α subunit of phosphorylase kinase, and the Type II regulatory subunit of cyclic AMP-dependent protein kinase. In all cases these dephosphorylation events would tend to terminate the response to cyclic AMP. Along with Ca^{2+} activation of cyclic nucleotide phosphodiesterase (Section 8.2.1) this would be a mechanism by which messengers increasing cytosolic Ca^{2+} could terminate the effects of messengers acting through cyclic AMP.

Protein phosphatase-2C

This cytosolic enzyme is a monomeric protein with two isoforms of 42 and 44 kDa: surprisingly these are not related in sequence to the catalytic subunits of protein phosphatases-1, -2A or -2B. It is completely dependent on Mg^{2+} for activity, and is not sensitive to okadaic acid. Although it has a relatively broad specificity, for most substrates it is a minor phosphatase activity when compared in crude extracts with protein phosphatase-2A. However the existence of okadaic acid-insensitive phosphoproteins in intact cells suggests that it may play a role in dephosphorylation of at least some substrates.

8.4.2 *Protein (tyrosine) phosphatases*

A number of protein phosphatases which are completely specific for tyrosine residues have been described in mammalian cells. The best characterized is protein (tyrosine) phosphatase (PTP) 1B from human placenta, a small monomeric protein which shows no homology with any protein (serine/threonine) phosphatase. Its specific activity is two orders of magnitude higher than that of any known protein tyrosine kinase, and since it has no known means of regulation, it is at first sight surprising that any tyrosine phosphorylation occurs *in vivo*. Possibly the high activity of this phosphatase ensures that tyrosine phosphorylation is restricted to proteins that form direct interactions with tyrosine kinases, as already proposed for the PDGF receptor (Section 7.4).

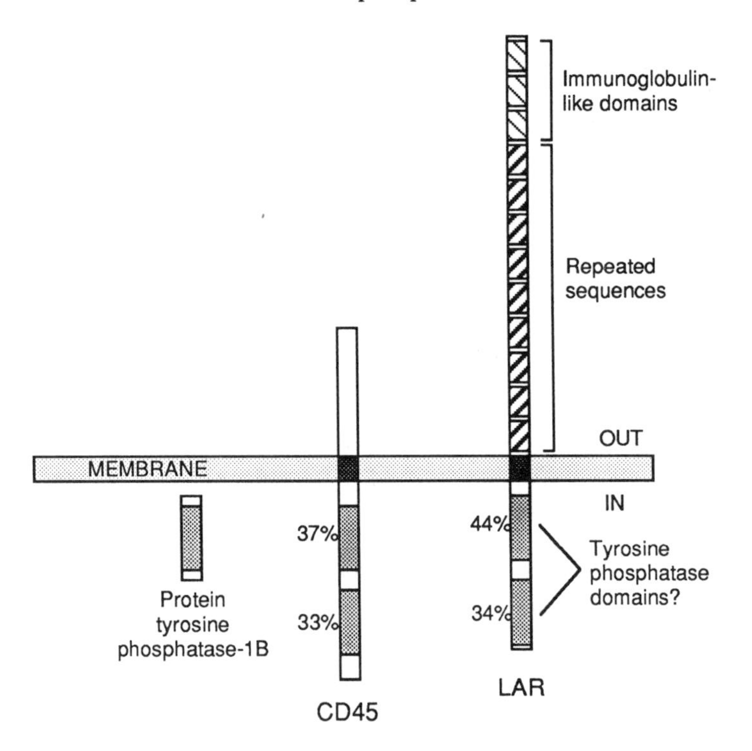

Fig. 8.41 The domain structures of soluble and membrane-bound protein (tyrosine) phosphatases. Figures next to the duplicated regions in the cytoplasmic domains of CD45 and LAR are the percentage identity with the catalytic region of protein (tyrosine) phosphatase-1B.

Intriguingly, at least two integral membrane proteins show homology with PTP1B. These are CD45, originally identified as a surface antigen on lymphocytes, and a related protein currently termed LAR. Both contain a single predicted transmembrane domain, and a cytoplasmic region with a tandem repeat of a sequence related to the core sequence of PTP1B. CD45 does indeed express protein tyrosine phosphatase activity, and it is particularly effective at inactivation of MAP kinase by tyrosine dephosphorylation (Section 8.3), although it is not yet certain that it performs this role *in vivo*. The predicted domain structure of CD45 is reminiscent of that of the EGF receptor (Figure 8.41), and by analogy it has been speculated that CD45 may be a messenger-activated protein tyrosine phosphatase. LAR is related to CD45 in the cytoplasmic domain (Figure 8.41). However, the ectoplasmic domain is quite different, being homologous to the ectoplasmic domain of the neural cell adhesion molecule N-CAM. This is intriguing because protein (tyrosine) kinases are involved in activating cell proliferation, and cell–cell contact inhibits proliferation in normal cells (Chapter 9). It is conceivable that LAR is involved in mediating this contact inhibition.

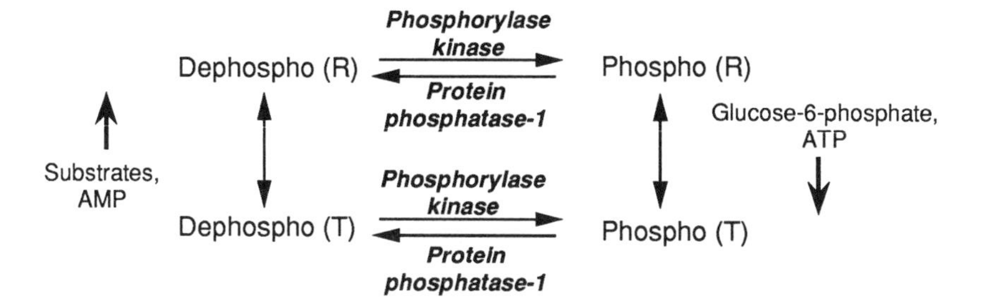

Fig. 8.42 The four forms of phosphorylase.

8.5 *How does phosphorylation modulate protein function?*

The only eukaryotic protein regulated by phosphorylation for which detailed tertiary structural information is available is skeletal muscle phosphorylase, the key enzyme of glycogen breakdown (Section 8.2.2). The regulatory properties of the enzyme are consistent with the existence of two conformational states, an active R state and an inactive T state, which are in equilibrium. Each of these can also exist in a phosphorylated (a) and dephosphorylated (b) form, making four species in all (Figure 8.42). The crystal structures of all of these forms have now been solved at high resolution. Phosphorylase is a homodimer which *in vivo* is largely bound to glycogen particles via binding sites which are distinct from the catalytic site (Figure 8.43). The two subunits contact each other mainly via antiparallel interactions between the "towers" formed by helix α7 in each subunit, and an interaction between the "cap" at the junction of helices α1 and α2 on one subunit, and helix α2 on the other subunit. The T↔R transition involves a 10° rotation of the two subunits with respect to each other, such that the "towers" on opposing subunits move apart, while the "caps" and helices α2 move together. The allosteric activator AMP forms interactions with the base of helix α2 and the "cap" on the other subunit. These interactions are more favourable in the R state, which is therefore stabilized by AMP. Conversely, the allosteric inhibitors glucose-6-phosphate and ATP bind at the same site as AMP, but preferentially bind to the inactive T state.

The site phosphorylated by phosphorylase kinase is serine-14, lying within the N-terminal 20 residues leading to the start of helix α1. In the dephosphorylated form of the T state, these 20 residues are poorly ordered and adopt an irregular, extended conformation involving only intrasubunit interactions. They also lie on the opposite side to the glycogen-binding sites, and therefore *in vivo* would be accessible for modification by phosphorylase kinase (a protein about six times bigger than phosphorylase!). In the phosphorylated form, the N-terminus folds into a partial helix and is reoriented by 120°, so that it comes to rest in a groove between the two subunits. The negatively charged phosphate group on serine-14 forms interactions with two positively charged arginine side chains, one

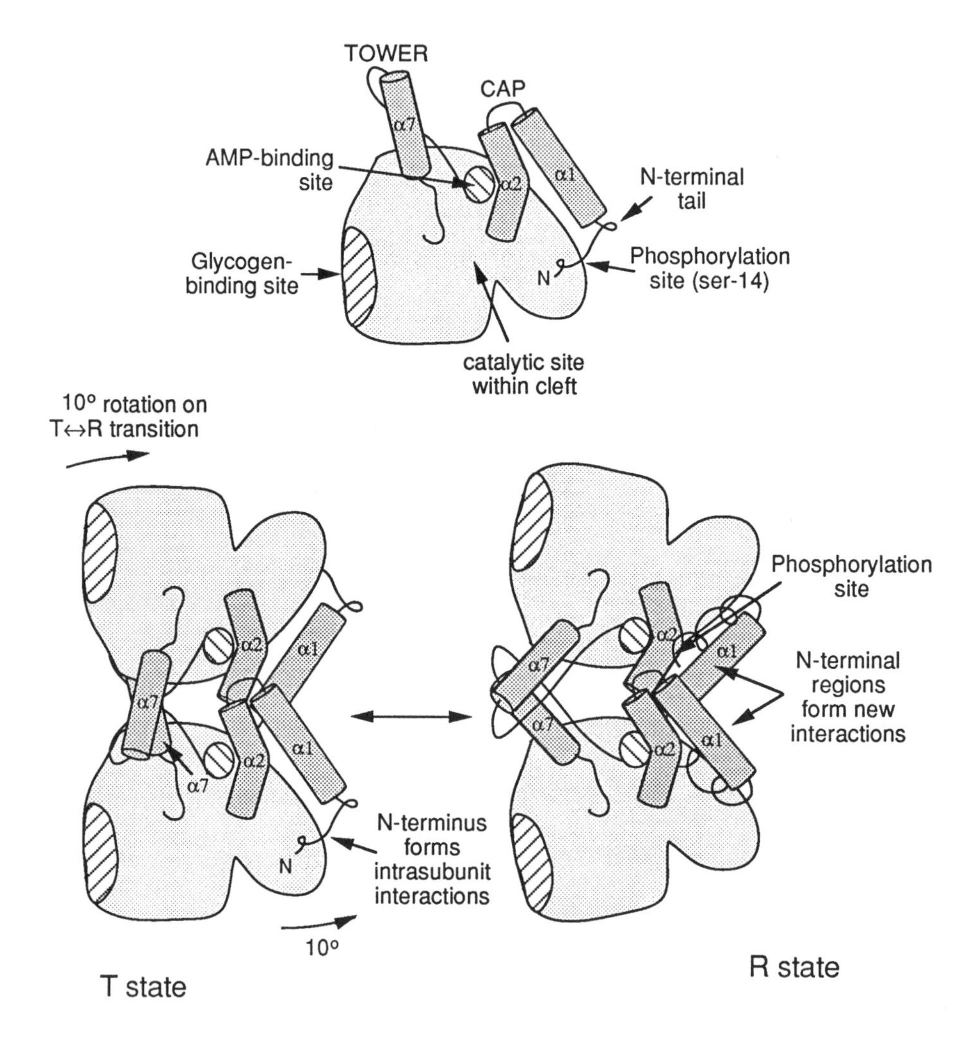

Fig. 8.43 Schematic structure of a single subunit of phosphorylase (top), and native dimers in the dephospho-T and -R states (bottom). Based on X-ray crystallographic results discussed in Barford, D. and Johnson, L.N. (1989) Nature **340**, 609–616. Helices α1, α2 and α7 are depicted as cylinders: the remainder of the structure (which includes β-sheets and several other α-helices) is not shown in detail. The phospho-T state structure is basically similar to the dephospho-T state shown, except that the 20 N-terminal residues form a stable interaction, similar to that shown here for the dephospho R state.

on helix $\alpha 2$ of the same subunit, and the other on the "cap" of the other subunit. Although this involves different residues to those which bind AMP, the effect is the same: helix $\alpha 2$ and the cap are pulled together and the R state is stabilized. Reflecting this, the equilibrium constant for the R$\leftrightarrow$T transition (in the absence of substrates or allosteric effectors) is ~3 for the phospho-form and ~2000 for the dephospho-form. This explains why the phosphorylated enzyme is almost independent of AMP (Figure 8.19), since in the presence of substrates it would already be largely in the R state.

How are these effects transmitted to the active site? The relative rotation of the two subunits on the T$\rightarrow$R transition causes a 20° tilt of the tower ($\alpha 7$) helices. These helices are attached to segments of polypeptide which extends down into the active site. The movement transmitted down this segment increases the affinity of binding of the substrate, phosphate, to lysine and arginine residues within the active site. Thus the effect of binding of AMP, or of phosphorylation, at locations more than 30 Å away from the active site, is converted into an enzyme activation.

8.6 *Advantages of protein phosphorylation for signal transduction*

By now it should be abundantly clear that reversible phosphorylation of proteins is the major device by which the binding of first messengers to cell surface receptors is converted into an end-effect inside the cell. It is pertinent to ask why such covalent modification of target proteins has been adopted. The alternative mechanism, used in a few cases for Ca^{2+}-mediated processes, is that an intracellular messenger binds directly to a target protein, or to a binding protein (e.g. calmodulin) which itself binds to a target protein. Why is the latter mechanism not more widely used? At least three possible advantages of the covalent mechanism can be discussed:

1) *Amplification*

In the direct-binding mechanism, the intracellular messenger must be produced in concentrations comparable with those of the target protein. Production of the intracellular messenger involves considerable expenditure of energy. For example, in skeletal muscle the direct activation of troponin C requires a very active Ca^{2+}-transporting ATPase to pump the cation into the sarcoplasmic reticulum. By contrast, concentrations of cyclic AMP around 1 $\mu mol.l^{-1}$ are able to activate phosphorylase (present at ~100 $\mu mol.l^{-1}$) because the signal is amplified by two rounds of protein phosphorylation catalysed by cyclic AMP-dependent protein kinase and phosphorylase kinase. The penalty for this saving in energy may be a slight loss in speed of response. It may be significant that smooth muscle, whose contraction is initiated by protein phosphorylation, contracts more slowly than skeletal or cardiac muscle.

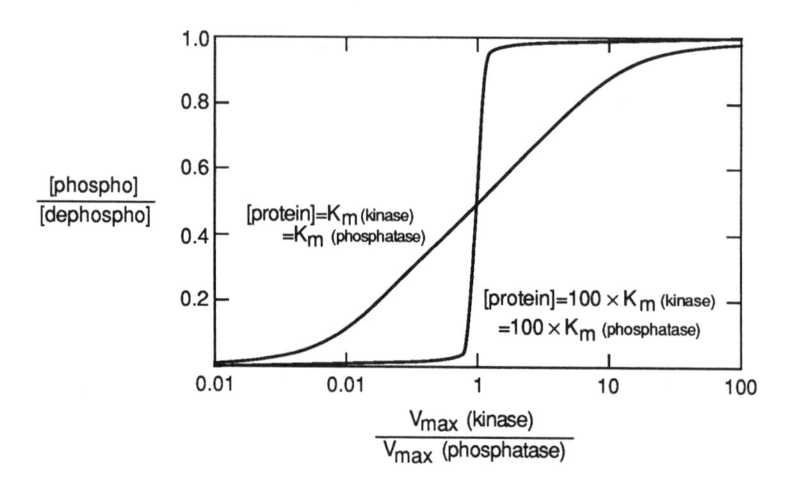

Fig. 8.44 Theoretical curve showing how a very small change in V_{max} results in a dramatic change in phosphorylation state when both kinase and phosphatase are saturated with their substrate (total protein substrate concentration = 100 × K_m for both the kinase and phosphatase). A much less sensitive system results if the kinase and phosphatase are not saturated (total protein substrate concentration = K_m). After Ledbetter and Koshland (1981) Proc. Natl. Acad. Sci. USA **78**, 6840-6844.

2) *Sensitivity*

Sensitivity is distinct from amplification: it refers to the phenomenon that a small percentage change in the concentration of the initial signal can produce a large percentage change in the end product. For a direct ligand-binding mechanism involving a single binding site, the percentage change in the amount of ligand bound cannot be greater than the percentage change in the ligand concentration, and may well be less if the system is approaching saturation. The situation can be improved by having two or more binding sites which display positive cooperativity (Box 7.1), as in the case of Ca^{2+} binding to calmodulin (Section 8.2.1). Nevertheless, with a Hill coefficient of 2 (as for calmodulin) the maximum possible change in ligand bound is the square of the change in ligand concentration.

With covalent modification systems, much higher sensitivity (**ultra-sensitivity**) is theoretically possible (Figure 8.44). This is especially the case if the kinase and/or the phosphatase are saturated with respect to their protein substrates. In this case, a small increase in the V_{max} of the kinase would result in an elevated rate of conversion of dephospho- to phospho-protein, and this would continue unabated until the concentration of dephospho-protein dropped to the point where it was no longer saturating. The rate of the opposing phosphatase reaction would also not initially be affected by the conversion if it was saturated

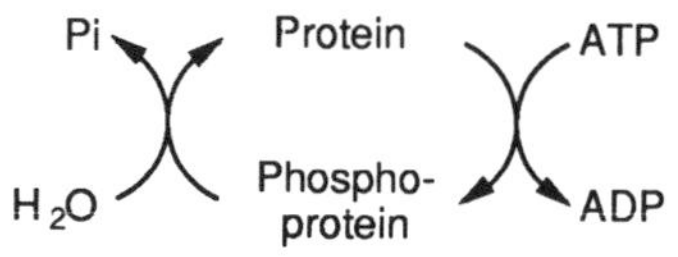

Pi Protein ATP

H_2O Phospho-protein ADP

Fig. 8.45 A protein kinase/phosphatase cycle.

with respect to its phosphoprotein substrate. Under these circumstances a covalent modification can act as a very sensitive switch, or trigger, mechanism (Figure 8.44).

Even greater sensitivity is possible if the kinase and phosphatase activities are simultaneously regulated in opposite directions. At least one mechanism has been shown for this, since activation of cyclic AMP-dependent protein kinase also inhibits protein phosphatase-1, by phosphorylation of the glycogen-binding subunit and inhibitor-1 (Section 8.4.1). Theoretical calculations have shown that the combination of these two effects can produce almost unlimited sensitivity.

3) *Flexibility*

An additional advantage of the covalent modification is that the intracellular messengers, instead of binding just to the target proteins, can interact with many different proteins, including one or more kinases, phosphatases and inhibitor proteins. As well as contributing to the sensitivity discussed above, this may allow much more complex interplay between regulatory networks than would be possible with simple ligand-binding mechanisms operating on a single target protein.

The above arguments (especially the second) assume that the protein kinase and phosphatase which act on a system are simultaneously active. This might at first sight seem to be wasteful, since the net result would be hydrolysis of ATP (Figure 8.45). However most substrate proteins are present in cells at concentrations of 10^{-6} mol.l^{-1} or less, and it can be calculated that the expenditure of ATP for protein phosphorylation is negligible compared with that involved in intermediary metabolism. In any case, the ATP consumed in protein phosphorylation reactions is not wasted, but is the small price the system pays to enable a very small initial signal to enable switching between different conformational states of a protein, some of which would be energetically unfavourable without an input of energy. Tony Hunter has pointed out that there is an analogy between protein kinase/phosphatase cycles in living systems, and the transistors in electronic circuits: both can switch the system from one state to another, and can also amplify signals. However, they both require a small input of energy in order to achieve this.

SUMMARY

The signal transduction mechanisms discussed in Chapter 7 produce three types of signal in the target cell: a change in membrane polarization, an increase in the concentration of a second messenger such as Ca^{2+}, cyclic AMP, cyclic GMP or diacylglycerol, or activation of a protein (tyrosine) kinase. In the vast majority of cases effects further downstream are produced by the action of protein (serine/threonine) kinases and phosphatases.

Depolarization of neurones causes opening of voltage-gated ion channels and the initiation of a particular frequency of action potentials. In secretory cells such as adrenal chromaffin cells, depolarization leads to influx of Ca^{2+} through voltage-gated channels, and hence triggers secretion. In skeletal and cardiac muscle, Ca^{2+} is released from the sarcoplasmic reticulum when the signal is transmitted into the cell as an action potential passing along the transverse tubules. A signal then passes to the sarcoplasmic reticulum, possibly via a direct protein–protein interaction, causing opening of Ca^{2+} channels.

The second messenger Ca^{2+} activates many enzymes through interactions mediated by a family of related Ca^{2+}-binding proteins. Examples include stimulation of contraction of striated muscle via Ca^{2+}/troponin C, and activation of several enzymes by Ca^{2+}/calmodulin, including a Ca^{2+}-transporting ATPase, adenylate cyclase, phosphodiesterase and at least four protein kinases. Specific Ca^{2+}/calmodulin-dependent protein kinases and the pathways they regulate include the myosin light chain kinases (smooth and striated muscle contraction), phosphorylase kinase (glycogen breakdown), and elongation factor-2 kinase (protein synthesis). The calmodulin-dependent multiprotein kinase phosphorylates many proteins and may regulate many different processes in vivo. Increases in cytosolic Ca^{2+} also produce concomitant rises in mitochondrial Ca^{2+}, which activates three dehydrogenases involved in carbohydrate oxidation and the tricarboxylic acid cycle.

The effects of cyclic AMP and cyclic GMP are almost entirely mediated by related cyclic AMP- and cyclic GMP-dependent protein kinases. Physiological substrates for the former kinase include key proteins involved in glycogen metabolism, glycolysis/gluconeogenesis, lipolysis, amino acid breakdown, heart muscle contraction, short-term memory in sea snails, and gene expression.

The effects of diacylglycerol are mediated by the Ca^{2+}- and phospholipid-dependent protein kinase, protein kinase C. Physiological substrates for this kinase are probably membrane associated, and include several receptors such as the epidermal growth factor receptor. In this case protein kinase C appears to mediate inhibition of the receptor tyrosine kinase activity.

Protein (tyrosine) kinases produce many of their downstream effects by modulating the activities of protein (serine/threonine) kinases or phosphatases, such as ribosomal protein S6 kinase. This appears to occur via complex cascades of protein phosphorylation. The links between protein (tyrosine) and protein (serine/threonine) kinases (e.g. MAP kinase, raf-1) are still being elucidated.

Eukaryotic cells contain four major cytosolic protein (serine/threonine) phosphatases, i.e. protein phosphatases-1, -2A, -2B and -2C. Protein phosphatase-1 is regulated by targetting subunits such as the glycogen-binding "G" subunit, and by inhibitor proteins, some of which are regulated in turn by phosphorylation by cyclic AMP-dependent protein kinase. Protein phosphatase-2B is a Ca^{2+}/calmodulin-dependent enzyme. The protein (tyrosine) phosphatases are very active in cells, which may explain why significant tyrosine phosphorylation only occurs if the substrate protein forms a stable complex with the protein (tyrosine) kinase. Some protein (tyrosine) phosphatases are soluble, while others form the cytoplasmic domains of transmembrane proteins which are putative receptors or cell adhesion molecules.

In the one case where structural studies have been performed on a protein regulated by phosphorylation (phosphorylase), the phosphorylated serine forms a novel interaction with two arginine residues at the interface of the two subunits, thus stabilizing the active conformation. Via small changes in quarternary structure, this difference in interaction between the two subunits is transmitted to the active site, which is located more than 30 Å from the phosphorylation site.

Protein phosphorylation is therefore the most widespread device which mediates the downstream effects of first messengers. It can provide a large amplification of the signal, allows a very sensitive response, and also enables great flexibility to be built into the control system. The ATP utilized in phosphorylation systems is not wasted, but is the small price paid in order to enable a very small initial signal to cause switching between different conformational states of a protein.

FURTHER READING

1. "The calcium signal" – introductory review on role of Ca^{2+} in muscle contraction and cell signalling. Carafoli, E. and Penniston, J.T. (1985) Scientific American, November, pp. 50–58.
2. "Primary structure of the receptor for calcium channel blockers from skeletal muscle" – dihydropyridine receptor. Tanabe, T., Takeshima, H., Mikami, A., Flockerzi, V., Takahashi, H., Kangawa, K., Kojima, M., Matsuo, H., Hirose, T. and Numa, S. (1987) Nature **328**, 313–318.
3. "Primary structure and expression from complementary DNA of skeletal muscle ryanodine receptor". Takeshima, H., Nishimura, S., Matsumoto, T., Ishida, H., Kangawa, K., Minamino, N., Matsuo, H., Ueda, M., Hanaoka, M., Hirose, T. and Numa, S. (1989) Nature **339**, 439–445.
4. "Ultrastructure of the calcium release channel of sarcoplasmic reticulum". Saito, A., Inui, M., Radermacher, M., Frank, J. and Fleischer, S. (1988) J. Cell Biol. **107**, 211–219.

5. "Structure of the calcium regulatory muscle protein troponin-C at 2.8 Å resolution". Herzberg, O. and James, M.N.G. (1985) Nature **313**, 653–659.
6. "Modulation of cAMP effects by Ca^{2+}/calmodulin" – review on calmodulin-dependent enzymes. Pallen, C.J., Sharma, R.K. and Wang, J.H. (1985) BioEssays **2**, 113–117.
7. "Adenylyl cyclase amino acid sequence: possible channel- or transporter-like structure" – structure of calmodulin-stimulated adenylate cyclase. Krupinski, J., Coussen, F., Bakalyar, H.A., Tang, W.J., Feinstein, P.G., Orth, K., Slaughter, C., Reed, R.R. and Gilman, A.G. (1989) Science **244**, 1558–1564.
8. "Protein kinases: a diverse family of related proteins" – review. Taylor, S.S. (1987) BioEssays **7**, 24–29.
9. "Protein serine/threonine kinases" – review. Edelman, A.M., Blumenthal, D.K. and Krebs, E.G. (1987) Ann. Rev. Biochem. **56**, 567–613.
10. "Pseudosubstrates turn off protein kinases" – minireview. Hardie, D.G. (1988) Nature **335**, 592–593.
11. "Multifunctional Ca^{2+}/calmodulin-dependent protein kinase: domain structure and regulation" – review. Schulman, H. and Lou, L.L. (1989) Trends Biochem. Sci. **14**, 62–66.
12. "Ca^{2+} as a second messenger within mitochondria" – review. McCormack, J.G. and Denton, R.M. (1986) Trends Biochem. Sci. **11**, 258–262.
13. "Rapid and reversible translocation of the catalytic subunit of cAMP-dependent protein kinase type II from the Golgi complex to the nucleus". Nigg, E.A., Hilz, H., Eppenberger, H.M. and Dutly, F. (1985) EMBO J. **4**, 2801–2806.
14. "Cyclic AMP stimulates somatostatin gene transcription by phosphorylation of CREB at serine 133". Gonzalez, G.A. and Montminy, M. (1989) Cell **59**, 675–680.
15. "The protein kinase C family: heterogeneity and its implications" – review. Kikkawa, U., Kishimoto, A. and Nishizuka, Y. (1989) Ann. Rev. Biochem. **58**, 31–44.
16. "Requirement for integration of signals from two distinct phosphorylation pathways for activation of MAP kinase". Anderson, N.G., Maller, J.L., Tonks, N.K. and Sturgill, T.W. (1990) Nature **343**, 651–653.
17. "Insulin receptor signalling. Activation of multiple serine kinases" – review. Czech, M.P., Klarlund, J.K., Yagaloff, K.A., Bradford, A.P. and Lewis, R.E. (1988) J. Biol. Chem. **263**, 11017-11021.
18. "The structure and regulation of protein phosphatases" – review. Cohen, P. (1989) Ann. Rev. Biochem. **58**, 453–508.
19. The molecular mechanism by which insulin stimulates glycogen synthesis in mammalian skeletal muscle" – activation of protein phosphatase-1 via phosphorylation of the G subunit. Dent, P., Lavoinne, A., Nakielny, S., Caudwell, F.B., Watt, P. and Cohen, P. (1990) Nature **348**, 302–308.

20. "Protein tyrosine dephosphorylation and signal transduction" – review. Tonks, N.K. and Charbonneau, H. (1989) Trends Biochem. Sci. **14**, 497–500.
21. "The allosteric transition of glycogen phosphorylase". Barford, D. and Johnson, L.N. (1989) Nature **340**, 609–616.
22. "Amplification and adaptation in regulatory and sensory systems" – review of regulatory cascade mechanisms. Koshland, D.E., Goldbeter, A. and Stock, J.B. (1982) Science **217**, 220–225.
23. "A thousand and one protein kinases" – review. Hunter, T. (1987) Cell **50**, 823–829.

NINE

REGULATION OF CELL PROLIFERATION AND CANCER

As discussed briefly in Chapter 1, there is a fundamental difference between unicellular and multicellular organisms in the factors which determine their rate of cell growth and division. Growth of unicellular organisms is usually limited by the availability of nutrients, and in rich medium they divide exponentially. By contrast, cells in a multicellular organism are normally bathed continually in a nutrient-rich medium (the extracellular fluid), yet their division must be strictly controlled. In healthy animals, cells tend to grow in organized layers or sheets, and cell proliferation generally occurs at well-defined locations (e.g. at the base of the dermis in skin) where there are **stem cells** capable of proliferation. In adult animals, proliferation generally only occurs to replace cells which have a limited lifetime (e.g. skin, gut and blood), and to repair wounds. Clearly there has to be a balance between production of new cells and loss of cells due to death or terminal differentiation. Cancer is caused by a breakdown in the systems which control this balance. A tumour is thought to be usually derived from a single aberrant stem cell, and it is interesting to note that a 1 g tumour, containing perhaps 10^9 cells, could result from only 30 division cycles. If it stops at that stage, a single tumour may be harmless or **benign**. However, **malignant** tumours go a stage further and **metastasize**: cells become detached and invade other parts of the body, where they form new tumours, with devastating results. **Leukaemias** are caused by uncontrolled division of leukocytes (white blood cells), and are also very serious. Since cancer is now responsible for nearly 20% of deaths in Western countries, in recent years there has been a considerable amount of time and money invested in attempts to understand how proliferation of eukaryotic cells is controlled.

9.1 *Growth of normal and abnormal cells in culture*

9.1.1 *Growth requirements of cultured cells*

The initial approach to the problem of when and how cells from multicellular organisms divide was to attempt to grow them outside the body, the technique of **tissue culture**. The aim was to determine the extracellular factors necessary to

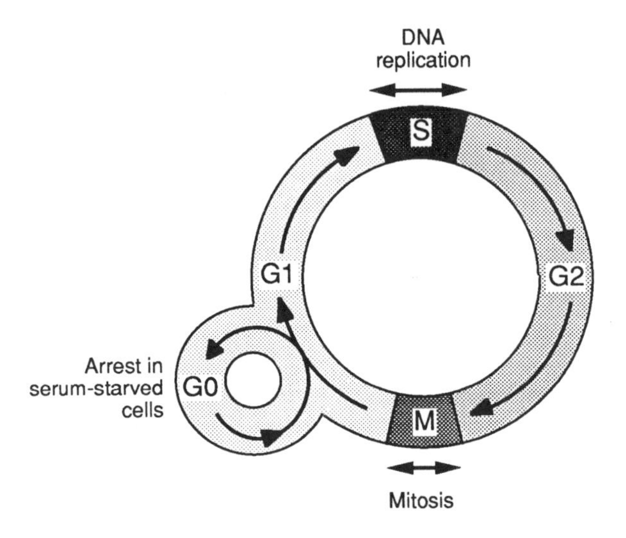

Fig. 9.1 The 4 phases of the cell cycle.

promote growth. By the early 1960s, the field had developed to the stage where many cell types could be maintained in a defined medium of ions, low molecular weight nutrients (e.g. glucose, essential amino acids) and vitamins. However, they would only **divide** in the presence of a crude mixture of macromolecules, usually given in the form of serum, especially from foetal animals (e.g. foetal calf serum). Painstaking fractionation of serum has now revealed that this is largely because serum provides protein **growth factors**.

Because of the relative ease with which they can be grown in culture, the most widely used cells for studies of growth control have been **fibroblasts**, which are responsible for secretion of the extracellular matrix in connective tissue. Fibroblasts from many sources have been established in culture, with a popular example being the mouse 3T3 cell line (Section 3.2.2 gives a definition of the term cell line). The normal division cycle of cells can be divided into four phases (Figure 9.1), defined by the periods of mitosis and DNA replication (M and S phase), with G1 and G2 being the "gaps" between them. If 3T3 cells are starved of serum or growth factors, they arrest in a **quiescent** state just beyond M phase known as G0. If restimulated with growth factors, after a period longer than a normal G1 phase, they recommence DNA replication (i.e. S phase). The effect of growth-stimulating agents is therefore usually monitored as incorporation of radioactive thymidine into DNA, measured 12–30 hours after stimulation of quiescent cells (Figure 9.2).

The most important growth factor for fibroblasts in serum is **platelet-derived growth factor** (PDGF, Section 2.3.2), which is released by platelets when whole blood is clotted to make serum. However, a large number of other growth factors

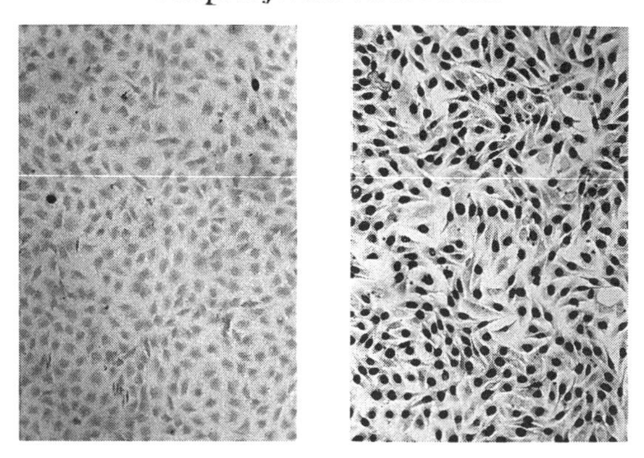

Fig. 9.2 DNA synthesis in proliferating 3T3 cells. The cells had been incubated with [^{3}H]thymidine, and sections made and coated with photographic emulsion to detect radioactive DNA. In this light micrograph the nuclei of cells which have been treated with foetal calf serum and have undergone DNA replication appear dark (right). Cells incubated without serum (left) remain quiescent. Courtesy of Theresa Higgins and Enrique Rozengurt.

Table 9.1 Combinations of growth factors stimulating DNA replication in 3T3 cells. Responses range fom zero (–) to a maximal amount equivalent to that obtained in response to 10% serum (++++). After Rozengurt (1986) Science **234**, 161-166. EGF, epidermal growth factor, PGE_1, prostaglandin E_1.

	PDGF	Bombesin	Vasopressin	Insulin	EGF	PGE_1
PDGF	+++	++++	+++	++++	++++	+++
Bombesin	++++	++	++	++++	+++	++
Vasopressin	+++	++	–	+++	++	++
Insulin	++++	++++	+++	–	+++	+++
EGF	++++	+++	++	+++	–	+
PGE_1	+++	++	++	+++	+	–

been found in other sources, and some of those which stimulate DNA replication in 3T3 cells are shown in Table 9.1. How do these agents mediate their effects? An important generalization is that they appear to use the same intracellular signals (e.g. cyclic AMP, diacylglycerol, inositol trisphosphate/Ca^{2+}, and tyrosine

Table 9.2 Synergistic combinations of pharmacological agents stimulating DNA replication in 3T3 cells (OAG=1-oleoyl-2-acetyl-glycerol). Responses range fom zero (–) to a maximal amount equivalent to that obtained in response to 10% serum (++++). Since no pharmacological agents are known which directly mimic the activation of a tyrosine kinase receptor, insulin has also been included in the Table. After Rozengurt, E. (1986) Science **234**, 161–166.

	Insulin	Phorbol ester	OAG	Cholera toxin	cAMP analogues
Insulin	–	+++	+++	+++	+++
Phorbol ester	+++	–	–	++	++
OAG	+++	–	–	++	++
Cholera toxin	+++	++	++	–	–
cAMP analogues	+++	++	++	–	–

phosphorylation) as the first messengers discussed elsewhere in this book. Indeed many of the growth factors listed in Table 9.1 are hormones or mediators which we have already met in earlier chapters. In support of this view, pharmacological agents which mimic the intracellular signals can stimulate growth and division of 3T3 (Table 9.2) and other cells, as well as the more acute effects discussed in Chapter 8.

A potential problem with this viewpoint is that one might therefore expect any hormone or mediator to stimulate growth and proliferation of any cell. That this does not happen may be explained in part by two observations:

1) The first messenger(s) normally need to be present in the medium **for several hours** before a division cycle can be initiated, whereas the acute effects occur within minutes.
2) Often, combinations of growth factors or pharmacological agents acting through two or more signal transduction pathways are required to initiate proliferation (Tables 9.1 and 9.2). For example, cholera toxin (which increases cyclic AMP and hence activates cyclic AMP-dependent protein kinase) has little effect on DNA replication in 3T3 cells on its own. However, if given in combination with insulin or epidermal growth factor (which stimulate tyrosine phosphorylation), phorbol ester (which stimulates protein kinase C), or vasopressin (which stimulates protein kinase C and Ca^{2+} release via phosphoinositide breakdown), cholera toxin greatly potentiates the effects of

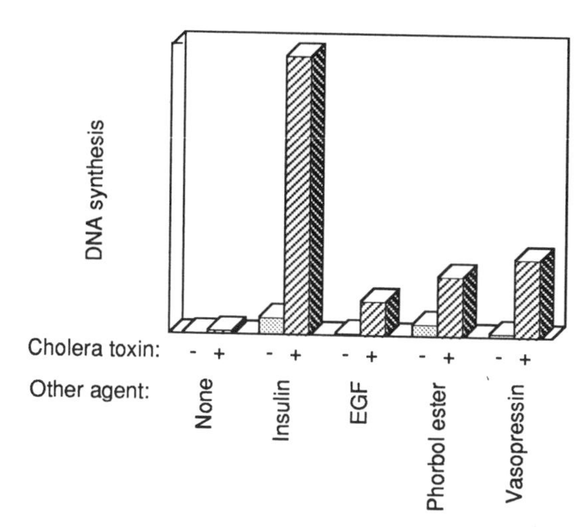

Fig. 9.3 Synergistic effect of cholera toxin and other mitogenic agents on DNA synthesis. Original data from Rozengurt, Legg, Strang and Courtenay-Luck (1981) Proc. Natl. Acad. Sci. USA **78**, 4392-4396.

the other agents (Figure 9.3). One reason that platelet-derived growth factor is so effective at stimulating DNA replication in 3T3 cells is that it not only activates tyrosine phosphorylation and phosphoinositide breakdown, but also stimulates synthesis of prostaglandins which elevate cyclic AMP.

9.1.2 *Transformation – a tissue culture model for cancer*

If fibroblasts are removed from an animal and grown in culture (a so-called **primary culture**), they normally go through a finite number of divisions (up to about 100) and then die. It appears that normal cells are programmed to die. However, occasionally one or more cells spontaneously undergo genetic changes (often associated with the appearance of extra copies of some chromosomes) which allow them to divide indefinitely in culture: they have become **immortal**, after which they are referred to as a **cell line**. For reasons that are not clear, immortalization occurs with high frequency in mouse embryo fibroblasts, and a very useful model system derived from this source is the 3T3 cell line. Because of their immortality, 3T3 cells cannot strictly be regarded as normal cells, but their appearance and general growth characteristics are similar to those of fibroblasts in primary culture. However, 3T3 cells readily undergo further changes in response to mutagenic agents or certain viruses, after which they closely resemble cancer cells. This process, known as **transformation**, is a model system for the initiation of cancer *in vivo*. Transformed fibroblasts differ from "normal" cells (i.e. primary cells or cell lines) in several ways:

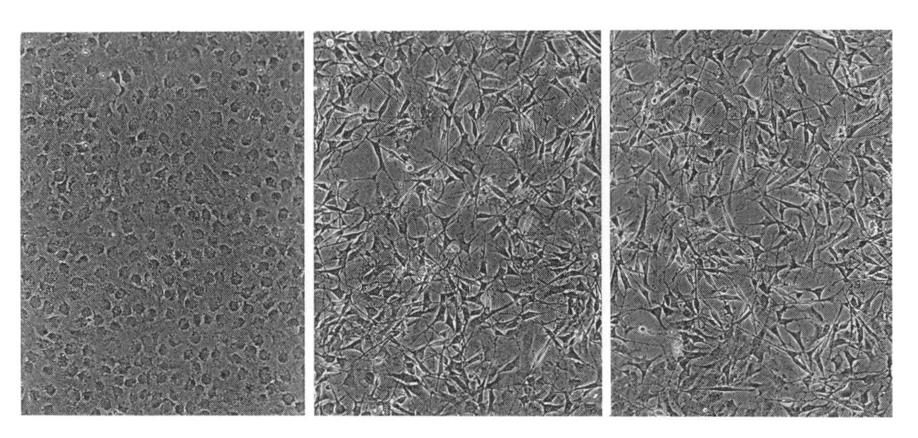

Fig. 9.4 Phase contrast light micrographs of normal (left) and transformed NIH-3T3 cells showing their different morphology. The cells had been microinjected with control buffer (left) or with an oncogene product (centre: H–*ras*; right: K–*ras*) 40 hours previously. Courtesy of Hugh Paterson.

1) *Anchorage dependence*
Normal fibroblasts only grow when attached to a solid surface via adhesive proteins (e.g. fibronectin) which they themselves secrete. *In vivo* this surface would be the extracellular matrix, but in culture they attach to surface charges on the culture dish. By contrast, transformed cells do not require a surface and can grow in semi-liquid medium, e.g. soft agar.
2) *Morphology and motility*
Normal fibroblasts are flattened and elongated, and inside the cell there are well-organized actin fibres leading to the sites (adhesion plaques) where they are attached to the substrate via the adhesive proteins. Transformed cells exhibit marked changes in shape (Figure 9.4) due to rearrangement of their cytoskeleton. They often also become less rigidly attached to the substrate and more motile.
3) *Contact inhibition*
Normal fibroblasts (particularly in the case of cell lines) often only grow in layers one cell thick (monolayers). They stop dividing when they grow to touch each other (contact inhibition). Transformed cells can continue dividing when they come into contact, so that if there are a few transformed cells in a field of normal cells they form mini-tumours called foci, which are visible to the naked eye (Figure 9.5).
4) *Growth factor dependence*
Normal cells only grow in the presence of growth factors, usually provided routinely by the presence of 10% foetal calf serum in the medium. Transformed cells can often grow in the presence of reduced concentrations of growth factors.

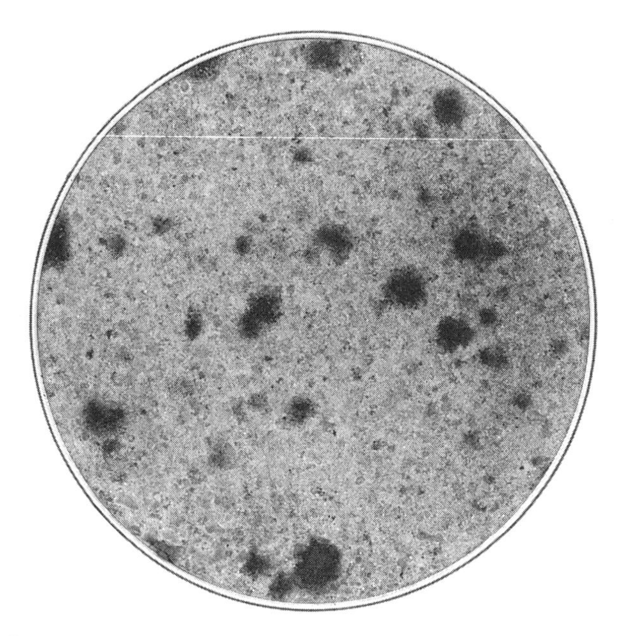

Fig. 9.5 A 10 cm petri dish of NIH-3T3 cells 16 days after transfection with a human oncogene form of H-*ras* DNA, showing clumps or **foci** of transformed cells visible to the naked eye. Courtesy of Hugh Paterson.

5) *Tumour formation in vivo*
Transformed cells, but not normal cells, will give rise to tumours when injected into host animals. In order to avoid immunological rejection of the tumour, it may be necessary to use host animals which have a deficiency in cell-mediated immunity, such as the "nude" mouse.

Some of the characteristics described above may be interdependent. For example, untransformed 3T3 cells may only grow when flattened out in a monolayer because of their requirement for high concentrations of growth factor. When rounded up and in suspension, they have a low surface area, and may simply be unable to express enough growth factor receptors on their surface to trigger division.

9.2 *Discovery of oncogenes*

It has been known for a long time that cancer can be initiated by agents which cause mutation of DNA, such as mutagenic chemicals, ionizing radiation, ultraviolet light and cigarette smoke. Since a tumour is normally of clonal origin, i.e. it arises from multiple divisions from a single cell, the idea arose that cancer was due to a

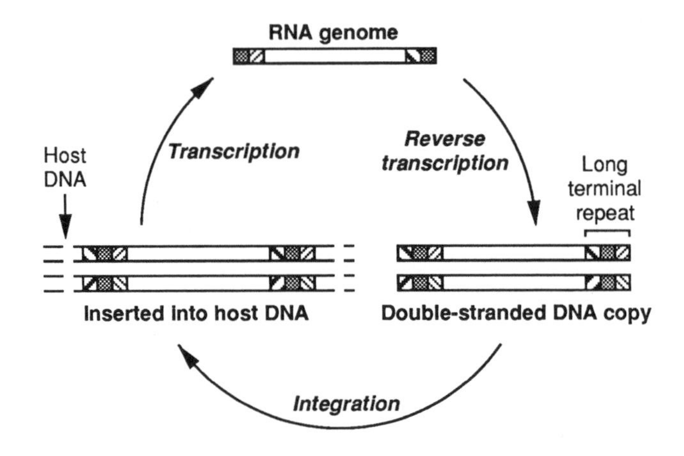

Fig. 9.6 Simplified version of the life cycle of a retrovirus.

mutation in that progenitor cell. From classical genetics we tend to think of mutations in multicellular organisms as only being important if they occur in germ cells (sperm or eggs), because only germ cell mutations are passed on to the next generation. Mutations do of course occur in non-germ cells (**somatic cells**), but a harmful mutation in a somatic cell is usually not noticed, because that cell merely dies and neighbouring cells take over its functions. However things would be different if the mutation caused the cell to divide continuously. It is now becoming clear that cancer can be caused by mutation, or altered expression, of key genes whose products are involved in regulation of cell growth. The mutated, cancer-causing genes are known as **oncogenes**.

9.2.1 *Discovery of viral oncogenes – the src gene*

One way in which harmful genes can get into cells is via infection by a virus. Important breakthroughs in understanding cancer came from the study of **acutely transforming retroviruses**, also called RNA tumour viruses. The best known of these is the **Rous sarcoma virus**, discovered by Peyton Rous in 1911. This virus causes sarcomas (cancer of connective tissue) in chickens, and also transforms fibroblasts in culture, and it is said to be acutely transforming because these endpoints invariably occur with rapid onset. Being a retrovirus, its hereditary material is carried in the form of single-stranded RNA rather than DNA, and in order to replicate the RNA has first to be copied into double-stranded DNA via the viral enzyme reverse transcriptase (Figure 9.6). Multiple copies of RNA are then produced by the host transcription machinery from the DNA template, a process that absolutely requires prior insertion of the viral DNA into the host genome.

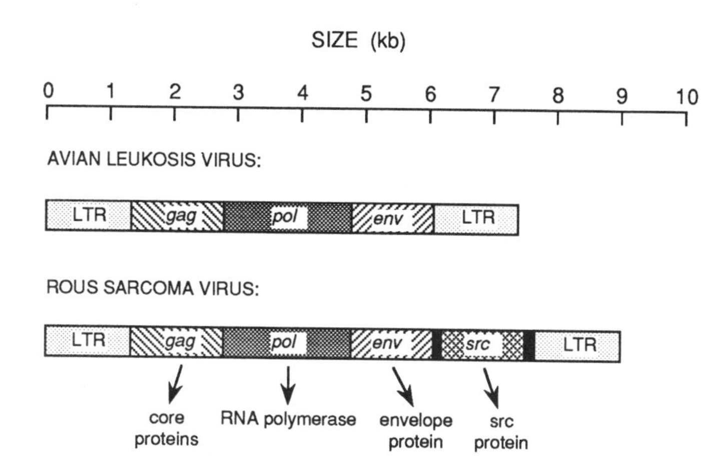

Fig. 9.7 Comparison of the genomes of a non-transforming retrovirus (avian leukosis virus) and an acutely transforming retrovirus which was derived from it (Rous sarcoma virus). LTR, long terminal repeat.

These RNA copies then form the genomes of new viruses as well as acting as mRNA for translation of viral proteins.

Among the great advantages of retroviruses for the study of cancer is their very simple genomes (Figure 9.7). The genome of all retroviruses is a linear molecule, and consists of sequences which are repeated at each end and which generate the **long terminal repeat** (involved in the mechanism of insertion into host DNA and in regulation of transcription) plus only three genes:

1) *gag*, which encodes a "polyprotein" which is proteolytically cleaved into the proteins of the viral core;
2) *pol*, which encodes the reverse transcriptase;
3) *env*, which encodes the viral envelope protein.

Unlike those of other retroviruses, the genome of the Rous sarcoma virus contains a fourth gene, *src* (Figure 9.7). Deletion of this gene showed that it was not required for virus replication, but that it was absolutely required for tumour formation or transformation. The converse experiment could also be carried out by transfection of cloned *src* DNA into cultured cells. **Transfection** is the incorporation of exogenous, naked DNA into intact cells, and is often carried out using NIH-3T3 cells, a sub-line of 3T3 cells that is particularly adept at taking up DNA. When pure *src* DNA is transfected into NIH-3T3 cells they are transformed. Taken together with the gene deletion experiments, this shows that the *src* gene is both necessary and sufficient to initiate this process. This was a very exciting finding because it suggested that a single gene product might give rise to cancer.

What is the function of the *src* gene? The gene encodes a polypeptide of 60 kDa, known originally as pp60src. When this polypeptide was precipitated from extracts of virus-infected cells using specific antibody, and incubated with radioactive ATP, a surprising finding was made: the antibody molecule became phosphorylated on **tyrosine** residues. Although phosphorylation of the antibody probably has no physiological relevance, this observation provided good evidence that pp60src was a tyrosine-specific protein kinase. Subsequently it was shown that a region of pp60src was homologous with the cytoplasmic domains of receptors for several growth factors such as epidermal growth factor, insulin and platelet-derived growth factor (Section 7.4), which also display tyrosine kinase activity. Thus a viral gene which produces cancer is related to normal host genes which regulate growth.

Why does the Rous sarcoma virus carry the *src* gene? From studies using pp60src DNA and antibody probes it became clear that normal, uninfected cells contain a closely related gene coding for a very similar 60 kDa protein (tyrosine) kinase. This is termed **cellular** or *c-src* to distinguish it from **viral** or *v-src*. Recall that one stage in the retroviral life cycle involves insertion of viral DNA into the host genome. It seems likely that at some time in the past a recombination event caused the "capture" of a host *src* gene which then became part of the viral genome: it is now known that incorporation of a random fragment of host DNA is a relatively common event in retroviral replication. Since that time mutations have occurred in the viral *src* gene, which now shows subtle differences from the host gene. One difference, discussed further in Section 9.3.2, is that the protein kinase activity of *v-src* is much higher than that of *c-src*. Unlike *v-src*, the *c-src* gene does not cause transformation when transfected into NIH-3T3 cells, and it is referred to as a **proto-oncogene**, i.e. a normal cellular gene which can be converted into an oncogene by mutation.

A different class of transforming virus is exemplified by the SV40 and polyoma viruses. Unlike the retroviruses, these have DNA genomes, and in so-called **permissive** host cells they merely replicate and kill the cells. However, in certain **non-permissive** host cells viral DNA replication is blocked, and what happens instead is that in a small proportion of infected cells, the viral DNA inserts into the host genome and the cell becomes transformed. Transformation has been shown to be due to viral proteins called tumour or T antigens. Unlike the *src* gene, the T antigen genes appear to be genuine viral genes which have no obvious host equivalent. Intriguingly, however, the "middle" T antigen of polyoma binds to and activates the host *c-src* protein kinase, and this may at least partially explain the transformation produced by the virus.

Thus two viruses (Rous sarcoma virus and polyoma) which at first sight cause cancer in quite distinct ways, appear to do so via oncogenes whose protein products either mimic a normal host gene product (*c-src*), or activate it.

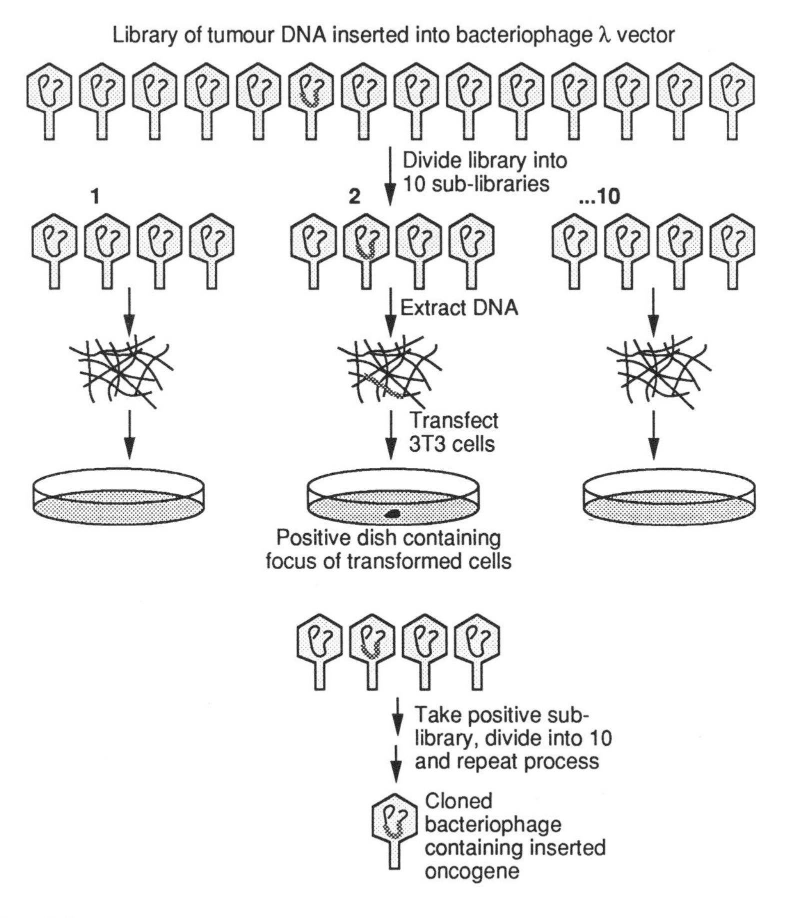

Fig. 9.8 Isolation of an oncogene from a human tumour cell line, by repeated screening of smaller and smaller portions of a tumour DNA library constructed using a bacteriophage λ vector.

9.2.2 *Oncogenes of non-viral origin*

Although oncogenic viruses provide simple model systems to study, most clinically important human cancers do not appear to be caused by viruses, and are probably the result of mutations induced by environmental radiation or carcinogens. Transfection of DNA into NIH-3T3 cells provides a method of isolating the oncogenes from human tumours of any origin: Figure 9.8 illustrates one method. DNA from the tumour cells is broken into fragments of a size which would be expected to carry only a few genes. These are amplified via insertion into a λ vector to produce a gene "library". The library is divided into ten sub-libraries, and DNA from each of these is used to transfect 3T3 cells. Only the cells which take up oncogene DNA will be transformed, and the process can then be repeated by subdivision of the positive library. Eventually the oncogene responsible for transformation can be cloned. In certain human tumours (e.g. cancers of the large bowel and pancreas), transformation of the 3T3 cells by tumour DNA is usually due to alteration of one of three closely related genes, *K-ras, H-ras* and *N-ras*. They are called *ras* genes because the first two appear to have given rise to the oncogenes encoded by the *Kirsten* and *Harvey* strains of the rat sarcoma virus, which causes connective tissue tumours in rodents. Comparison of the sequences of normal and oncogene forms of *ras*, from both viruses and human cancers, shows that the latter have usually arisen by single point mutations. The function of the *ras* proteins and the effects of these mutations will be discussed further in Section 9.3.3.

9.3 *Functions of oncogenes and their normal counterparts*

At least 50 oncogenes have now been discovered, and many of these were originally discovered as genes "captured" by acutely transforming retroviruses. As well as helping us to understand cancer, the study of these oncogenes has greatly illuminated the normal control of cell growth and division. This is because a common theme in the mechanism of action of the oncogene products is that they subvert a normal cellular control pathway. We will now consider the various levels at which this can occur, starting on the outside of the cell and working inwards.

9.3.1 *Oncogenes can mimic growth factors*

Human platelet-derived growth factor (PDGF – Section 2.3.2) exists as three isoforms, which are disulphide-linked dimers of related 12 kDa subunits (A–A, B–B or A–B). Amino acid sequencing of the B chain produced the exciting observation that it was almost identical to the mature product of the *sis* oncogene of simian sarcoma virus. This suggests that the virus transforms cells by an **autocrine** mechanism, i.e. infected cells continually produce a growth factor which stimulates receptors on the same cell (Figure 9.9). In agreement with this mechanism, virus infection closely mimics the effect of continual addition of PDGF:

1) the *sis* gene product and PDGF compete for the same binding sites;
2) antibodies against PDGF prevent proliferation of primary fibroblasts after infection by simian sarcoma virus;
3) virus infection produces anchorage-independence, but only to the same limited level produced by PDGF.

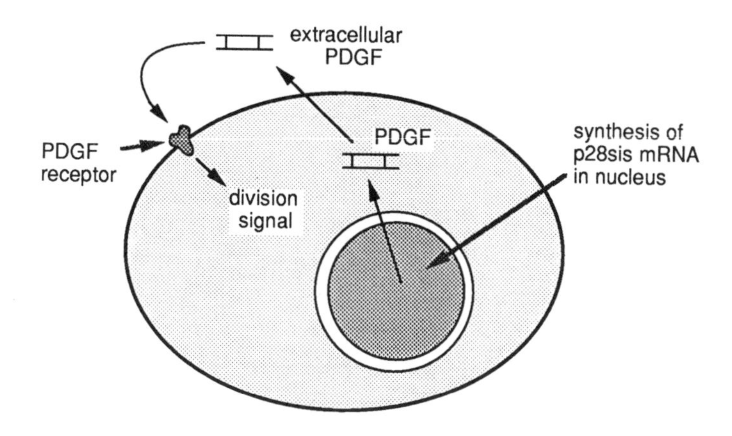

Fig. 9.9 The autocrine mechanism for transformation by the *sis* oncogene.

Could this autocrine mechanism be involved in other forms of cancer, particularly human cancers? A large amount of circumstantial evidence indicates that it is:

1) several human tumour cell lines (e.g. derived from bone or brain tumours) secrete PDGF B–B homodimer into the medium;
2) fibroblasts transformed with the DNA tumour virus SV40 secrete a PDGF-like growth factor into the medium: in this case the virus does not encode a growth factor, but switches on a host gene;
3) cells transformed by Moloney murine leukaemia virus secrete so-called transforming growth factors, TGF-α and -β: despite their name these are now known to be normal host gene products. TGF-α is related to epidermal growth factor and acts by binding to the EGF receptor;
4) at least some cell lines derived from human "small cell" lung cancer have been reported to secrete a bombesin-like peptide into the medium: bombesin is known to be a growth factor for fibroblasts (Table 9.2).

9.3.2 *Oncogenes can mimic growth factor receptors*

The erb-B oncogene

When the first amino acid sequencing on the epidermal growth factor receptor was carried out, the exciting discovery was made that it was very similar to the sequence of one of the two oncogenes (*v-erb* B) of the avian erythroblastosis virus. The EGF receptor gene is therefore a proto-oncogene which could be termed *c-erb* B. However, the viral and cellular forms differ in that the viral protein is truncated

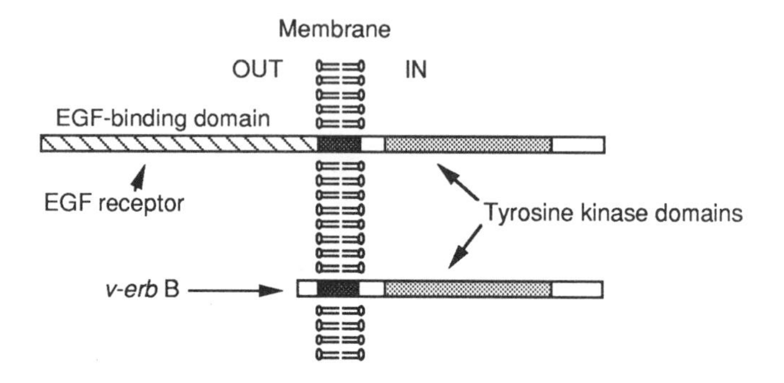

Fig. 9.10 Comparison of the domain structures of the EGF receptor and *v-erb* B.

and almost entirely lacks the ectoplasmic domain (Figure 9.10). Since this domain binds the messenger, this raised the intriguing possibility that *v–erb* B was a constitutively active form of the EGF receptor, an idea which has now been confirmed.

The v-erb A gene

Although *v-erb* B causes transformation of some cells on its own, with erythroblasts (the precursors of red blood cells) a full effect is only achieved if a second gene from the avian erythroblastosis virus, *v-erb* A, is also transfected. Sequencing of *v-erb* A indicated that it showed some homology with the nuclear receptors for steroid hormones, particularly in the region of the latter which is thought to bind DNA (Chapter 10). However, when the normal cellular *c-erb* A was cloned and expressed it was found to bind the thyroid hormones (T3 and T4) rather than steroids. At that time thyroid hormone receptors had been very poorly characterized, and this was another example where the study of disruption of cell metabolism by a virus illuminated how normal cellular regulation occurs.

Unlike *c-erb* A, *v-erb* A protein does not bind thyroid hormones and it was at first suggested, by analogy with *v-erb* B, that *v-erb* A may be a constitutively active receptor. However, it is now thought that thyroid hormones promote differentiation of erythroblasts into erythrocytes (red blood cells), thus abolishing their ability to proliferate. Possibly *v-erb* A is an **inactive** form of thyroid hormone receptor which competes for the nuclear binding sites (Chapter 10) with the normal form, thus preventing differentiation and allowing *v-erb* B to transform the cells.

The neu oncogene

One strain of newborn rats exposed to the mutagen, ethylnitrosourea, develop tumours of the nervous system called **neuroblastomas**. An oncogene invariably associated with these tumours, *neu*, was identified by serial transfections of NIH-3T3 cells as described in Section 9.2.2. The predicted amino acid sequence of the *neu* gene product suggested that it was closely related to, but distinct from, the EGF receptor. The *neu* gene from normal rats does not cause transformation when transfected into NIH-3T3 cells. Sequence analysis of the *neu* gene from several different neuroblastomas showed that a single thymine to adenine mutation had occurred, leading to the replacement of a valine in the transmembrane helix by a glutamic acid.

Although the *neu* gene product is assumed to be a growth factor receptor, the growth factor which regulates it is not known. Since the single val→glu mutation converts a proto-oncogene into an oncogene, it is assumed that it causes the receptor to become constitutively active.

The mas oncogene

Not all oncogenes are readily detected by the NIH-3T3 transfection assay. An alternative is to transfect tumour DNA into 3T3 cells and then to look for genesis of tumours *in vivo* by injecting the cells into "nude" mice. The *mas* oncogene, which is only weakly transforming in the *in vitro* assay, was isolated from a human tumour cell line in this way. Hydropathy plots of the sequence of the *mas* protein suggested that it would form seven transmembrane helices, similar to receptors of the "seven-pass" variety (Section 6.3.2). This was no coincidence, because expression of the gene in *Xenopus* oocytes revealed that it was an angiotensin II receptor which was coupled to phosphoinositidase C. The *mas* oncogene has only been isolated once, and it has been suggested that the screening method used would detect mutations generated during transfection into the 3T3 cells. Nevertheless these results demonstrate that abnormal growth can result from alterations to receptors other than those of the tyrosine kinase type.

Other links between growth factor receptors and oncogenes

As discussed in Section 6.3.3, the sequences of the retroviral *v-kit* and *v-fms* oncogenes are related to that of the PDGF receptor. Normal cells contain *c-kit* and *c-fms* genes distinct from the PDGF receptor, so these oncogenes were not derived from the PDGF receptor gene itself. The *c-fms* gene product has now been identified as the receptor for **colony stimulating factor-1**, a growth factor for macrophages.

What is the function of c-src?

Although *c-src* and *v-src* are protein (tyrosine) kinases, they do not contain a transmembrane domain and, by themselves, clearly could not act as receptors for first messengers. The tyrosine kinase activity of the viral protein against model substrates is about ten-fold higher than that of the cellular protein. When the amino

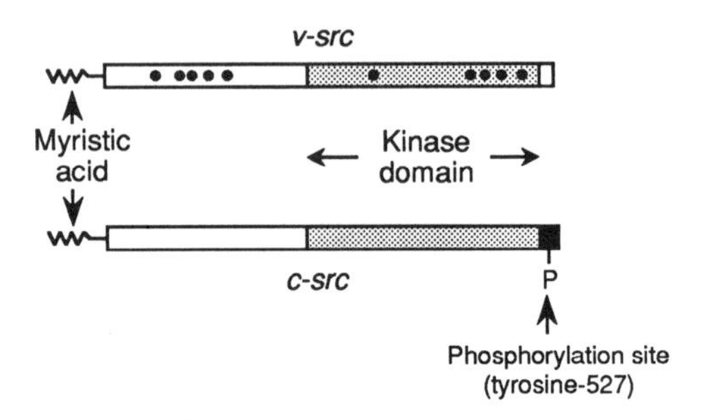

Fig. 9.11 Comparison of the domain structures of *v-src* and *c-src*. The black dots show amino acid replacements in the N-terminal and catalytic domains of *v-src*. The extreme C-termini are completely different (see text).

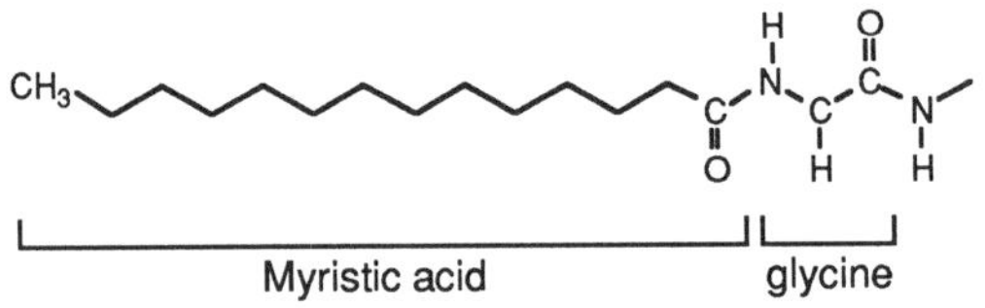

Fig. 9.12 Structure of the N-terminus of a myristoylated protein.

acid sequences are compared, there are minor differences throughout their length, but the most significant difference is at the C-terminus where the last 19 residues of *c-src* have been replaced by 12 unrelated amino acids in *v-src*. The significance of this is that *c-src* is normally found to be phosphorylated *in vivo* on a tyrosine residue within this segment (Figure 9.11). Dephosphorylation of this site causes a 5-fold activation of the tyrosine kinase activity of *c-src* so that it approaches that of *v-src*. The middle T antigen of polyoma virus (Section 9.2.1) activates *c-src* by binding to it and promoting net dephosphorylation of this site.

Another interesting feature of both *v-src* and *c-src* is that they are modified at the N-terminus by myristoylation, the attachment of a C_{14} fatty acid (Figure 9.12), and both proteins are found to be associated with membrane fractions. This is probably due to insertion of the acyl chain of the myristic acid into the membrane, because if the N-terminal glycine of *v-src* is altered by site-directed mutagenesis, the protein is not myristoylated and does not associate with the membrane. Although the protein kinase activity is apparently normal, this mutant does **not** cause transformation of chick embryo fibroblasts. Thus the membrane localization of the protein is very important in transformation.

The normal function of *c-src* remains unclear. A variant produced by alternative mRNA splicing is expressed at high levels in postmitotic, differentiated

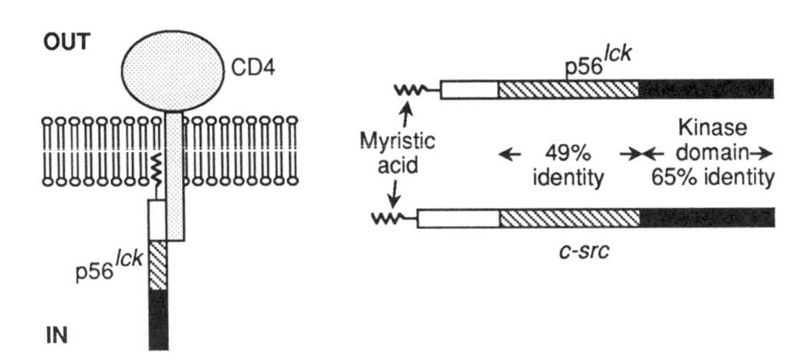

Fig. 9.13 Association of $p56^{lck}$ with CD4, and comparison of domain structures of $p56^{lck}$ and *c-src*.

cells such as neurones, suggesting that its function may not solely be in regulation of cell proliferation. Although *c-src* in itself is not a receptor, it remains possible that to fulfil its normal function it associates with the cytoplasmic domain of a transmembrane receptor. A possible precedent for this is seen in the case of the protein (tyrosine) kinase $p56^{lck}$, which is expressed in T lymphocytes and forms a complex with the cytoplasmic tail of the transmembrane glycoprotein CD4 (Figure 9.13). CD4 forms part of a receptor through which antigen-presenting cells bind to and activate the T cells. *C-src* and $p56^{lck}$ are closely related, myristoylated proteins, but differ in their N-terminal domains (Figure 9.13), which is the region of $p56^{lck}$ responsible for association with CD4.

9.3.3 *Some oncogenes are mutated GTP-binding proteins*

As discussed in Section 9.2.1, some human tumours analysed by the NIH-3T3 cell transfection assay contain point mutations in one of the three *ras* genes. The *ras* protein sequences show a weak relationship with the α (GTP-binding) subunit of the G proteins which couple "seven-pass" receptors to adenylate cyclase (Chapter 7). This raised the intriguing possibility that *ras* might be a G protein coupling an unknown receptor to some unknown effector which produces a growth signal. Consistent with this idea, expression of *ras* DNA in bacteria showed that the proteins bind GTP and have an intrinsic GTPase activity. The tertiary structure of H-*ras* has now been determined by X-ray crystallography (Figure 9.14). The phosphates of GDP/GTP bind in a pocket and forms interactions with loops L1, L2 and L4, including residues around glycines-12 and -13 on L1 and glutamine-61 on L4. Mutations in *ras* genes derived from numerous different human tumours have been analysed, and they invariably occur at glycine-12, glycine-13 or glutamine-61. The commonest mutations are at glycine-12, and site-directed mutagenesis has shown that any change to this residue, other than to proline, converts *ras* into an

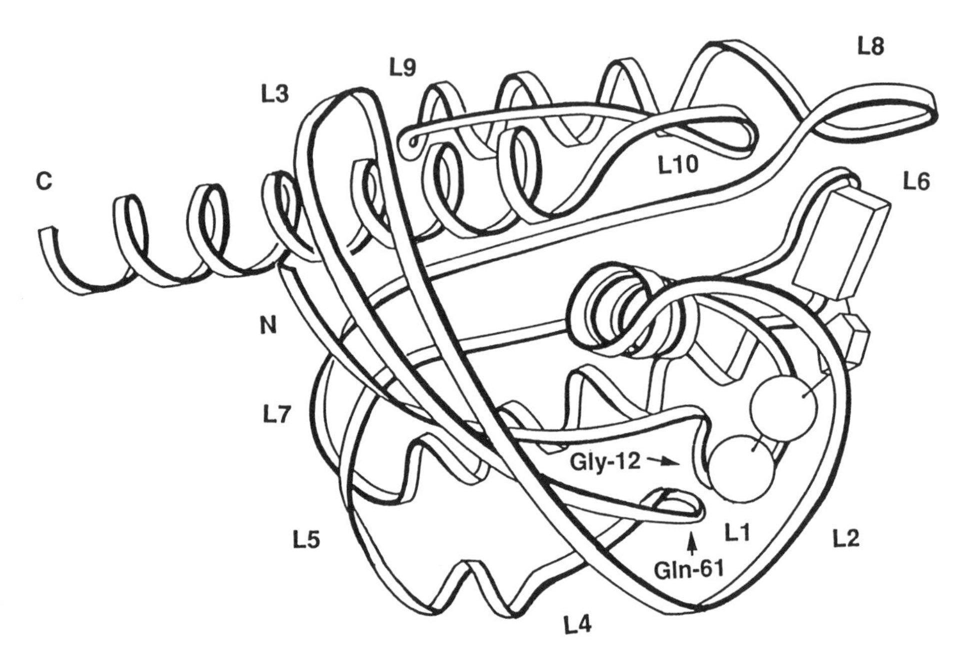

Fig. 9.14 Ribbon model of *ras* in the presence of GDP. The two phosphate groups of GDP are represented by spheres. The locations of glycine (gly)-12 and glutamine (gln)-61 are marked.

oncogene.

The intrinsic GTPase activities of oncogenic forms of *ras* are usually between 2- and 10-fold lower than those of normal *ras*. However, in the presence of another cellular protein called **GTPase activator protein**, the GTPase activity of normal, but not oncogenic, forms of *ras* is stimulated up to 100-fold. Continuing the analogy with Gs and adenylate cyclase, the very low GTPase activity of oncogenic forms of *ras*, observed even in the presence of GTPase activator protein, would mean that they would remain permanently in the active *ras*.GTP complex and constitutively activate the production of the growth signal (Figure 9.15). This hypothesis is supported by observations that normal *ras* protein will mimic the effect of oncogenic *ras* when injected into cells together with a non-hydrolysable GTP analogue.

These ideas also received strong support when it was found that the budding yeast *Saccharomyces cerevisiae* contained two proteins, RAS1 and RAS2, that were homologous to mammalian *ras*, and that these proteins coupled adenylate cyclase to a receptor which was activated by high levels of glucose and other nutrients (Figure 9.16). In *S. cerevisiae*, cyclic AMP is absolutely required for normal mitotic division, and mutations in the catalytic subunit of cyclic AMP-dependent protein kinase become blocked early in G1 phase. The yeast RAS

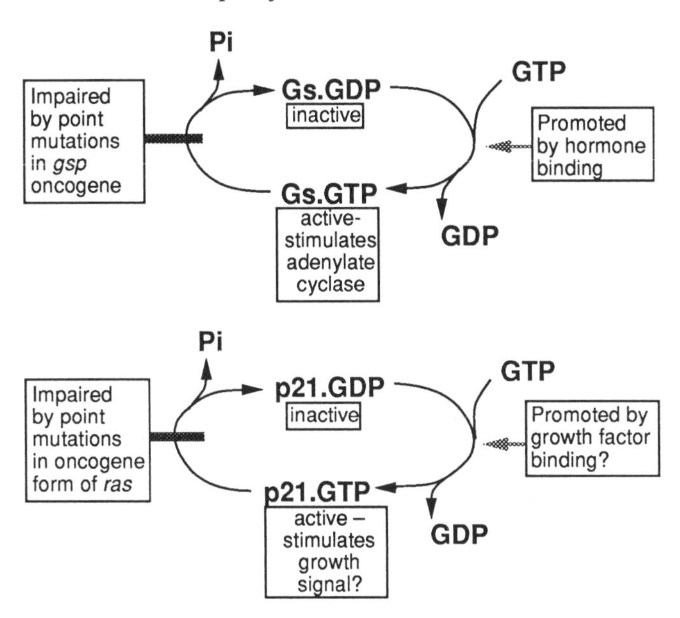

Fig. 9.15 Comparison of the known role of Gs in regulation of adenylate cyclase, and the proposed role of *ras* in control of cell growth and proliferation. The steps in the cycle that are impaired in oncogenic mutants of *ras* and Gs (*gsp* oncogene) are indicated.

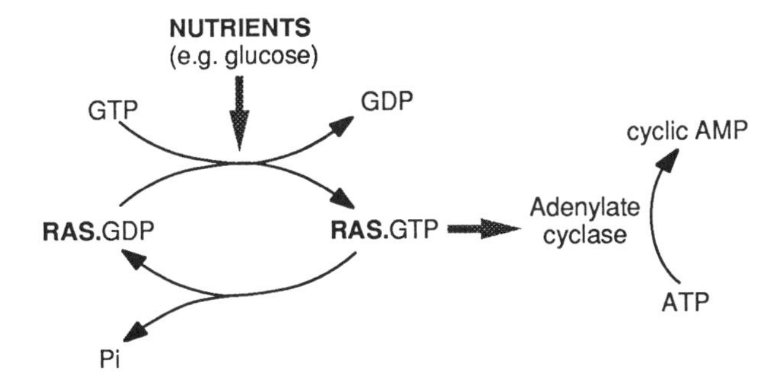

Fig. 9.16 Role of RAS gene products in control of the mitotic cell cycle in the budding yeast *Saccharomyces cerevisiae.*

proteins and mammalian *ras* are homologous in their N-terminal regions, and hybrids in which the N-terminal region of yeast RAS are replaced by the equivalent region of the mammalian protein are functional in yeast (Figure 9.17). Intriguingly, yeast cells containing a single RAS gene in which glycine-19 (equivalent to glycine-12 in mammalian *ras*) is mutated, carry out normal mitotic

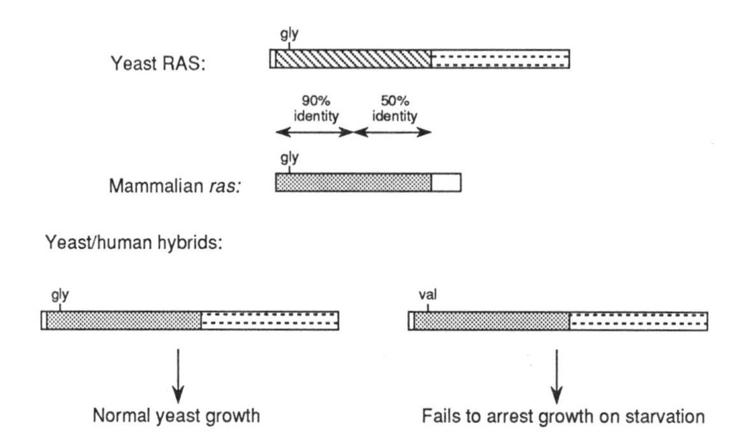

Fig. 9.17 Comparison of *ras* proteins from mammals and the budding yeast *Saccharomyces cerevisiae* (top). The GTP-binding regions containing glycine-12 are particularly well conserved. Yeast/human hybrids (bottom left) are functional in yeast, but the phenotype is abnormal if the human sequence contains an oncogenic mutation (glycine-12→valine).

division in rich medium. However, when starved of nutrients, they behave abnormally in that there is no decrease in cyclic AMP, no cessation of growth, and no switch to meiosis and spore formation. This is a kind of yeast equivalent of cancer.

Despite these analogies with yeast, it is clear that Gs, rather than *ras*, couples receptors to adenylate cyclase in mammals. The putative growth signal to which *ras* couples is still unknown, although one possibility is that GTPase activator protein is at least a component of the effector which produces it. This is supported by findings that point mutations around threonine-35 in loop L4 abolish both interaction with GTPase activator protein *in vitro,* and the transforming activity of oncogenic *ras* in intact cells. Intriguingly, sequencing of GTPase activator protein shows that it has one region homologous with adenylate cyclase from *S. cerevisiae*, which of course is the effector for yeast RAS. Whatever the identity of the ultimate growth signal, it appears that protein (tyrosine) kinases can interact with this pathway, since GTPase activator protein is phosphorylated by the PDGF receptor (Section 7.4) and by soluble kinases such as *v-src*. Moreover, microinjection of anti-*ras* antibody into cells inhibits the ability of PDGF to stimulate DNA synthesis, and inhibits transformation by *v-src*.

Ras is not the only GTP-binding protein found to be mutated in human tumours. In some but not all mammalian cell types, cyclic AMP is a positive signal for cell division, as it is in yeast. An example is provided by the cells in the pituitary gland which produce somatotropin. Stimulation by somatotropin

```
                   201
Wild-type:  -Arg-Cys-Arg-Val-Leu-

 Tumour 1:  -Arg-Cys-Cys-Val-Leu-

 Tumour 2:  -Arg-Cys-His-Val-Leu-


                   227
Wild-type:  -Gly-Gly-Gln-Arg-Asp-

 Tumour 3:  -Gly-Gly-Arg-Arg-Asp-
```

Fig. 9.18 Location of point mutations in Gsα in the *gsp* oncogene.

releasing hormone (from the hypothalamus) increases cyclic AMP in these cells, and the result is a stimulation of their growth as well as release of somatotropin. It has been found that several different human tumours of these cells contain point mutations in the α subunit of Gs which reduce its GTPase activity and cause constitutive cyclic AMP production and growth. Intriguingly, one of the sites of mutation, found in two tumours, occurs at the residue (arginine-201) at which ADP-ribosylation by cholera toxin inactivates the GTPase activity (Section 7.2.4). Another occurs at glutamine-227, which is in a position homologous with glutamine-61 in *ras*, one of the sites at which conversion to an oncogene occurs in that protein (Figure 9.18).

9.3.4 *Some oncogenes encode transcriptional activator proteins*

A number of cellular proto-oncogenes which have been "captured" by RNA tumour viruses (e.g. *fos, jun, myc*) encode proteins which are normally only expressed transiently in the nucleus in cells exposed to growth factors and other agents. Thus these viruses may cause transformation by bringing about high levels of expression of nuclear regulatory proteins that are normally only produced transiently when cells are stimulated to divide. Two of these proteins, *fos* and *jun*, are related to each other, and show limited similarity in sequence (Figure 9.19) with GCN4, a yeast transcription factor which binds to promoter regions on DNA and activates gene expression. This finding suggested that *fos* and *jun* might also be transcriptional activators.

If this idea is correct, what genes are activated by *fos* and *jun*? A number of genes have been found to be expressed in response to phorbol esters which activate protein kinase C, such as TPA (see Box 8.3). A region of the promoters of these genes contains the consensus sequence TGACTCA, which has been termed a **TPA response element**. The transcriptional activators which bind to this element are

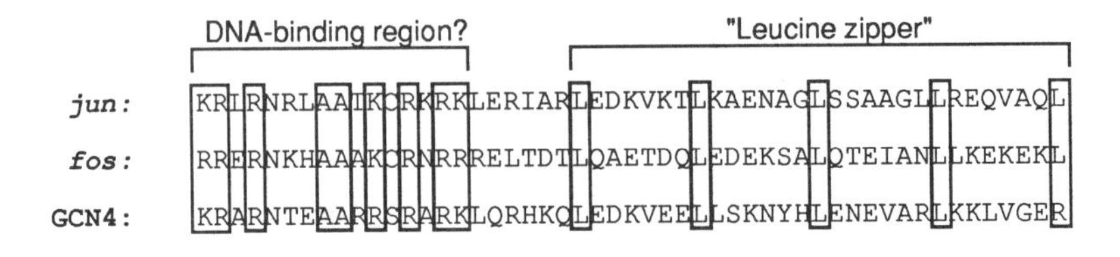

Fig. 9.19 Sequence comparison of the DNA-binding and leucine zipper regions of *jun*, *fos* and GCN4.

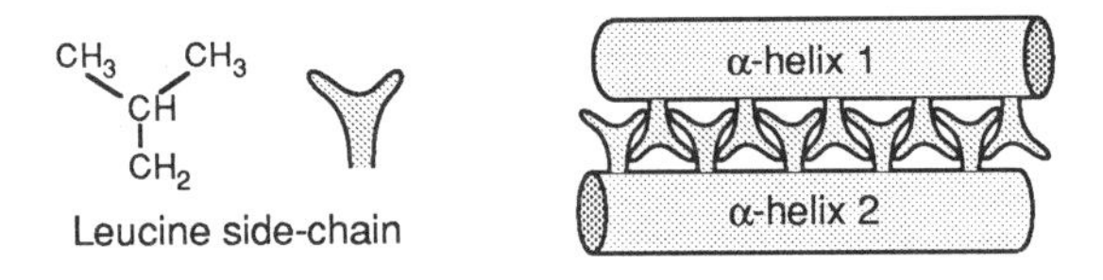

Fig. 9.20 Model for mechanism of dimerization via a leucine zipper.

known collectively as AP-1 proteins, and it is now clear that a heterodimeric complex between *fos* and *jun* represents at least one type of AP-1 protein. The sequences of *fos*, *jun* and GCN4 contain basic regions which are thought to bind to DNA, and just C-terminal to these basic regions are regions where the hydrophobic amino acid leucine occurs as every seventh residue (Figure 9.19). The significance of this spacing is that α-helices contain ~3.5 residues per turn, so that if this region formed a helix there would be a row of leucines down one face. The helices from *fos* and *jun* could then come together to form a heterodimer in which the hydrophobic leucine side chains from one polypeptide could interdigitate with those of the other polypeptide: a structure called for obvious reasons a **leucine zipper** (Figure 9.20). GCN4 also contains a leucine zipper sequence, but unlike the mammalian proteins appears to form readily a homodimer.

Formation of the leucine zipper would leave the basic regions of *fos* and *jun* free to interact with the TPA response element. The latter is a palindromic sequence, i.e. it base-pairs with itself, and one obvious possibility is that the basic elements on the *fos* and *jun* components of the heterodimer bind to the two symmetrical halves of this sequence (Figure 9.21). Although this model contains speculative elements, it is well supported by studies using site-directed mutagenesis. Thus alteration of two of the conserved leucine residues in *fos* prevent both its association with *jun* and binding to the TPA response element.

Although the *myc* family of nuclear proto-oncogenes are not obviously related to *jun* or *fos*, they do contain potential leucine zipper sequences and, in positions equivalent to those of the basic, DNA-binding sequences in *jun* and *fos*, a sequence reminiscent of that around the DNA binding site of bacteriophage λ cII protein

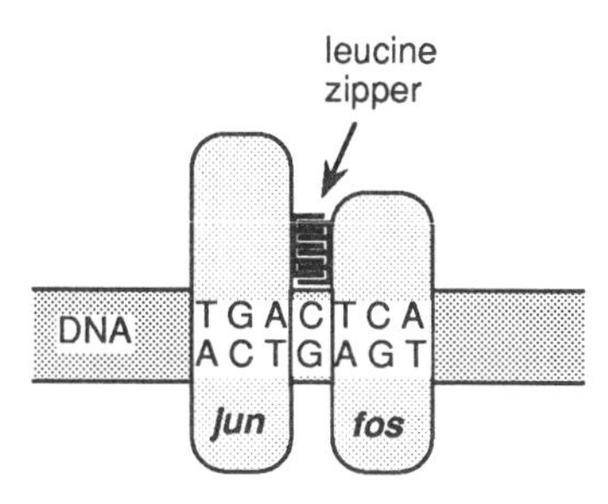

Fig. 9.21 Model for the association of a *fos–jun* heterodimer with a palindromic DNA sequence.

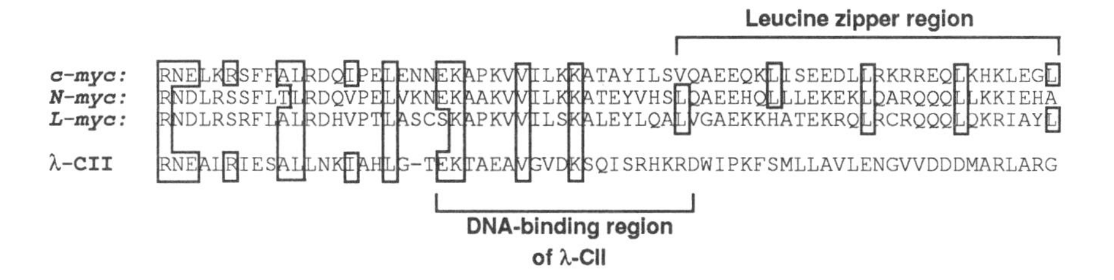

Fig. 9.22 Comparison of the proposed DNA-binding regions of the *myc* protein family and that of bacteriophage λ CII protein. The proposed leucine zipper regions of the *myc* family are also shown.

(Figure 9.22). The latter is a homodimeric regulatory protein which binds to a specific palindromic sequence on the bacteriophage DNA, via **helix-turn-helix** structural motifs which are present in each half of the homodimer. Thus it is likely that *myc* is also a transcription factor which forms dimeric complexes via leucine zippers, but which may bind to DNA via a different type of sequence from *fos* and *jun*.

The *myc* gene is particularly interesting in that certain leukaemias are associated with chromosomal rearrangements which involve a break point near the gene. Chromosomes can be visualized at metaphase using certain stains which produce characteristic banding patterns which a cytogeneticist can recognize. Chromosomal abnormalities are consistently observed in Burkitt's lymphoma, a tumour of B lymphocytes which occurs with relatively high frequency in children in parts of Africa. These abnormalities involve an exchange of part of the long arm of chromosome 8 with specific parts of chromosomes 2, 14 or 22 (Figure 9.23). The break points in these chromosomes are always the same, and are close to the *myc* gene on chromosome 8, and to immunoglobulin genes on chromosomes 2, 14 and 22. Recombination of DNA is, of course, part of the mechanism whereby diverse

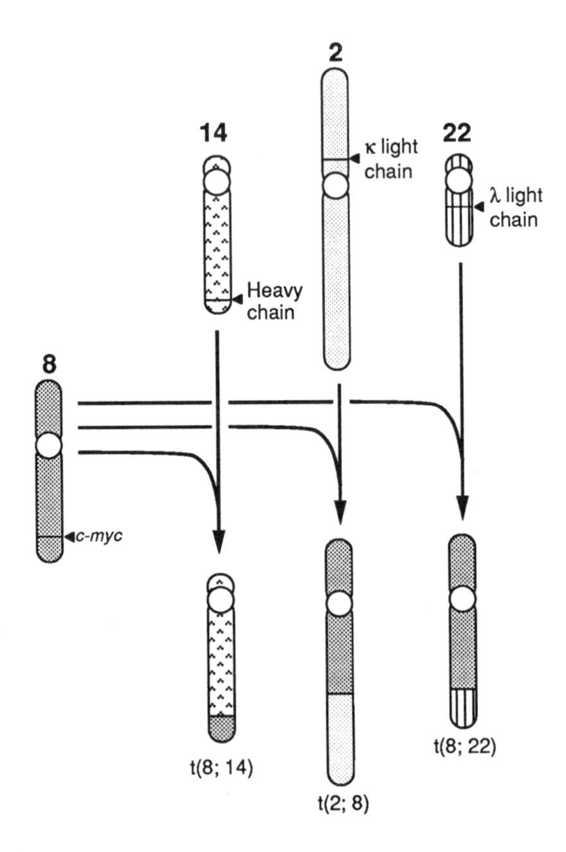

Fig. 9.23 Chromosomal rearrangements between chromosome 8 (*myc* gene locus) and chromosomes 14, 2 and 22 (immunoglobulin gene loci) found in Burkitt's lymphoma.

antibodies are produced by lymphocytes, and it is likely that aberrant recombination events remove the normal constraints on expression of the *myc* gene, leading to abnormal production of the *myc* protein.

9.4 *The anti-oncogenes – tumour suppressors*

Acquisition of **growth stimulating functions**, such as mutant genes encoding *ras* or the EGF receptor, clearly contributes to the excessive cell proliferation in some cancers. However, a variety of evidence indicates that losses of **growth inhibitory functions** are also critical events, and may be much more common. If, for example, tumour cells are fused with normal cells, the hybrid cells usually acquire the growth characteristics of the normal cells, indicating that the latter

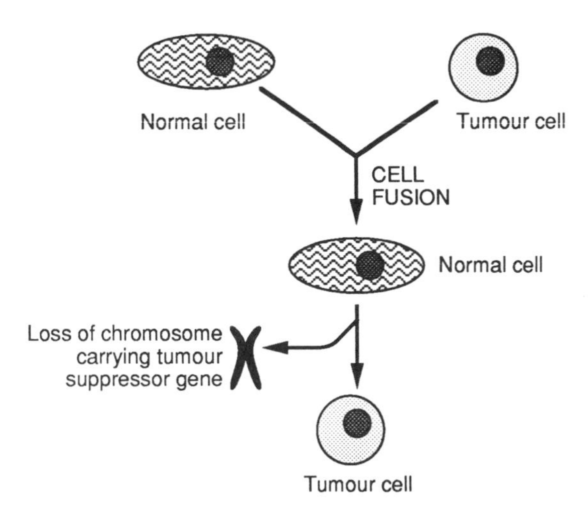

Fig. 9.24 Evidence for tumour suppressors from cell fusion experiments.

provide an inhibitory factor which is absent from the tumour cells (Figure 9.24). The chromosomes of hybrid cells are usually unstable, and the hybrids will often revert back to the tumour phenotype due to the loss of a particular chromosome. This indicates that that chromosome contains a gene which encodes an inhibitor of abnormal growth or division – an **anti-oncogene** or **tumour suppressor gene**. Since conversion of a proto-oncogene to an oncogene activates a growth pathway, such mutations are dominant and are expressed even in the presence of the normal gene. By contrast, the loss of function of a tumour suppressor will not normally be evident if a copy of the original gene is still present, so these mutations are usually recessive.

What is the normal function of tumour suppressor genes? The protein phosphatases which reverse the actions of protein kinases (Section 8.4) represent good candidates. This is suggested by the finding that okadaic acid, an inhibitor of protein phosphatases-1 and -2A, is a potent tumour promoter, and by findings that vanadate ions, which inhibit protein (tyrosine) phosphatases, can produce growth characteristics reminiscent of transformation in some cell lines. However mutations in protein phosphatases have not yet been found in human tumours. The best characterized human tumour suppressor is the product of the **retinoblastoma** gene. Retinoblastoma is a tumour of the retina of the eye which occurs in childhood. Although relatively rare, it has received a great deal of interest because it has been shown that in the **familial** form of the disease, susceptibility to the tumour can be inherited as a simple Mendelian trait. Genetic analysis suggested that affected children received a mutant copy of the retinoblastoma (*Rb*) gene from one parent: a somatic mutation in a retinal cell then produced the tumour, which fortunately can be treated by surgical removal of the eye. Cytogenetic analyses showed that carriers of retinoblastoma sometimes had a specific segment of one copy of chromosome 13 deleted in all somatic cells. Since the defect in these cases was therefore likely to involve the complete **loss** of the *Rb* gene on chromosome 13, it seemed likely that the gene encoded a tumour suppressor, and that the somatic mutation in the tumour cells was the loss of the normal chromosomal copy of the gene which had been inherited from the non-carrier parent. The rarer **sporadic** form of the disease is not inherited, but is due to two somatic mutations affecting both copies of chromosome 13.

Since the loss of a growth-suppressing function is recessive, tumour suppressor genes cannot be detected so readily using transfection into NIH-3T3 cells, because the latter will contain a normal copy of the same gene. However, the *Rb* gene has been cloned by a painstaking process involving hybridization between cloned segments of chromosome 13 and DNA from a battery of retinoblastoma tumours. Eventually a small segment of chromosome 13 DNA was found, containing a gene which was at least partially deleted in a large proportion of the tumours tested. The gene encodes a protein of 105 kDa which is localized in the nucleus. Although the function of this protein remains unknown, it is suggested that it is a regulator of transcription which either switches off growth-activating genes, or switches on other growth suppressor genes. The gene has now been found to be deleted in cancers other than in the retina, such as those of breast and bone.

An intriguing twist to the retinoblastoma story came when it was found that the 105 kDa *Rb* protein forms a complex with oncogene products of several DNA tumour viruses, including the large T antigen of SV40 (Section 9.2.1), and oncogene products of human adenoviruses and papillomaviruses (the latter being associated with cervical cancer). These results suggest the viral oncogene products may act, at least in part, by inactivating this tumour suppressor protein.

Another nuclear protein (termed p53 because of its polypeptide molecular weight) was originally identified by its ability, like *Rb*, to form a complex with the SV40 large T antigen. The p53 gene is deleted in several tumour cell lines, and it inhibits transformation by other oncogenes when transfected into fibroblasts. It thus appears to be another member of the class of tumour suppressor or anti-oncogenes. However, as well as being lost by deletion, it has been shown that this particular anti-oncogene can be converted into a **positive** oncogene by point mutations. It is thought that these mutations inactivate the growth-inhibiting properties of p53, but that the mutant polypeptide competes at the same binding sites as the wild-type protein, thus inhibiting its action. Although these mutations involve a **loss** of function, they are dominant in character, and hence are referred to as **dominant negative** mutations. The *v-erb* A oncogene (Section 9.3.2) may be another example of this.

9.5 *Mutation of more than one gene may be necessary to produce cancer*

It is a sobering thought that single mutations (point mutations, deletions, or chromosomal rearrangements) can convert proto-oncogenes into oncogenes. However, the position may not be quite as bleak as it first appears. It is true that a single point mutation in the *ras* or *neu* genes can cause transformation when they are transfected into NIH-3T3 cells. However, recall (Section 9.2.2) that the latter are an immortal cell line which have already undergone genetic changes. It has been shown that **at least two** oncogenes acting on cellular control at different levels, e.g. *ras* and *myc*, are required to elicit transformation of **primary** fibroblasts in culture. In addition, some human cancer cells have been found to contain alterations in both of these genes. Thus cancer probably requires at least two, and probably more, mutation events (in the case of cancer of the large bowel, at least four different oncogenic mutations have been found in different tumours). It must also be remembered that although we may understand transformation in tissue culture fairly well, there are still many unanswered questions about development of cancer *in vivo*. For example, how do tumours acquire a blood supply, how does spread to secondary sites (metastasis) occur, and how do tumour cells evade the immune system?

SUMMARY

Cells from multicellular organisms will normally grow and divide only in the presence of growth factors. Growth factors act by the same signal transduction mechanisms as other first messengers discussed in previous chapters, but may need to be present for prolonged periods and to activate two or more signal transduction pathways for proliferation to occur. Cancer is caused by mutations in the DNA of a single cell which allows it to divide in an uncontrolled manner. One class of mutations converts proto-oncogenes, which encode proteins involved in normal regulation into cancer-causing oncogenes. Transformation of 3T3 cells by transfected DNA is the usual assay for an oncogene. Transformed cells show anchorage-independence, loss of contact inhibition, and lower dependence on growth factors, and cause tumours when injected into immunodeficient mice.

Oncogenes can act at several levels, and the proto-oncogenes from which they are derived can encode growth factors, growth factor receptors, G proteins, or transcription factors which are normally expressed transiently in response to growth factors. Generally the mutations convert the proto-oncogene from a form which is strictly regulated to a form which is constitutively active. However, in the case of the transcription factors, over-expression (e.g. due to a chromosomal rearrangement in Burkitt's lymphoma) may be sufficient to cause transformation.

Many human tumours are probably caused by the loss of activity of tumour suppressors, rather than acquisition of an oncogene. Since tumour suppressor mutations are usually recessive, they are harder to detect, but two examples found recently are the retinoblastoma and p53 gene products. Proteins encoded by DNA tumour viruses often form stable interactions with these tumour suppressor proteins.

Although cells in culture can be transformed by a single point mutation, it appears likely that several mutations are necessary for tumour formation and spread of secondary cancers (metastasis) in vivo.

FURTHER READING

1. "Oncogenes" – book containing several useful chapters. Edited by Glover, D.M. and Hames, B.D. (1989) Oxford University Press, Oxford.
2. "Early signals in the mitogenic response" – review of control of proliferation in 3T3 cells. Rozengurt, E. (1986) Science **234**, 161–166.
3. "Cell transformation by the *src* oncogene" – review. Jove, R. and Hanafusa, H. (1987) Ann. Rev. Cell Biol. **3**, 31–56.
4. "Growth factors as transforming proteins" – review. Heldin, C.H. and Westermark, B. (1989) Eur. J. Biochem. **184**, 487–496.
5. "Human epidermal growth factor receptor cDNA sequence and aberrant expression of the amplified gene in A431 epidermoid carcinoma cells" – homology with *v-erb* B oncogene. Ullrich, A., Coussens, L., Hayflick, J.S., Dull, T.J., Gray, A., Tam, A.W., Lee, J., Yarden, Y., Libermann, T.A., Schlessinger, J., Downward, J., Mayres, E.L.V., Whittle, N., Waterfield, M. and Seeburg, P.H. (1984) Nature **309**, 418–425.
6. "The *c-erb* A gene encodes a thyroid hormone receptor". Weinberger, C., Thompson, C.C., Ong, E.S., Lebo, R., Gruol, D.J. and Evans, R.M. (1986) Nature **324**, 641–646.
7. "Multiple independent activations of the *neu* oncogene by a point mutation altering the transmembrane domain of p185". Bargmann, C.I., Hung, M.C. and Weinberg, R.A. (1986) Cell **45**, 649–657.
8. "The *mas* oncogene encodes an angiotensin receptor". Jackson, T.R., Blair, L.A.C., Marshall, J., Goedert, M. and Hanley, M.R. (1988) Nature **335**, 437–440.
9. "The *lck* tyrosine protein kinase interacts with the cytoplasmic tail of the CD4 glycoprotein through its unique amino-terminal domain". Shaw, A.S., Samrein, K.E., Hammond, C., Stern, D.F., Sefton, B.M. and Rose, J.K. (1989) Cell **59**, 627–636.
10. "A molecular basis of cancer" – review on the discovery of the *ras* oncogene. Weinberg, R.A. (1983) Scientific American, November, pp. 102–116.

11. "Structure of *ras* protein" – three-dimensional structure of the GTP-bound form. Tong, L., Milburn, M.V., de Vos, A.M. and Kim, S.H. (1989) Science **245**, 244–244.
12. "Structure of the guanine-nucleotide-binding domain of the Ha-*ras* oncogene product p21 in the triphosphate conformation". Pai, E.F., Kabsch, W., Krengel, U., Holmes, K.C., John, J. and Wittinghofer, A. (1989) Nature **341**, 209–214.
13. "Cloning of bovine GAP and its interaction with oncogenic *ras* p21". Vogel, U.S., Dixon, R.A.F., Schaber, M.D., Diehl, R.E., Marshall, M.S., Scolnick, E.M., Sigal, I.S. and Gibbs, J.B. (1988) Nature **335**, 90–93.
14. "GTPase inhibiting mutations activate the α chain of Gs and stimulate adenylate cyclase in human pituitary tumours". Landis, C.A., Masters, S.B., Spada, A., Pace, A.M., Bourne, H.R., and Vallar, L. (1989) Nature **340**, 692–696.
15. "The leucine zipper: a hypothetical structure common to a new class of DNA binding proteins". Landschulz, W.H., Johnson, P.F. and McKnight, S.L. (1988) Science **240**, 1759–1764.
16. "The role of the leucine zipper in the fos–jun interaction". Kouzarides, T. and Ziff, E. (1988) Nature **336**, 646–651.
17. "Positive and negative controls on cell growth" – review on tumour suppressors. Weinberg, R.A. (1989) Biochemistry **28**, 8263–8269.

TEN

SIGNAL TRANSDUCTION BY NUCLEAR RECEPTORS

The steroid and thyroid hormones, and the morphogen retinoic acid, differ from all other first messengers described in this book in that they act at intracellular receptors, rather than receptors expressed at the cell surface. All of these three classes of messenger are small lipophilic compounds, and they readily permeate the plasma membrane. An additional difference from other first messengers is that the primary action of these types is to regulate gene expression at the level of transcription. While messengers acting at cell surface receptors can certainly affect transcription (e.g. Section 8.2.4), this only occurs as a secondary response to changes initiated in the cytoplasm.

The effect of steroid hormones on synthesis of proteins can be analysed by two-dimensional separation in gels after labelling cells briefly with a radioactive amino acid such as [^{35}S]methionine. Figure 10.1 shows an experiment in the proteins have been separated in one dimension by isoelectric focusing (IEF, which separates according to isoelectric point), followed by electrophoresis in the presence of sodium dodecyl sulphate (SDS, which separates according to molecular weight) in the second dimension. Newly synthesized proteins are detected as dark spots after autoradiography. It can be seen that the steroid dexamethasone leads to specific induction of about 12 proteins (squares), and repression of two proteins (circles), these changes representing about 1% of all newly synthesized proteins detectable on the gel. Since this analysis probably only detects around 10% of cell proteins, the true figure for the number of glucocorticoid-regulated genes may be an order of magnitude higher.

Activation and repression of genes can be more directly visualized in larvae of insects such as *Drosophila melanogaster*. In the giant cells of these larvae, DNA replication has occurred in the absence of normal chromatid separation by mitosis, and up to 10^9 copies of the DNA remain in perfect alignment, so that individual transcription units (containing one or, at most, a few genes) appear as bands visible in the light microscope. Genes active in transcription appear as expanded bands known as puffs, in which the DNA is less condensed. When larval cells are treated with the steroid hormone 20-hydroxyecdysone, which initiates metamorphosis into the adult, several bands corresponding to "early" genes (e.g. 74EF, 75B) are rapidly but transiently converted into puffs: one is thus able to directly visualize gene activation and deactivation (Figure 10.2). Over the next few hours, the "late" genes (e.g. 78D, 71EF) are also switched on transiently in a

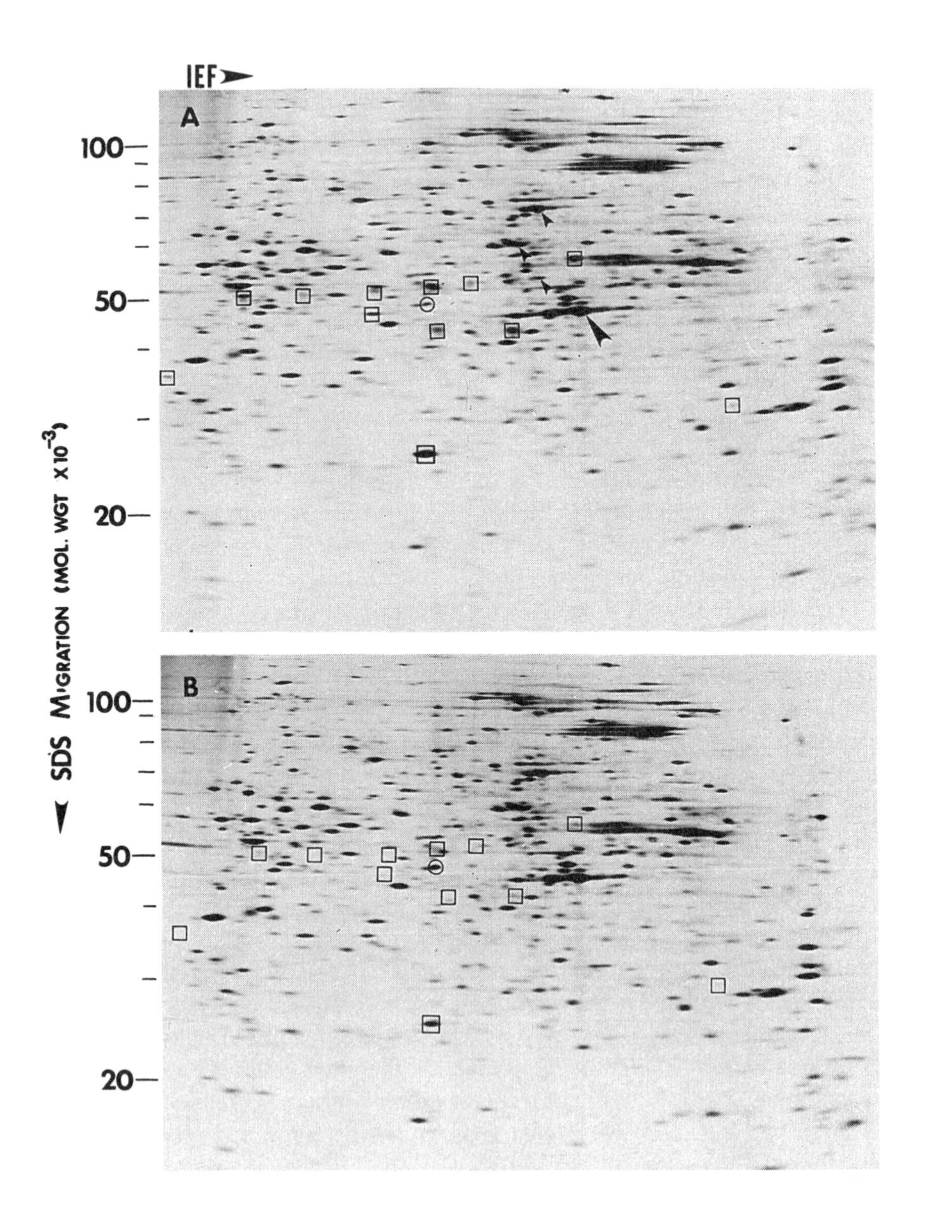

Fig. 10.1 Analysis of newly synthesized proteins in rat liver tumour cells by two-dimensional electrophoresis. Cells were labelled with [^{35}S]methionine in the presence (top) or absence (bottom) of dexamethasone. Proteins whose synthesis is increased (squares) or decreased (circles) by steroid are marked. Reproduced from Ivarie and O'Farrell (1978) Cell **13**, 41–55, courtesy of Patrick O'Farrell.

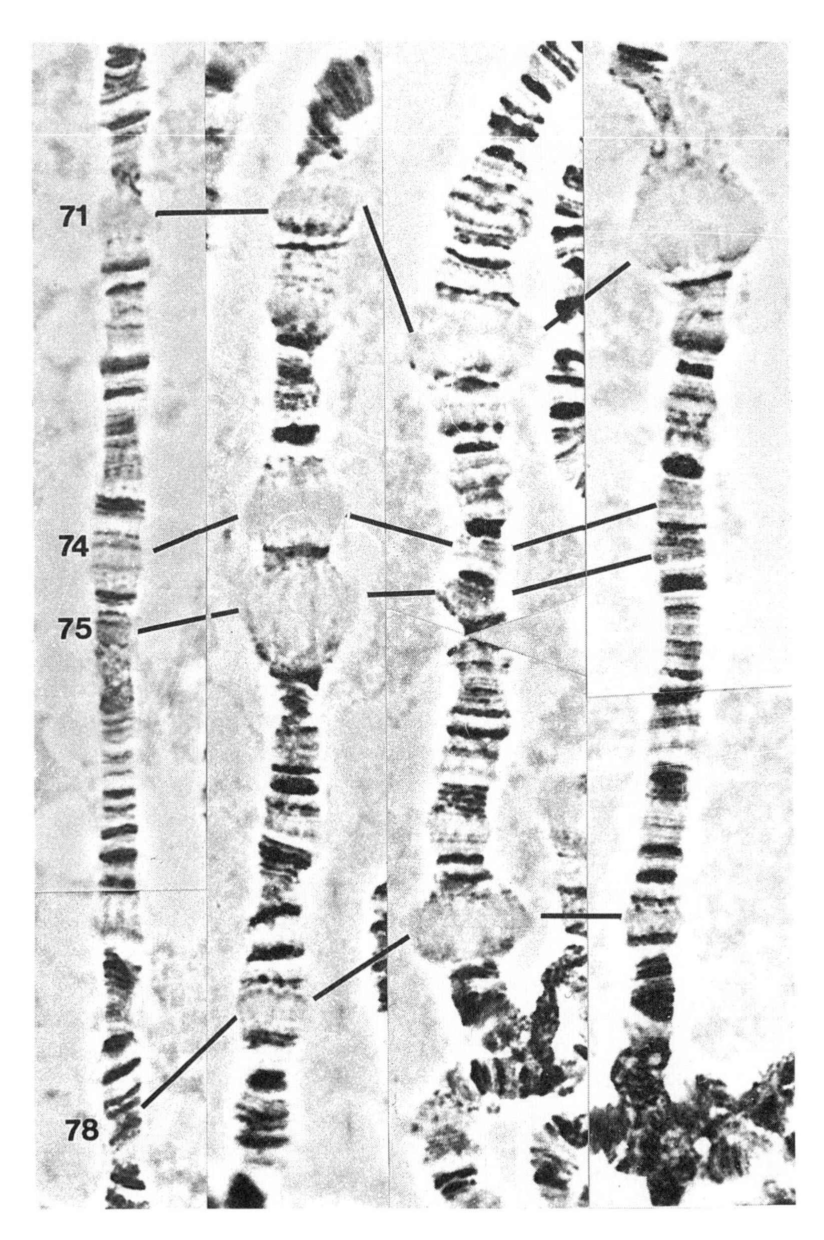

Fig. 10.2 Sequence of gene activation and deactivation induced by the steroid hormone 20-hydroxyecdysone in giant chromosomes of *Drosophila* larvae, visualized as "puffing" and "depuffing" of bands on chromosome 3L. Bands 74 and 75 are "early puffs" where the genes are activated within minutes. 78 and 71 are "late puffs" which are activated at different times over the following few hours. Courtesy of Michael Ashburner.

defined order: unlike that of the early genes, activation of these genes is blocked by inhibitors of protein synthesis. The products of at least some of the early genes are DNA-binding proteins, and current evidence favours a model in which the ecdysone–receptor comple: **inhibits** expression of the late genes, but in which this is relieved by the presence of early gene products.

Further progress in understanding the mechanism of action of steroid hormones required the analysis of regulation of specific genes. Some genes which are regulated by steroid and thyroid hormones, which have been utilized as model systems for study, are listed below:

1) *Regulation by glucocorticoids*
   a) **Tyrosine aminotransferase**, involved in tyrosine breakdown, is induced by glucocorticoids in liver and liver tumour cell lines. One of the induced proteins in the experiment shown in Fig. 10.1 is tyrosine aminotransferase.
   b) **Metallothioneins** are proteins which bind heavy metals and thus reduce their toxicity. Metallothionein expression is induced both by glucocorticoids and by the heavy metals themselves, and metallothionein promoters have been a particularly well-studied system.
   c) **Mouse mammary tumour virus** (MMTV) is a retrovirus which causes mammary (breast) tumours with low frequency in mice. As in all retroviruses (Section 9.2.1), the RNA genome of this virus is copied into DNA which is then integrated into a host chromosome via a mechanism involving the **long terminal repeat** sequences at each end of the viral genome. Once incorporated into the host genome, the viral genes are activated by glucocorticoids, and this is now known to be due to binding of the activated glucocorticoid receptor to sequences in the long terminal repeats. Because of the relative ease with which they could be isolated, the long terminal repeats of MMTV have occupied an important position in studies of steroid hormone action.
   d) **Pro-opiomelanocortin**, the precursor for adrenocorticotropin and certain other peptide messengers (Section 3.2.2) is **repressed** by glucocorticoids. Since the main role of adrenocorticotropin is to stimulate synthesis of glucocorticoids in adrenal cortex, this is an important feedback mechanism.
2) *Regulation by vitamin D*
   Ca^{2+}-binding protein, a cytosolic protein which may be involved in Ca^{2+} uptake from the gut, is induced by the active hydroxylated form of vitamin D_3 in intestinal cells.
3) *Regulation by progesterone*
   Ovalbumin, **conalbumin** and **ovomucoids**, the major proteins of egg white, are induced by progesterone in the oviduct of birds.
4) *Regulation by oestrogens*
   Vitellogenins, the major proteins of egg yolk, are induced by oestrogens in the liver of birds.

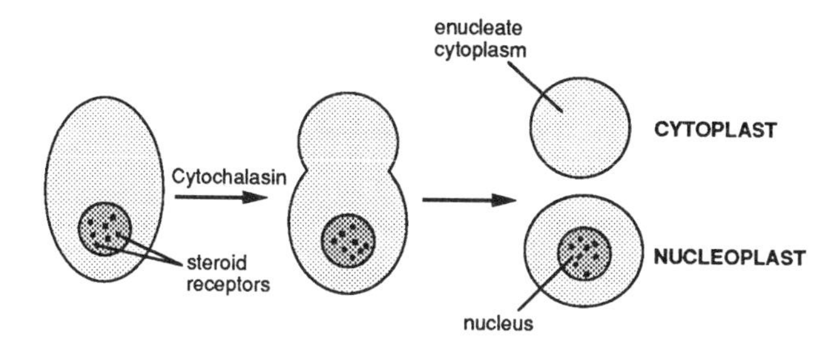

Fig. 10.3 Enucleation by the drug cytochalasin B, which induces cell division without nuclear division.

5) *Regulation by thyroid hormone (T3)*
   a) **Fatty acid synthase** and **malic enzyme** are two examples of enzymes involved in fatty acid synthesis which are induced by T3 in liver.
   b) **Thyrotropin**, the pituitary trophic hormone, is repressed by T3. Since thyrotropin stimulates T3 release (Section 2.1.2), this is another example of a feedback mechanism.

While most of these genes are activated by their respective hormone, it should be noted that some (e.g. pro-opiomelanocortin and thyrotropin) are repressed. Any model for the mechanism of action of these hormones must take this into account.

10.1 *Nuclear receptors bind directly to target genes*

Purification of the receptors for steroid and thyroid hormones was a difficult task because they are present in particularly small amounts in target cells, and must be purified by factors of 10^5 or 10^6 to achieve homogeneity. However, using affinity chromatography methods it eventually proved possible to purify the receptors for several steroid hormones, i.e. glucocorticoids, oestrogen, progesterone and vitamin D_3. In general the receptors are single polypeptides with molecular masses in the range 50 to 100 kDa, although the active forms are now thought to be homodimers. They all show affinity for DNA, but a key feature is that their affinity for DNA is greatly increased by binding of the hormone.

Early experiments, in which receptors were analysed by binding of radioactive steroid to isolated subcellular fractions, indicated that in unstimulated cells they were present in the cytoplasm, but that after stimulation of cells they were found in the nucleus. However, with purification of the receptors and the development of monoclonal antibodies, immunohistochemical evidence suggested that some, if not all, steroid hormone receptors were present in the nucleus even in the unoccupied state. Convincing support for this idea came from studies with enucleated cells (Figure 10.3). If certain cells are treated with the drug cytochalasin B, they divide

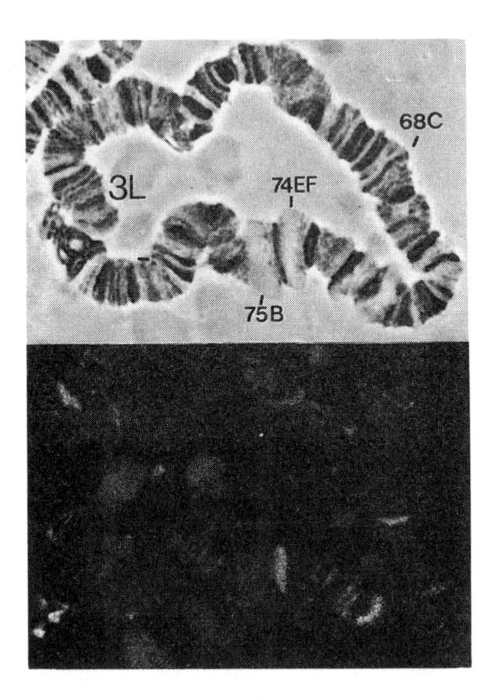

Fig. 10.4 Phase contrast (top) and fluorescence (bottom) micrographs of part of chromosome 3L of *Drosophila melanogaster* after stimulation with ecdysone. The steroid was crosslinked to the chromosomal binding sites using ultraviolet light, and detected using fluorescent-labelled antibody. Regions of intense fluorescence correspond with chromosome puffs which are induced by ecdysone. Courtesy of Olaf Pongs.

into **nucleoplasts**, containing nuclei and some cytoplasm, and **cytoplasts**, containing cytoplasm and no nuclei. This is a very effective method of separating nuclei and cytoplasm without breaking open the cells. With some cells it can even be achieved by simple centrifugation, without the use of a drug. In either case, centrifugation can be used to separate the nucleoplasts and cytoplasts, and the receptors for glucocorticoids, oestrogens and progesterone were found only in the nucleoplasts, irrespective of whether the cells had been pre-treated with hormone. It now seems clear that the recovery of unoccupied receptors mainly in the "cytoplasmic" fraction of homogenized cells is an artefact caused by leakage out of nuclei, although the receptors must be present in the cytosol, at least transiently, during synthesis. The hormone-bound receptors may remain associated with the nuclear fraction because they have a higher affinity for DNA. In the case of the thyroid hormones, the evidence has always favoured a nuclear location for both occupied and unoccupied receptors.

Since hormone-activated receptors bind with reasonably high affinity to any

GLUCOCORTICOID RESPONSE ELEMENT:

Mouse mammary tumour virus:	ATGGTTACAAACTGTTCTTAA
Somatotropin:	TTGGGCACAATGTGTCCTGAG
Metallothionein IIA:	CCCGGTACACTGTGTCCTCCC
Tyrosine aminotransferase 2:	TGCTGTACAGGATGTTCTAGC

CONSENSUS SEQUENCE:

```
5' GGTACA---TGTTCT 3'
   ||||||   ||||||
3' CCATGT---ACAAGA 5'
```

OESTROGEN RESPONSE ELEMENT:

Vitellogenin:	TCCTGGTCAGCGTGACCGGAG
Ovalbumin:	TTCAGGTAACAATGTGTTTTC
Prolactin:	TTTTTGTCACTATGTCCTAGA

CONSENSUS SEQUENCE:

```
5' GGTCA---TGACC 3'
   |||||   |||||
3' CCAGT---ACTGG 5'
```

THYROID RESPONSE ELEMENT:

Growth hormone:	AAGATCAGGGACGTGACCGC

CONSENSUS SEQUENCE:

```
GATCA------TGACC
|||||      |||||
CTAGT------ACTGG
```

Fig. 10.5 Hormone response elements and derived consensus sequences in upstream regions of genes regulated by steroid and thyroid hormones.

random DNA, it was at first difficult to demonstrate specific binding sites associated with gene regulation. However, early indications that occupied receptors bound specifically to regulated genes came from experiments in which ecdysterone was crosslinked to chromosomes in the giant cells of *Drosophila* larvae using ultraviolet light, and then detected using fluorescent antibodies. The most intense fluorescence was seen in the ecdysone-induced puffs (Figure 10.4). A more detailed analysis of receptor binding sites became possible when a number of genes regulated by steroid or thyroid hormones were cloned. Experiments in which cloned genes containing specific deletions were transfected into cells revealed short sequences (**hormone response elements**) which were required for gene activation by the hormones (Figure 10.5). That these are binding sites for the activated receptors was confirmed by "footprinting", in which it was shown that the presence of the activated receptor protected these sequences against nuclease digestion.

The hormone response elements are around 15 base pairs long and, like the binding sites for many other bacterial and eukaryotic DNA-binding proteins (e.g. *fos* and *jun* – see Section 9.3.4), contain sequences which are at least partially palindromic, i.e. that are nearly symmetrical in the double-stranded form (Figure 10.5). This suggests that the receptor may interact with the element as a homodimer which itself has two-fold symmetry, a hypothesis which has now been

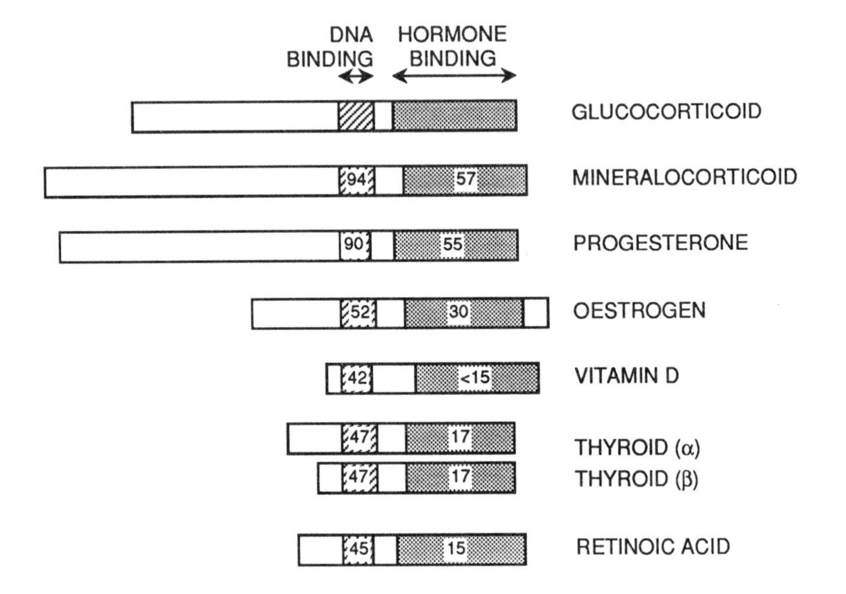

Fig. 10.6 Aligned domain structures of steroid, thyroid and retinoic acid receptors. Figures refer to the percentage sequence identity of the boxed region with the equivalent region on the glucocorticoid receptor.

directly confirmed for the oestrogen receptor. Comparison of the hormone response elements from several genes has suggested "consensus sequences" required for optimal binding of glucocorticoid, oestrogen and thyroid hormone receptors. The consensus sequences for these three classes of hormone are surprisingly similar (Figure 10.5), and it appears that quite subtle changes in sequence can markedly affect binding specificity.

10.2 *Anatomy of nuclear receptors*

Purification of several steroid hormone receptors allowed cloning and sequencing of cDNAs encoding these receptors. Using these cDNAs as probes for screening, other steroid hormone receptors have been cloned by low stringency hybridization of cDNA libraries. As discussed in Section 9.3.2, a thyroid hormone receptor had already been accidentally "cloned" by the avian erythroblastosis virus as the *v-erb* A gene, which had been acquired from the avian host. This fortuitous discovery facilitated cloning of normal thyroid hormone receptor DNA from mammalian cDNA libraries. cDNA encoding a retinoic acid receptor was cloned using the bold, but correct, assumption that it would be related to steroid hormone receptors. A cDNA library from a mammary tumour cell line known to contain retinoic acid

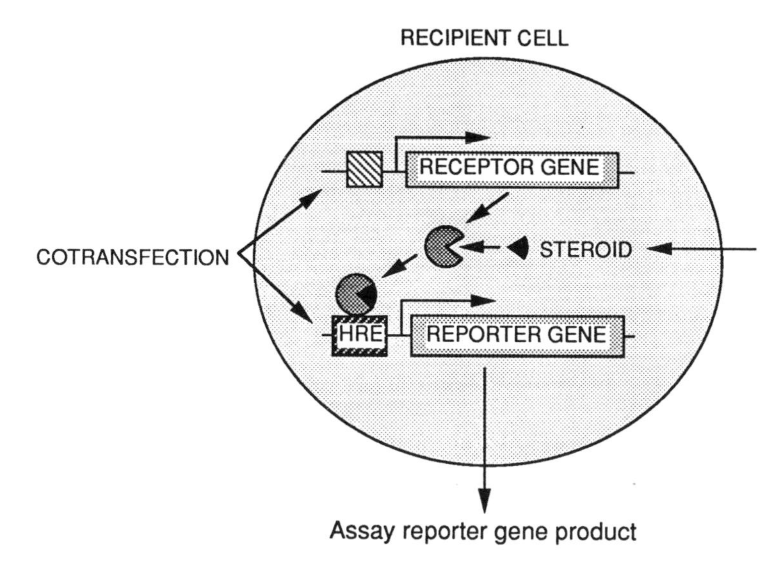

Fig. 10.7 The principle of a hormone receptor DNA/hormone response element-cotransfection assay. HRE = hormone response element.

receptors was screened with an oligonucleotide probe based on sequences conserved in steroid hormone receptors. One cloned cDNA was shown to encode a product which bound retinoic acid with high affinity when expressed in HeLa cells. At least three receptors for retinoic acid (α, β and γ), and two for thyroid hormones (α and β) have now been found by further screening of cDNA libraries.

Comparison of the sequences of these receptors shows that they are indeed related, and that the steroid, thyroid and retinoic acid receptors form a "superfamily" of nuclear receptors (Figure 10.6). The sequences are particularly well conserved in a central region which is now known to represent the DNA-binding domain (see below). The C-terminal region is the messenger-binding domain, and is also well conserved between receptors binding the same class of messenger (e.g. those for glucocorticoids, mineralocorticoids and progesterone, which are >50% identical in this region). Analysis of the functions of different regions of the receptor has been carried out by elegant experiments in which receptor cDNAs coupled to strong viral promoters are cotransfected into cultured cells with a reporter gene coupled to a hormone response element. One particularly convenient and sensitive reporter gene is that for the luciferase from fireflies, since luciferase catalyses the chemiluminescent emission of light in cell extracts (Figure 10.7).

Using these techniques, it could be readily confirmed that the conserved central region was the DNA-binding domain. This was demonstrated in a particularly satisfying way by experiments involving DNA constructs in which the putative DNA-binding domain of the oestrogen receptor was replaced with that of the

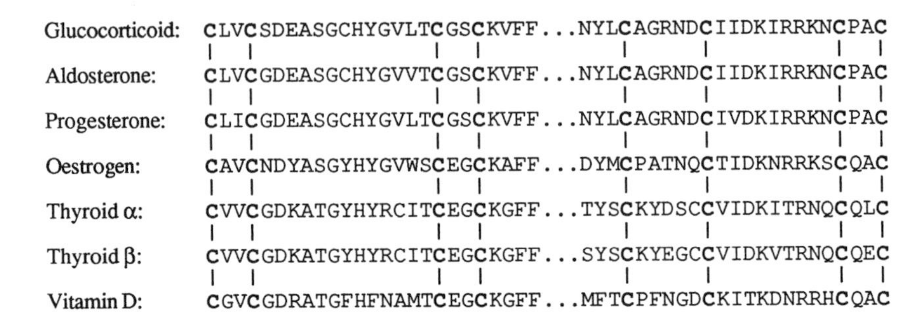

Fig. 10.8 Aligned sequences of the DNA-binding domains in steroid, thyroid and retinoic acid receptors showing the conserved cysteine (C) residues. See the Appendix for the single letter amino acid code.

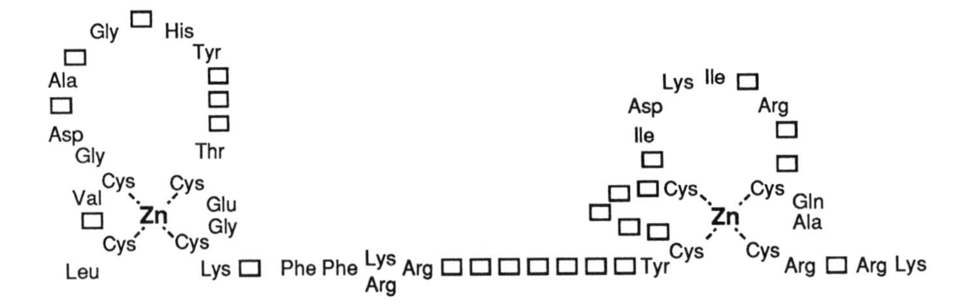

Fig. 10.9 Probable zinc coordinating side chains in steroid hormone receptors. Residues which are not highly conserved between different receptors are shown as boxes.

glucocorticoid receptor: in the transfected cells a reporter gene coupled to a glucocorticoid response element was activated in response to oestrogen. A prominent feature of the DNA-binding regions is a pattern of eight conserved cysteine residues (Figure 10.8), and site-directed mutagenesis has shown that all of these are essential for DNA binding. This pattern of 2×4 cysteine residues is similar to a motif found in the yeast transcription factor GAL4, and a related motif containing 2 cysteines and 2 histidines which is repeated several times in TFIIIA, a transcription factor required for transcription of 5S RNA by RNA polymerase III. In the latter case it has been shown that the motif forms a small domain termed a **zinc finger**, in which two cysteines and two histidines form a tetrahedral coordination sphere around an atom of zinc (cf. Figure 10.9). Figure 10.10 shows a speculative model, based on experiments with TFIIIA-DNA complexes, in which two adjacent zinc fingers bind in the major groove of DNA on opposite sides of the double helix. There is now good evidence that the DNA-binding domains of the

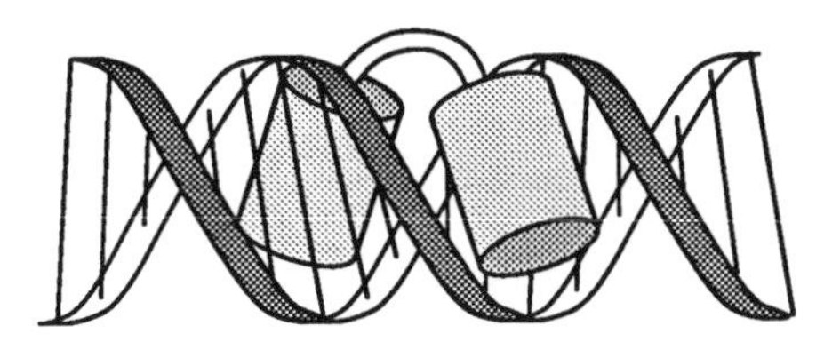

Fig. 10.10 Model for the binding of two zinc fingers to DNA. After Klug and Rhodes (1987) Trends Biochem. Sci. **12**, 464-469.

steroid hormone receptors contain two zinc fingers, with the metal ion coordinated to four cysteine residues in each (Figure 10.9). More recent experiments involving hybrid receptors suggest that it is the first (N-terminal) zinc finger and the residues immediately following it that determine the specificity of binding to the hormone-response element. This first zinc finger presumably forms specific interactions with the bases exposed in the major groove of the DNA.

As expected from the sequence comparisons between receptors, deletion of the C-terminal regions leads to loss of hormone binding. More interestingly, such deletions can result in a protein which is still partially active in binding DNA and activation of transcription, at least for the glucocorticoid receptor. In other cases, deletion experiments suggest that sequences in the C-terminal domains (and possibly in the variable N-terminal regions) are also necessary for transcriptional activation. The interpretation of such results is however complicated by the fact that the C-terminal domain seems to be involved in dimerization and nuclear localization of the receptors. Since dimerization may be required for binding to the palindromic hormone response elements, any mutations which affected this process would abolish transcriptional activation, as would the failure to localize in the nucleus.

10.3 *Mechanism of gene activation*

How does binding of the receptor to the hormone response element activate transcription? Since steroid hormones can activate different genes in different cell types, despite the presence of the same receptor protein, it is clear that other proteins which are expressed in a tissue-specific manner are required for gene activation. Interaction of the activated receptors with other transcription factors can be directly demonstrated using DNA constructs in which a hormone response element and a binding site for another factor are coupled to the promoter for a reporter gene. The construct is transfected into recipient cells and the activity of the reporter gene is measured. This has shown that binding sites for a number of transcription factors, such as nuclear factor-1, Sp1 and "CACCC-factor", can act

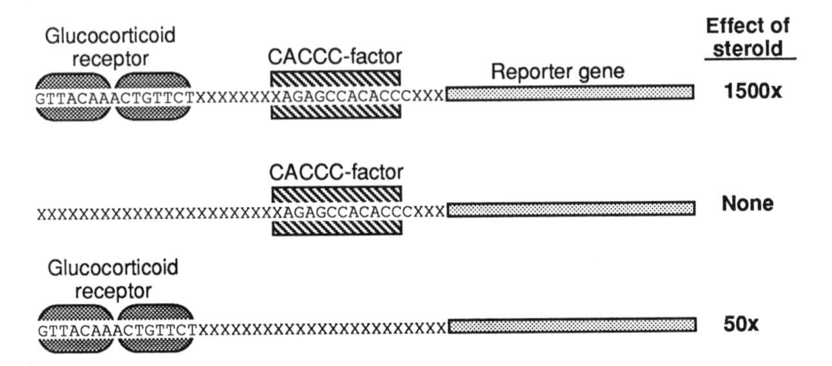

Fig. 10.11 Example of DNA constructs used to demonstrate synergistic interactions between steroid hormone response elements and sites for other transcription factors. The presence of a binding site for the "CACCC factor" increases the degree of glucocorticoid induction of the reporter gene from 50-fold to 1500-fold.

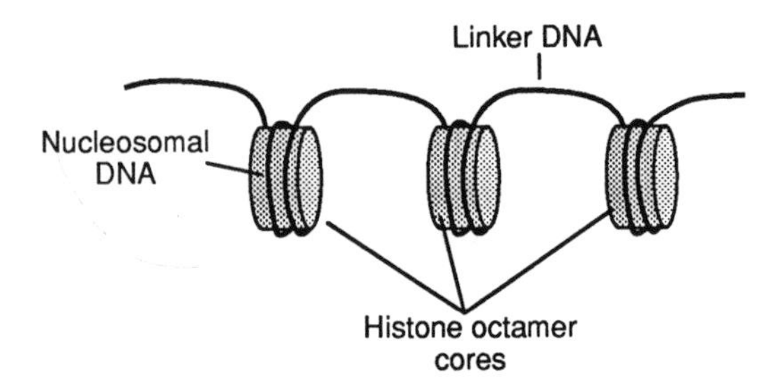

Fig. 10.12 Model for the packaging of DNA into nucleosome particles. The single thick line represents a DNA double helix.

synergistically with a hormone response element in promoting hormone-dependent gene expression, although the best response of all is obtained using **two** hormone response elements (Figure 10.11). This latter finding may explain why the long terminal repeat of mouse mammary tumour virus, which appears to contain no less than **four** glucocorticoid response elements, is so effective at making neighbouring genes responsive to glucocorticoid.

It is not yet clear whether the synergistic effects between steroid hormone receptors and other transcriptional activators involve direct protein–protein interactions. Experiments in which the spacing between the binding sites is varied have shown that this parameter is important, but that there is no single critical spacing. An alternative model is that the binding of the activated receptor may alter the chromatin structure around the gene, and thus **indirectly** allow binding of other transcription factor(s). DNA in mammalian cells (Figure 10.12) is normally

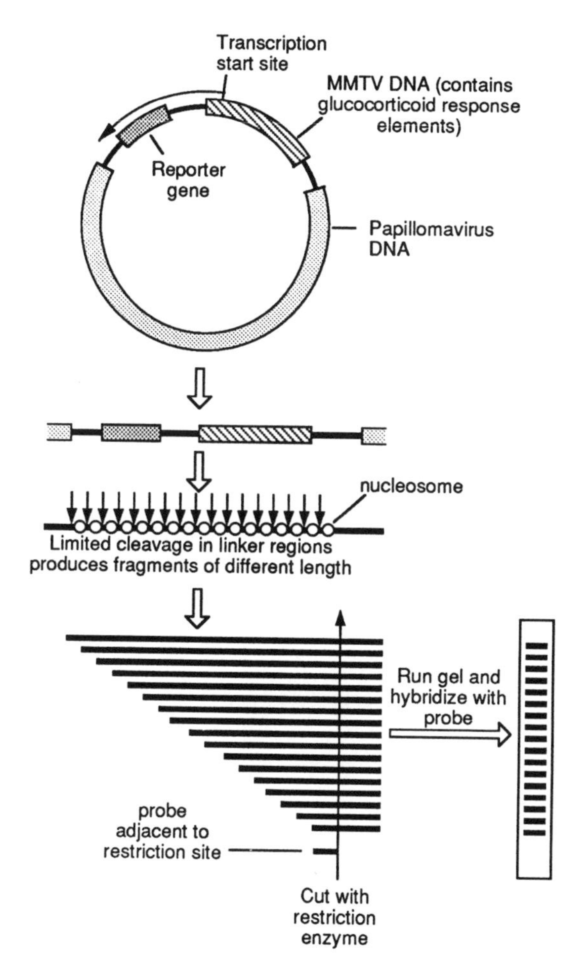

Fig. 10.13 Minichromosome constructs used to test the effects of glucocorticoid response elements on nucleosome positioning. The sites of cleavage between nucleosomes can be mapped accurately, by means of a hybridization probe based on a sequence adjacent to a known restriction site.

packaged into **nucleosome** particles, each of which is formed by the wrapping of two turns of double helix around a protein core made from histones. Mild treatment with certain nucleases or chemicals which degrade DNA results in cleavage only in the linker regions (which are ~50 base pairs long), with the DNA in the nucleosome particles (~150 base pairs) being protected by association with the histones. Due to the great length and complexity of mammalian DNA, it is difficult to study the chromatin structure around individual genes in native chromosomes. However, experiments have been carried out (Figure 10.13) using

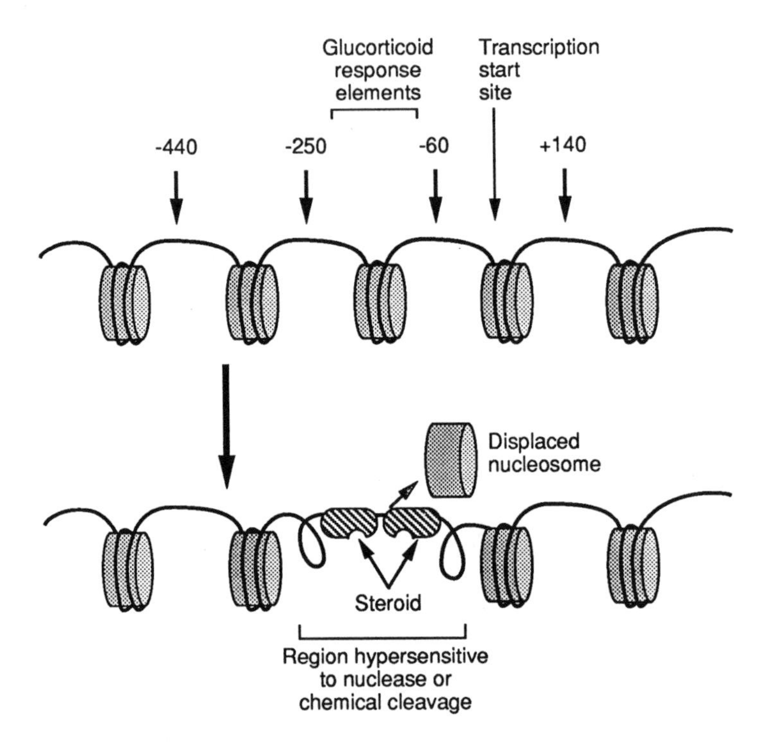

Fig. 10.14 Effect of dexamethasone on protection of DNA by nucleosomes in minichromosomes containing the long terminal repeat of mouse mammary tumour virus. Binding of the hormone–receptor complex makes a region (–60 to –250) formerly occupied by a nucleosome sensitive to cleavage.

circular vectors (based on papillomavirus DNA), which acquire nucleosomes and replicate as "minichromosomes", independently of the host chromosomes, in mammalian cells. Constructs were made in which these vectors contained an inserted reporter gene, fused to a region of mouse mammary tumour virus (MMTV) containing two glucocorticoid response elements. In the absence of glucocorticoid, nuclease or chemical cleavage of minichromosomes occurred at specific locations, ~200 base pairs apart, on the 5' side of the transcription initiation site (at –60, –250, –450, –650, etc.). This shows that the nucleosomes were 'phased', rather than being bound at random locations, with respect to the initiation site. Experiments using various different constructs showed that this phasing was produced by the inserted MMTV DNA, rather than being a function of the vector. Intriguingly, in the presence of glucocorticoid, the region between –60 and –250 was no longer protected against chemical cleavage: this is exactly the region containing the hormone response elements. Thus the binding of the activated receptor appears to have displaced a nucleosome (Figure 10.15), or at least to have altered the nucleosome structure in this region. Further experiments showed that this also exposed a region from the initiation site to –60 which bound two other transcription factors, one of which was nuclear factor-1 (see above). Although both of these factors appear to be present in extracts of unstimulated cells, they do not bind to the chromatin unless the cells are first treated with glucocorticoid. Thus, binding of the activated hormone receptor may cause a local change in chromatin structure which enables the binding of other proteins essential for initiation of transcription. At least some of the latter may be tissue-specific, thus explaining the different responses to the same hormone in different tissues.

How are genes inhibited by hormone–receptor complexes? Although the sequences of the receptor binding sites in two or three repressed genes have been determined, they are very similar to hormone response elements at which activation occurs (Figure 10.5), and not enough have yet been studied to enable one to state whether a different consensus sequence is involved. An alternative possibility is that the binding site may be identical to that for activated genes, but its location is such that the bound receptor blocks, rather than enhances, the binding of other transcriptional activators. Support for this has come from studies of the gene encoding the α subunit of the glycoprotein hormones (Section 3.3.1), which is repressed by glucocorticoids. It was found that the binding sites for the glucocorticoid receptor and a transcriptional activator of the gene (CREB: see Section 8.2.4) were overlapping. The critical feature may therefore be the location of the binding site for the receptor relative to those of other transcription factors.

SUMMARY

The steroid and thyroid hormones, and the morphogen retinoic acid, are small lipophilic molecules which can diffuse through the plasma membrane and bind to soluble intracellular receptors. The receptors for all of these messengers form a family of related proteins which, once synthesized, are now thought to be localized exclusively in the nucleus. These receptors contain a variable N-terminal domain, a conserved, central DNA-binding domain, and a C-terminal messenger-binding domain. The central domain, which folds into two "zinc-finger" domains, binds to specific DNA sequences known as hormone response elements, which are located in the upstream region of regulated genes. Beyond this point, our understanding of the mechanism of gene regulation is still incomplete. A hormone response element is a partially palindromic sequence, and the hormone-bound receptor probably binds as a homodimer. The receptors may activate genes by causing local changes in the packaging of the DNA into nucleosomes. The expression of the gene is certainly dependent on the presence of additional transcription factors, some of which may be tissue-specific. Whether the gene is activated or repressed probably depends on the location of the hormone response element relative to the binding sites for other transcription factors.

FURTHER READING

1. "The glucocorticoid domain: steroid-mediated changes in the rate of synthesis of rat hepatoma proteins". Ivarie, R.D. and O'Farrell, P.H. (1978) Cell **13**, 41-55
2. "Localization of ecdysterone on polytene chromosomes of *Drosophila melanogaster*". Gronemeyer, H. and Pongs, O. (1980) Proc. Natl. Acad. Sci. USA **77**, 2108–2112.
3. "The steroid and thyroid hormone receptor superfamily" – review. Evans, R.M. (1988) Science **240**, 889–895.
4. "Gene regulation by steroid hormones" – review. Beato, M. (1989) Cell **56**, 335–344.
5. "A human retinoic acid receptor which belongs to the family of nuclear reeptors". Petkovich, M., Brand, N.J., Krust, A. and Chambon, P. (1987) Nature **330**, 444–450.
6. "Zinc fingers: a novel protein motif for nucleic acid recognition" – review. Klug, A. and Rhodes, D. (1987) Trends Biochem. Sci. **12**, 464–469.
7. "The function and structure of the metal coordination sites within the glucocorticoid receptor DNA binding domain" – direct demonstration of coordination of two zinc atoms with four cysteine residues each. Freedman, L.P., Luisi, B.F., Korszun, Z.R., Basavappa, R., Sigler, P.B. and Yamamoto, K.R. (1988) Nature **334**, 543–546.
8. "Many transcription factors interact synergistically with steroid receptors". Schüle, R., Muller, M., Kaltschmidt, C. and Renkawitz, R. (1988) Science **242**, 1418–1420.
9. "Sequence-specific positioning of nucleosomes over the steroid-inducible MMTV promoter". Richard-Foy, H. and Hager, G.L. (1987) EMBO J. **6**, 2321–2328.

APPENDIX: ABBREVIATIONS FOR AMINO ACIDS

Single letter	Three letter	Amino acid	Nature of side chain
A	ala	Alanine	Small, uncharged
C	cys	Cysteine	Hydrophobic, forms crossbridges
D	asp	Aspartic acid	Acidic
E	glu	Glutamic acid	Acidic
F	phe	Phenylalanine	Hydrophobic
G	gly	Glycine	Small, uncharged
H	his	Histidine	Basic
I	ile	Isoleucine	Hydrophobic
K	lys	Lysine	Basic
L	leu	Leucine	Hydrophobic
M	met	Methionine	Hydrophobic
N	asn	Asparagine	Hydrophilic
P	pro	Proline	Cyclic: bends the peptide chain
Q	gln	Glutamine	Hydrophilic
R	arg	Arginine	Basic
S	ser	Serine	Hydrophilic
T	thr	Threonine	Hydrophilic
V	val	Valine	Hydrophobic
W	trp	Tryptophan	Hydrophobic
Y	tyr	Tyrosine	Hydrophobic

INDEX

The suffixes F, B and T indicate that the page numbers refer to Figures, Boxes and Tables respectively. Numeric (1-, 2-...) or Greek (α-, β-...) prefixes have been ignored during alphabetization.